BUSINESS SYSTEMS
ANALYSIS AND DESIGN

ANAHEIM PUBLISHING COMPANY
Specialist In Data Processing Textbooks

INTRODUCTION TO DATA PROCESSING

The Computers In Our Society, Logsdon & Logsdon
The Computers In Our Society Workbook, Logsdon & Logsdon

ASSEMBLER LANGUAGE

IBM System/360 Assembler Language, Cashman & Shelly
IBM System/360 Assembler Language Workbook, Cashman & Shelly
IBM System/360 Assembler Language Disk/Tape Advanced Concepts, Shelly & Cashman
DOS Job Control For Assembler Language Programmers, Shelly & Cashman

COBOL

Introduction To Computer Programming ANSI COBOL, Shelly & Cashman
ANSI COBOL Workbook, Testing & Debugging Techniques & Exercises, Shelly & Cashman
Advanced ANSI COBOL Disk/Tape Programming Efficiencies, Shelly & Cashman
Introduction To Computer Programming System/360 COBOL, Cashman
IBM System/360 COBOL Problem Text, Cashman
IBM System/360 COBOL Disk/Tape Advanced Concepts, Shelly & Cashman
DOS Job Control For COBOL Programmers, Shelly & Cashman

RPG

Introduction To Computer Programming RPG, Shelly & Cashman
Computer Programming RPG II, Shelly & Cashman
IBM System/360 RPG Programming, Volume 1-Introduction, Fletcher & Cashman
IBM System/360 RPG Programming, Volume 2-Advanced Concepts, Fletcher & Cashman

FORTRAN

Introduction To Computer Programming Basic FORTRAN IV-A Practical Approach, Keys

PL/I

Introduction To Computer Programming System/360 PL/I, Shelly & Cashman

SYSTEMS ANALYSIS AND DESIGN

Business Systems Analysis And Design, Shelly & Cashman

FLOWCHARTING

Introduction To Flowcharting and Computer Programming Logic, Shelly & Cashman
Basic Logic For Program Flowcharting and Table Search, Jones & Oliver

JOB CONTROL-OPERATING SYSTEMS

DOS Utilities Sort/Merge Multiprogramming, Shelly & Cashman
OS Job Control Language, Shelly & Cashman

CERTIFICATE IN DATA PROCESSING

Review Manual For Certificate In Data Processing, Cashman

UNIT RECORD TECHNIQUES

Introduction To Control Panel Wiring-548 Interpreter, Cashman & Keys
Control Panel Wiring-514 Reproducer, Cashman & Keys
Control Panel Wiring-85 Collator, Cashman & Keys
Control Panel Wiring-402 Accounting Machine, Cashman & Keys
Control Panel Wiring-407 Accounting Machine, Cashman & Keys

BUSINESS SYSTEMS ANALYSIS AND DESIGN

By:

Gary B. Shelly
Educational Consultant
Fullerton, California

&

Thomas J. Cashman, CDP, B.A., M.A.
Long Beach City College
Long Beach, California

ANAHEIM PUBLISHING COMPANY
1120 E. Ash, Fullerton, CA, 92631
(714) 879-7922

First Printing
November 1975
Second Printing
March 1976

Third Printing

June 1977

Fourth Printing

October 1977

Library of Congress Catalog Card Number 74-77398

Photo Credits:

Figure 5-3 Courtesy of IBM
Figure 5-6 Courtesy of IBM
Figure 5-7 Courtesy of IBM
Figure 5-8 Courtesy of IBM
Figure 6-2 Courtesy of IBM
Figure 5-4 Courtesy of Mohawk Data Sciences Corp.
Figure 5-6 Courtesy of Mohawk Data Sciences Corp.

ISBN 0-88236-043-4

Printed in the United States of America

Preface

For many years Industrial Engineers and Systems and Procedures Analysts have been an important part of the business environment. During the 1960's, computer systems for business applications became widely used and the need for business data processing systems analysts developed. These systems analysts were individuals that were primarily concerned with the conversion of manual business systems to "computerized" systems. With the continued growth in the number of computer systems installed during the 1970's, the need for individuals trained in systems analysis and design is even greater and instruction in business data processing systems analysis and design is essential.

Courses in systems analysis and design have been offered for many years at our colleges and universities. Often, however, these systems courses were management oriented with emphasis upon cost analysis and hardware acquisition, or they were oriented toward the needs of the systems and procedures analysts with emphasis upon systems theory, records and form management, procedure and process flowcharting, and work simplification and measurement. Neither approach has served to adequately meet the needs of the student majoring in business data processing or the student attempting to prepare for a career as a business data processing systems analyst. Far too often the instruction presented has failed to give the student even the remotest understanding of the sequence of events occuring when a systems project is approved by management and the systems analyst undertakes the job of converting the current system to a computerized system or developing a new computer system—yet the tasks which must be performed in a systems project are the very tasks with which the emerging student will be faced upon entering industry as a systems analyst trainee.

This text has been written to meet the need for instructional material to educate students in the art of business systems analysis and design. The text has been designed to provide the student with an understanding of the duties of the systems analyst together with an understanding of the specific methods and techniques for conducting a systems project—from the preliminary investigation of the project through the systems implementation and evaluation. Every effort has been made to introduce subject matter in the text that will assist the student in gaining an understanding of the techniques and procedures for conducting a systems project.

Because "systems" is a relatively new profession, there is a noticeable lack of standards within the systems profession relative to the approach to be taken when conducting a systems project. For this reason the authors have found it necessary to draw upon their combined backgrounds in both education and business data processing to define the specific duties and tasks which must be performed during the progress of the systems project. Thus, much of the material has been drawn from the authors' own experience in conducting systems projects, resulting in the academic subject matter within the text being widely supplemented with real life problems and situations that the student will encounter in industry.

In drawing upon industry experience the authors have defined a five phase approach to conducting a systems project. These phases are:

PHASE I   Initiation of the Systems Project and Preliminary Investigation
PHASE II   Detailed Systems Investigation and Analysis
PHASE III Systems Design
PHASE IV   Systems Development
PHASE V   Systems Implementation and Evaluation

One or more chapters of the text are devoted to an explanation of each of these phases of the systems project.

In the first portion of each chapter of the text, the methods and techniques used by the analyst in conducting each of the phases of the systems project are explained. Every effort has been made to be as exact as possible in the explanations, citing throughout the textbook numerous rules and specific steps to follow when conducting the systems project. To enhance the practical value of the text, at the end of each chapter a segment of a case study is presented involving the design of a simplified payroll system. The purpose of this section in each chapter is to illustrate, by means of a practical example, the concepts presented in each chapter and to provide the opportunity for the student to "live" through the actual problems involved as the systems project moves through the various phases of the study, as well as "seeing" a systems project evolve from initial problem definition through actual implementation.

At the conclusion of each chapter Review Questions are included to provide the student with a systematic review of the subject matter covered. One or more "controversial" Discussion Questions are also included to encourage the student to critically analyze issues and to point out that there are a variety of opinions relative to many aspects of data processing and systems analysis.

One of the most important learning experiences for the student is the discussion, analysis, and actual design of a system as reflected in the Ridgeway Company Case Study. At the end of each chapter a segment of the Ridgeway Company Case Study is presented to provide the student with exposure to the problems involved in conducting a "realistic" systems project. Upon completion of the textbook and completion of the Case Study the student will have received experience in the designing of an actual system. It should be noted that the Case Study used within the text is a modification of an actual systems project with which the authors have had experience.

Many systems textbooks which the authors have had the opportunity to review are so generalized that the material is extremely difficult to teach or the material is so difficult that the students cannot understand the basic concepts presented. Great emphasis has been placed by the authors upon developing a "teachable" textbook on systems analysis and design—a textbook that is interesting to read and that will develop the knowledge and skills required of the student that will contribute to his success upon entering the field of data processing systems.

The authors would like to express thanks to Sandi McMurray for typesetting the final manuscript, and to Kay Kreidel for her excellent job in laying out the text, performing all the art work and original drawings in the book, and designing the cover.

Gary B. Shelly
Thomas J. Cashman, CDP

Table of Contents

PHASE III   SYSTEMS DESIGN

PHASE IV   SYSTEMS DEVELOPMENT

PHASE V   SYSTEMS IMPLEMENTATION AND EVALUATION

INTRODUCTION TO SYSTEMS ANALYSIS AND DESIGN

CHAPTER 1

INTRODUCTION TO SYSTEMS ANALYSIS AND DESIGN

1

INTRODUCTION

The world in which we live is governed by "systems" of all types. We speak of the solar system in our universe, the highway system within the country, the payroll system within a business, and the nervous system within our bodies. According to the dictionary, a system is "a regularly interacting group of items forming a unified whole."

Business activity for many hundreds of years has been governed by systems of some type. Historically, one of the earliest forms of business activity was the "bartering system" in which the goods and services of one individual were exchanged for the goods and services of another individual. With this type of business activity there was little need for formalized record keeping procedures, as there was no credit, no accounts receivable or payable, and no payroll. Although formalized accounting concepts were developed over 500 years ago, business organizations prior to the Industrial Revolution of the 1800's often consisted of the sole proprietor or the skilled artisan, and the transactions related to a business were normally simple in nature and detailed records of daily operational procedures of a business were often not required. Even though it must be recognized that in any organized society, no matter how simple in structure, there must be some systematic approach to doing business, the need for formal, documented systems for controlling and directing business activity did not become necessary until business organizational structures became relatively complex.

With the Industrial Revolution of the 1800's came the expanded use of machines, the techniques of mass production, many new inventions, and the growth of world markets. Business organizations evolving from the Industrial Revolution had the result of separating the worker from control over his work, and the skilled artisan became a member of a larger group of specialists with a shift in emphasis from a single worker and his tools to techniques of mass production with hundreds of workers gathered together within a single business organization.

These concepts substantially complicated the structure of business organizations. As business units grew and gigantic corporations developed, the need for specific methods for managing the activities within an organization became necessary. Thus, the concept of a "system" for accomplishing a given task within a business emerged.

1

In the 1950's, with the application of the electronic computer to the solution of business problems, the need for a "systems approach" to controlling business activity became even more evident. With the electronic computer came the ability to perform calculations in millionths of a second and to process thousands of business transactions each day without human intervention! A formalized system was absolutely mandatory in order to properly handle these transactions.

SYSTEMS AND PROCEDURES

In the course of doing business, a company will perform many functions in a specified manner, such as writing orders for customers who buy their products, or writing payroll checks for the employees of the company. In most companies and businesses the manner in which these functions are performed depends upon the needs of the business and the requirements which must be satisfied for each function. The steps which are followed are normally tied to what may be termed a SYSTEM, with a system being made up of a sequence of PROCEDURES which are used to reach a given objective.

A "Procedure" may then be defined as "a series of logical steps by which all repetitive business actions are initiated, carried forward, controlled, and finalized," with a "System" being considered the "network of related procedures designed to perform some major business activity." Thus, a business system is merely an organized method for accomplishing definite goals, with the procedures of the system consisting of the logical steps by which the repetitive actions of the business are carried out.

Systems may take many forms and the procedures which are incorporated into a given system may be as diverse as any activity imaginable within a business. For example, there may be a "system" to build a complex rocket engine or a system to allow an employee to deduct money from his paycheck for the credit union. Although each of these activities differs from the other, they are both governed by a set of procedures which either the individual in the manufacturing plant must follow to build the engine or the employee must follow to deduct money for the credit union. See Figure 1-1.

Example of a "System"

Figure 1-1 Example of a "System" to Deduct Credit Union Dues

It is extremely important that procedures be understood, because it is the function of the systems analyst to define the procedures which will be placed in a given sequence to form the system.

THE BUSINESS ENTERPRISE

There are a variety of types of business organizations in operation in the economy today, but most businesses may be classified into two basic types. They include:

1. INDUSTRIAL ORGANIZATIONS - Industrial or production oriented business organizations are defined as organizations of people, material, and equipment organized to produce, sell, and distribute goods.

2. SERVICE ORGANIZATIONS - The service oriented business organization is based solely upon the sale and distribution of goods or services.

The distinctive characteristic of the industrial organization is the production of goods, while the activities of service organizations center only on the selling and distributing of goods or services. The industrial organization is normally much more complex in terms of organizational structure than the service organization.

Another type of organization which is of significance within our economy is the "governmental" organization, including city, county, state and federal governments. Governmental organizations commonly assume the characteristics of service organizations as governmental organizations are normally concerned with providing services of some type.

BUSINESS INFORMATION PROCESSING SYSTEMS

Regardless of the type, purpose, or organizational structure of businesses, the activity of the business is directed and controlled by some type of information processing system. An information processing system is one which generates data and information which must be analyzed and acted upon in order for the business to be successful.

The entire business information processing system of a company normally consists of a series of "subsystems", many of which are common to all business organizations. In small enterprises, these subsystems may be handled by a single individual, whereas in large organizations, management personnel may be assigned with related staff to each element within the subsystem. Figure 1-2 illustrates some of the elements which can possibly make up subsystems within a total company information processing system.

ELEMENTS OF AN INFORMATION PROCESSING SYSTEM

Personnel Subsystems
Candidate Selection
Employee Histories
Labor Distribution

Finance Subsystems
Payroll
General and Tax Accounting
Cost Accounting

Order Processing Subsystems
Order Input
Editing
Pricing
Billing
Credit and Collection

Purchasing Subsystems
Vendor Selection
Ordering
Receiving
Inspection
Invoice Matching and Payment

Production Subsystems
Finished Goods Inventory Control
Raw and In Process Inventory Control
Operations and Quality Planning
Scheduling
Dispatching
Production Reporting
Inspection
Packaging
Shipping

Marketing Subsystems
Sales Statistics
Advertising Performance
New Product Scheduling
Long Range Forecasting
Short Range Forecasting

Figure 1-2 Elements in an Information Processing System

Each of the subsystems illustrated in Figure 1-2 represents a business information system which generates data that must be processed and analyzed if the business is to be successful. A breakdown in any one of these subsystems can drastically affect the success of a company. It is indeed a tribute to modern business management that large organizations, some of which employ over a half a million employees, can effectively integrate such a complex set of subsystems to produce a successful, profitable business organization.

COMPANY STRUCTURE

Whenever a system is designed to perform any function within a company, it must be designed within the structure of the company. The structure of a company is the hierarchy of management and the responsibilities of each member of the company in relation to other members of a company.

A company structure is normally illustrated through the use of a ''management tree'' or organization chart wherein the various levels of management are depicted graphically. The example in Figure 1-3 is an example of an organization chart.

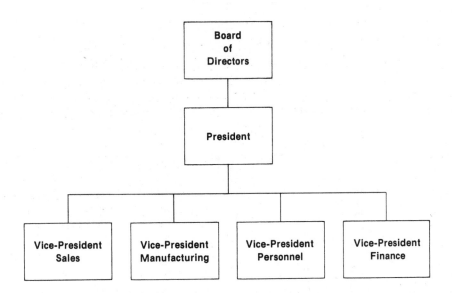

Figure 1-3 Example of Organizational Chart

The function of the organizational chart is to depict graphically who reports to whom in terms of the structure of the company. Thus, in the example above, the President of the company would report to the Board of Directors, and the Vice-President of Sales, the Vice-President of Manufacturing, the Vice-President of Personnel and the Vice-President of Finance of the company would report to the President.

It should be noted that the example presented in Figure 1-3 is an over-simplification of the organizational chart which would be found in most companies; that is, there would be more officers, and more people reporting to the president; or the vice-presidents depicted above might report to an executive vice-president, who would then report to the president. Thus, there could be many different structures, depending upon the size and organization of a given company.

The use of an organizational chart is not restricted to top-level management such as the president and vice-presidents as shown in Figure 1-3. It can be used to illustrate any relationships which are desired to be shown. The organizational chart in Figure 1-4 is used to depict the structure of the machining and tool departments of a company which is engaged in producing machined parts.

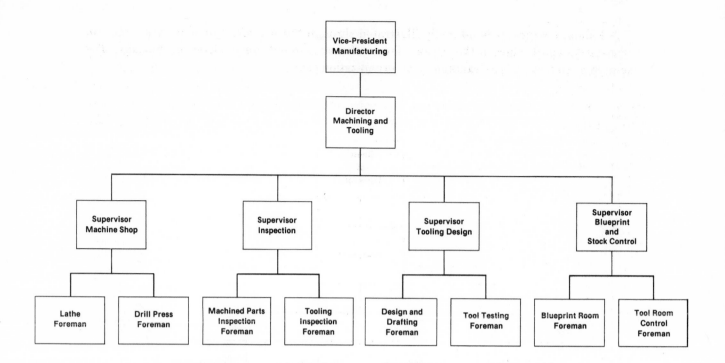

Figure 1-4 Example of Lower-Level Organizational Chart

As can be seen from the chart in Figure 1-4, the level of the organizational chart has been lowered from vice-presidents to foremen of various departments. In addition, the chart is more comprehensive than that shown in Figure 1-3 because there are more positions illustrated. An organizational chart may become as involved as desired in order to show the relationships between various positions within a company.

In addition, it can be noted that the organizational chart is useful in depicting the relationship of positions even if there is not a thorough understanding of the function of each position. Thus, if one were not familiar with what the function of a drill press foreman really was, he can still see the organizational arrangement and the persons to whom this foreman would report.

LINES OF AUTHORITY

Within every company structure there must exist some formal lines of authority and responsibility. Small companies are normally characterized by what is known as the LINE organizational structure. In the line organizational structure, authority flows from top to bottom with each function on the same level of authority considered to be an independent self-contained unit. Figure 1-3 illustrated a single line type organizational structure, as each vice-president reports directly to the president.

It was recognized in business organizations, however, that there was often a need for specialists to advise management in the efficient operation of the business. Thus, the concept of the LINE AND STAFF organizational structure arose. In line and staff organizational structures, the ''staff'' functions are essentially service departments which are set up to advise management and assist line personnal. Normally, members of the ''staff'' do not give orders directly but consult, recommend, and advise. It is the duty of the ''line'' personnel to take action and to implement.

An example of an organizational chart illustrating the inclusion of staff functions is contained in Figure 1-5.

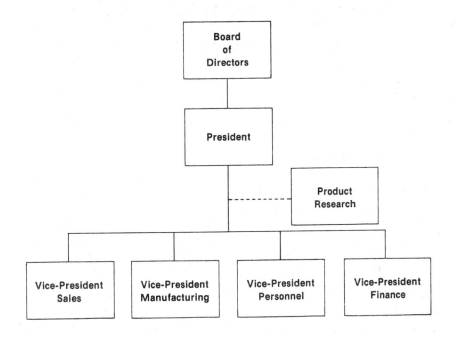

Figure 1-5 Organization Chart with a Staff Function

Note from the organizational chart illustrated in Figure 1-5 that the Product Research Group has been included. The function of the Product Research Group is to research the development of new products for the company and to make recommendations to management. In the line and staff organization, this group would have no authority to implement the results of its research. This would be the responsibility of the line personnel who have the authority for directing and implementing.

The understanding of the use of a company organizational chart is important, because when analyzing business systems, it is often necessary to determine which position would be able to supply information about a given problem, and it is through the use of an organizational chart that it may be determined who should be questioned or who might have useful information.

THE SYSTEMS AND PROCEDURES DEPARTMENT

In a competitive business environment, the success or failure of the business is often directly related to the operating efficiency of the business. The operating efficiency of a business, in turn, is commonly dependent upon two basic factors: (1) the speed at which employees can accomplish assigned tasks; (2) the efficiency of the system of activities which take place within the organization. As there is often a limit to the speed at which employees can work, an area that often offers an opportunity for improving the operating efficiency is the analysis and improvement of existing methods of doing business. For this reason, Systems and Procedures Departments have become increasingly important in many business organizations. The Systems and Procedures Department is commonly composed of one or more employees who are responsible for the analysis and improvement of the systems existing within a company. Typical duties of the Systems and Procedures Department include:

1. Work Measurement

2. Work Simplification

3. Forms Analysis and Control

4. Records Management

5. General Management Surveys and Systems Studies

In large organizations, the Systems and Procedures Department is commonly considered to perform a staff function reporting to top management. See Figure 1-6.

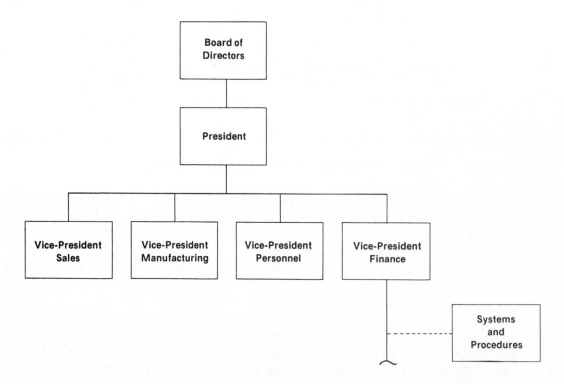

Figure 1-6 Systems and Procedure Department in an Organization Chart

In the past two decades with the advent of the electronic computer and its widespread use in business, two distinct types of systems specialists have appeared within the business organization—the general systems and procedures specialist, concerned with areas such as work measurement, work simplification, forms control, and general systems studies, and the data processing systems analyst who is concerned with the analysis of existing systems with the purpose of implementing these systems on electronic data processing equipment if feasible. The data processing systems analyst is commonly found within the data processing department of the business organization.

THE DATA PROCESSING DEPARTMENT

Although there are great differences in the organizational structures in the data processing departments of various organizations, three functions are commonly performed within the department. They are: (1) Systems Analysis and Design; (2) Programming; (3) Operations. The personnel within these groups commonly report to a director of data processing or the data processing manager. This structure is shown in Figure 1-7.

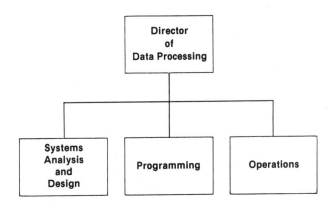

Figure 1-7 Functional Organization of Data Processing Department

The functions of the personnel in each of these areas is discussed in the following paragraphs:

1 - Director of Data Processing - This is the individual who is responsible for the administration and control of the data processing department. He will direct the people within the department in terms of jobs to be done, procedures to be followed, and money to be spent. One of the major tasks of the data processing manager is the budgeting of people and money to complete the jobs which top management within the company deem to be of importance.

2 - Systems Analysis and Design - This area of the data processing department is responsible for defining and developing the systems which will be implemented on the company's computer. The systems analysis and design section of the data processing department forms the liaison between user departments, that is, departments which will use the output from data processing systems, and the data processing department. In addition, most of the design of the systems is done in this area, and in many cases, much of the impetus for designing new systems comes from the systems analysis and design department.

3 - Programming - This group within the data processing department is responsible for all programming activities which take place within the company. This includes the business applications programming for areas such as payroll, accounts receivable, accounts payable, etc. In addition, the programming group is normally responsible for updating and maintaining systems which have already been programmed, for handling any changes or maintenance which must be performed on the operating system, and the members of the programming staff frequently act in an advisory role to the systems department on any programming problems which may occur in systems design work.

4 - Operations - The Operations area controls all of the day-to-day activities which take place within the data processing department, including such things as operation of the computer, preparation of input data, checking and disbursement of output data and materials, and scheduling of the data processing department.

The organizational chart in Figure 1-7 depicted the general departments. Each of these departments are further broken down into functioning units. An example of a more detailed organization chart of a data processing department is illustrated in Figure 1-8.

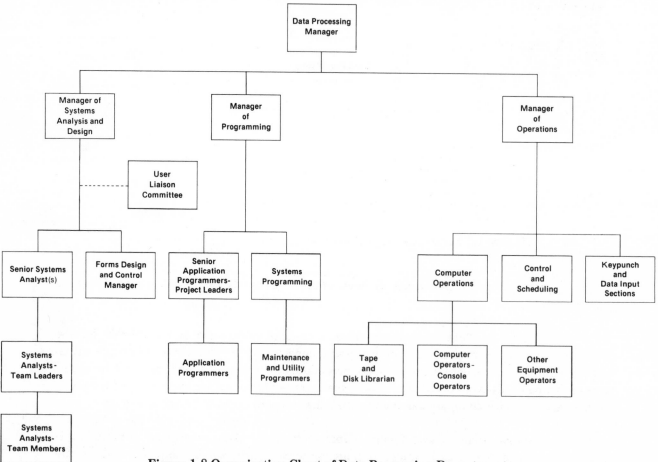

Figure 1-8 Organization Chart of Data Processing Department

It should be noted that in many organizations there is a merging of the systems analysis and programming responsibilities. Thus, the "systems analyst," sometimes called the "programmer analyst," is not only responsible for analyzing the existing systems and designing new systems, but also may have the responsibility for programming and implementing the system.

THE DATA PROCESSING DEPARTMENT AND COMPANY STRUCTURE

It is of importance to the systems analyst to realize where, within the structure of the company, the data processing department lies, and also where in the company the systems department lies. Unfortunately, it is difficult to state in all cases where the data processing department will be found within the company. This is because the historical evolution and importance of the data processing function within the company has resulted in the data processing department rising to increasingly higher levels in the typical organizational structure.

When the data processing department began to be a significant addition to the company structure in the middle or late 1950's, most of the applications which were processed were of an accounting or financial nature, such as payroll, accounts receivable, and accounts payable. Therefore, the control of the data processing department quite naturally was placed in the hands of the chief financial officer of the company, such as the Vice-President of Finance. It was felt that since the primary function of the data processing department was to handle financial matters, the applications of data processing should be controlled by those who were most familiar with company moneys. See Figure 1-9.

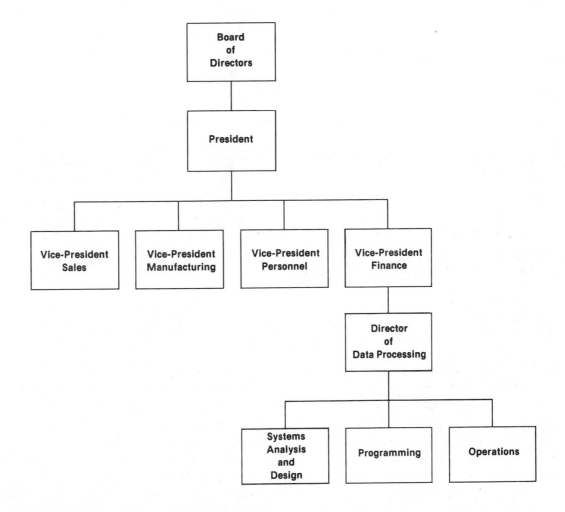

Figure 1-9 Organizational Chart in which the Director of Data Processing Reports to the Vice President of Finance

In the 1960's, with the advent of more sophisticated hardware and software, and with the realization that computers could be used for more than just adding and subtracting financial figures, there was a movement away from strict and absolute control of the data processing facilities by the financial arm of the company. In some large installations, the data processing "manager" was given the rank of a vice-president and reported directly to the president of the company. See Figure 1-10.

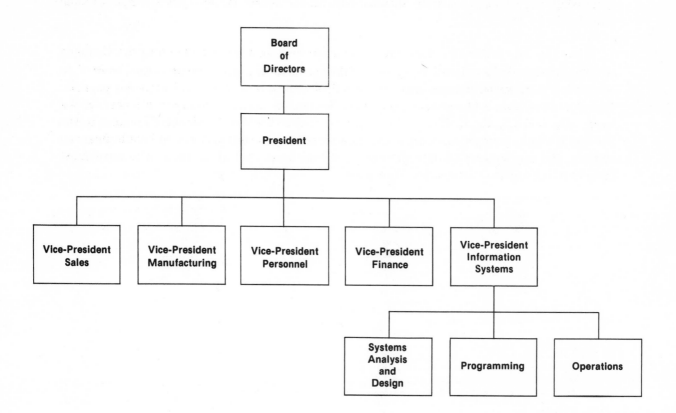

Figure 1-10 Organizational Chart in which the Director of Information Systems is a Company Vice-President

Thus, although the position of each data processing department will vary within a company structure, it is safe to say that the status of the business data processing department has risen considerably since the time of electro-mechanical accounting machines. In addition, a larger percentage of a company's money is being spent on data processing than was true 10 - 15 years ago, and top management has recognized that the management of the data processing and computer installation is one of the more important management positions in the company.

CONDUCTING A SYSTEMS PROJECT

As previously discussed, the data processing systems analyst is that employee of a company whose primary duties consist of analyzing existing systems and methods of operation for the purpose of converting those systems to computerized processing if feasible. The systems analysis function is commonly performed within a well defined, scientific problem-solving approach. This approach is applied to all SYSTEMS PROJECTS which are undertaken by a systems department. Any systems project can be broken down into a series of specific phases. These phases include:

Phase 1 - INITIATION OF THE SYSTEMS PROJECT AND PRELIMINARY INVESTIGATION

This phase of the systems project begins with a request, normally from someone outside the data processing department, to the systems department to conduct an investigation of some portion of the business. These requests can vary from a request to prepare a single sales report to a request to design a complete accounting system. When the request is received, the systems manager will normally review the request and determine if the request offers sufficient potential benefits to the company to warrant further review. If so, the manager will normally order that a preliminary investigation be initiated by the systems department. This phase of the systems project is sometimes called the ''problem definition'' phase.

The purpose of this preliminary investigation is to identify the true nature of the problem specified in the request for services and to gain an overall understanding of the procedures used in the current system. In order to accomplish this, the analysts will interview those persons who have a knowledge of the current system. In this preliminary investigation, the interviews will normally be with management level personnel who have a knowledge of current operating procedures and who are in a position to point out any problems which are occurring within the system. At this time, the analysts will also review available documentation in order to understand the methods which are **supposed** to be used in the system.

At the conclusion of the preliminary investigation, the systems department will submit a report to management specifying what they have found to be the problems within the system, and also what they suggest as further action. If warranted, the systems department will recommend that a detailed systems investigation and analysis of the current system take place for the purpose of developing recommendations relative to the solution of the problems which appear to exist in the current system.

Phase 2 - DETAILED SYSTEMS INVESTIGATION AND ANALYSIS

After approval by management for continued study of the systems project, the systems manager will assign systems personnel to conduct the full-scale investigation and systems analysis. In this "fact-gathering" portion of the systems project, the analysts will normally conduct further interviews with selected individuals, observe current activities related to the system, and obtain samples of all documents relating to the system being studied in an attempt to define all elements of the current system.

After the facts have been obtained and recorded in some systematic fashion, the analyst must then analyze and evaluate the existing system. From this analysis, several alternative plans should be developed which will solve the problems found in the current system. For example, recommendations could be presented suggesting the hiring of additional personnel, the redesign of existing forms, or perhaps the complete redesign of the existing system. The systems department will normally make a recommendation concerning which plan they feel is the most appropriate. Upon management approval of a plan, the systems project enters the next phase—that of Systems Design.

Phase 3 - SYSTEMS DESIGN

With a knowledge of the present operation, it is now possible to systematically design a new system or modify the existing system to meet the operational needs of the business. In the systems design phase, the outputs required are designed, the format and type of input to be used is determined, and the processing and type of file organization is established.

Prior to implementing the system, there is normally a formal systems presentation to management relative to the new system design and formal approval of the systems project.

Phase 4 - SYSTEMS DEVELOPMENT

The systems development phase includes the actual programming of the system, the system testing and documentation, and the development of formalized procedures for scheduling and maintenance of the system.

Phase 5 - SYSTEMS IMPLEMENTATION AND EVALUATION

After the system is fully tested and documented and schedules established, the system is ready to be implemented.

As a part of the complete systems project, provisions must be made to allow for a post-implementation evaluation at regular intervals. This evaluation should be aimed at determining if the system is operating as proposed, and if the economics and costs are as anticipated.

It can be seen that a systems project follows the commonly used scientific method of problem analysis and consists of an entire cycle of events from initial problem analysis to final evaluation.

CONDUCTING A SYSTEMS PROJECT - A CASE STUDY

To provide an overview of the sequence of events occurring in a systems project, a brief case study is introduced in this chapter and explained in the following paragraphs.

Case Study - Introduction

The Ames Medical Supply Company is a company which sells medical supplies to doctors throughout the Southern California area. The President of the company, Henry Ames, recently formed a separate department within the company to prepare and market the mailing list of doctors which the company has on file. This mailing list is sold to other non-competing companies which have products to be sold to doctors. The list now comprises approximately 3,000 individuals. To prepare mailing labels for the Ames Company and other companies desiring to purchase the list, the doctors' names and addresses have been recorded on edged-punched paper tape cards. These edged-punched cards are processed individually through an automatic word processing machine which reads the cards, and automatically, at 100 words per minute, prints the required mailing labels. This is illustrated in Figure 1-11.

Figure 1-11 Preparing Mailing Labels Using Edged-punched Cards

Case Study - Phase 1 - Initiation of the Systems Project and Preliminary Investigation

In the past, one of the big selling features in the marketing of the mailing list was that the list could be provided to customers within five days after receipt of an order. In the past several months, because of an increased volume of business, it has been necessary to have the employee in charge of preparing the mailing labels work considerable overtime, and even with the overtime, a number of cases have arisen in which it has taken over 10 days to deliver the mailing list after original receipt of the order. Because of the recent delays, several orders have been cancelled.

The president of the company has been apprised of the situation which has developed in the newly formed department and has contacted the systems department to find out if anything can be done about improving the efficiency of operations.

Upon receipt of a formal request to study the mailing list system, the manager of the systems department assigned one of his analysts to make a preliminary investigation of the current system. After interviewing the president of the company and other persons associated with the mailing list system the analyst confirmed the problems that reportedly existed in the current system. The analyst then prepared a report detailing the problems and indicated that an improvement of the existing mailing list system would be of benefit to the company. The analyst also recommended this system be considered for further investigation.

This recommendation was presented to management for their approval. Upon reviewing the report and asking a number of questions relative to the problems existing within the current system, management gave their approval for a detailed systems investigation and analysis of the mailing label system currently in operation. Thus, the next phase of the systems project is entered.

Case Study - Phase 2 - Detailed Systems Investigation and Analysis

When the second phase of the systems project begins, the manager of the systems department will assign persons within his department to gather facts and investigate the current system. He will also direct them to analyze the facts which they obtain and prepare a report for management giving alternative solutions to the problem.

As noted, this phase of the systems project is the ''data-gathering'' and ''fact-finding'' stage of the project. Through interviewing and analyzing the operation of preparing the mailing labels, the systems analyst determined that over an extended period of time an average of three labels per minute were being prepared using the edged-punched cards. Production records also revealed that an average of 6,000 labels per week were being produced. As the mailing list had grown to over 3,000 names, a single request for the list took approximately 2-1/2 days to prepare. Thus, if over two requests for mailing lists per week were received, overtime was necessary to provide the customer with the five-day service.

As a result of the information gathered in the investigation of the current system, the analyst concluded that the file of names comprising the mailing list could readily be adapted to punched cards and processed on a computer. It was estimated by the analyst that the list of 3,000 names could be prepared in approximately 10 minutes using the computer with its high speed printer operating in excess of 1,000 lines per minute. Other alternative solutions included the hiring of additional personnel or the purchase of additional word processing machines.

Figure 1-12 High Speed Printer

The systems analysts concluded that the most feasible solution to the mailing labels problem was the implementation of the system on the computer. Thus, they presented a report to management recommending that a system be designed using the computer which Ames had recently acquired. Management had previously planned to only put several billing applications on the computer in the beginning, but with this breakdown of the mailing label process and the presentation from the system analyst indicating that the processing time for mailing labels could be substantially reduced using the computer, management enthusiastically endorsed the project and gave approval to the systems department to design a computerized system for producing mailing labels.

Case Study - Phase 3 - Systems Design

The first task when designing a system is to determine what the output from the system is to be. In many cases, the output will consist of printed reports which are to be used either within the company or by persons outside the company. In the mailing label system, the only output of the system is to be a set of mailing labels to be used on envelopes containing advertising material.

Designing the Output

As a result of determining the needs of the mailing list system, the systems analyst in charge designed the mailing labels illustrated in Figure 1-13. Note that instead of producing a single label, using the computer it is possible to produce two labels at one time with one run through the computer. It was decided by the analyst to produce two labels in a single run.

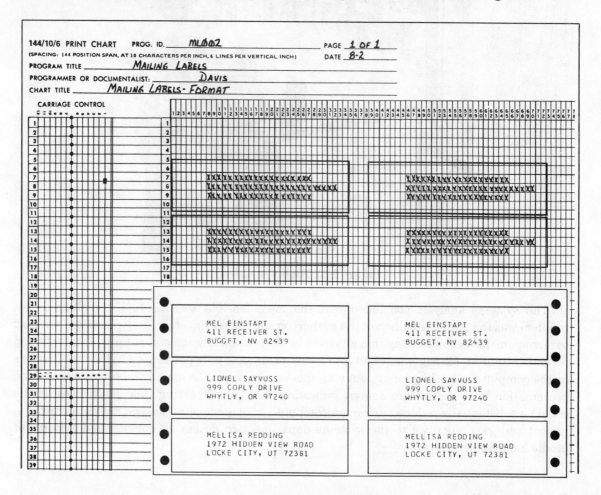

Figure 1-13 Example of Mailing Labels

Note from the example above that there is both the mailing label as it actually appears and the drawing of a printer spacing chart, which is a form which can be conveniently used for illustrating the columns and the spacing to be used on a printed report.

Since only the mailing labels are to be produced from the system, the definition of the output of the system is complete.

Designing the Input

Once it is determined what is desired from the system, it must be determined what information is necessary to produce the output. Since only the mailing labels are to be produced from the system, the only input required is the information which will be printed on the mailing labels, that is, the name and address of the person to whom the material will be sent.

It was decided to use punched cards as the input medium because of the relative ease of preparation and the ease of inserting new customers in the card file. The format of the input card to be used for the mailing list is illustrated in Figure 1-14. To define the size of the fields a Multiple-Card Layout Form is used.

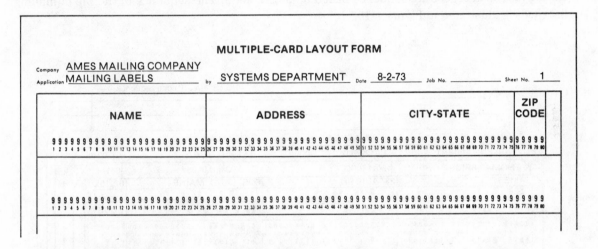

Figure 1-14 Example of Card Format

In the example above, it can be seen that the card will contain the name, the address, the city, the state and the ZIP code of the person to whom the material is to be sent. Since both the medium (card) and the format as illustrated above have been designed, and since there is no other input required for the system, the input design step is complete.

Designing the Files and Processing

After the desired output and the required input are designed, the processing which is to be accomplished by the system must be specified. In addition, any files, that is, groups of records which are to be saved and processed within the system, must also be designed. In this system the file is to consist of punched cards.

In the mailing list system, the processing which is to be accomplished is that the card file for the mailing labels is to be punched from the current mailing lists. In addition, as new entries are made to the mailing list, these punched cards must be added to the file of mailing list cards. Since it is desired that the labels be arranged in ZIP code sequence for bulk mailing, all of the cards must be in ascending sequence based upon ZIP code. The sequence of the file of mailing list cards is illustrated in Figure 1-15.

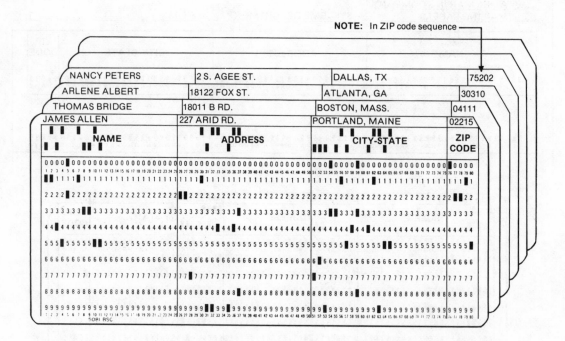

Figure 1-15 Example of Mailing List Card File

This file of cards is then to be read by the computer and the mailing labels, which are illustrated in Figure 1-13, will be created.

Documenting the System

After the output and input have been designed, the systems analyst must document the system in order to indicate the processing which is to take place within the system. One of the tools used by the systems analyst to perform this function is the systems flowchart. The systems flowchart for the mailing label system is illustrated in Figure 1-16.

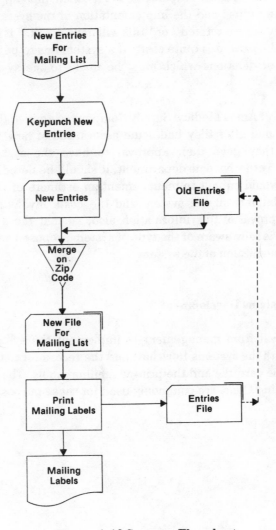

Figure 1-16 Systems Flowchart

As can be seen from Figure 1-16, special flowcharting symbols are used to depict the sequence of events which are to take place within the system processing. Thus, the first event to occur is that any new entries to the mailing list will be keypunched from source documents. Next, these entries are merged, in a ZIP code sequence, with the file containing all of the previously used cards. This new file is then used as input to a computer program which will print the mailing labels. The mailing list card file will then be stored to be used for the next cycle of labels.

Presentation of System to Management

After the design of the system has been completed, and before programming and implementation is begun, the system should again be reviewed by management for approval and should also be reviewed by the user so that there is no misunderstanding about the function of the system and how the system is to perform.

It is necessary at this point for those responsible to give their final approval of the system so that when the programming effort and the implementation of the system is completed, there are no cries of "this is not what we wanted" or "this will never work." It is quite important that this approval take place because it is quite costly if a system has to be redesigned or if there are significant additions or deletions which must be made to the system after it has been implemented.

After the management of Ames Medical Supply Company examined the system proposed by the systems department, and after they had some assurances of how fast the system could produce mailing labels, they gave their approval to have the system programmed and implemented as designed by the systems department. It should be noted here that at this time, the systems department would submit to management an estimate of the time which will be needed to program and implement the system and the costs involved. There should be a "buy-off," that is, an approval of this information also, so that the systems department is assured that management is fully aware of the type of system proposed and the costs associated with the design and implementation of the system.

Case Study - Phase 4 - Systems Development

After obtaining approval from management to implement the system, the analyst must supply the programmer with the systems flowchart and the record formats, that is, the location of the fields within both the card file and the printed mailing labels. The Multiple-Card Layout Form and the Printer Spacing Chart are commonly used for these purposes, and are illustrated in Figure 1-17.

Card Format

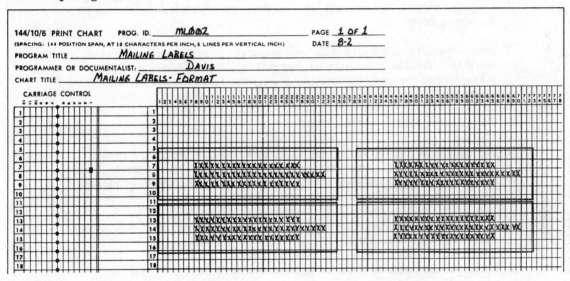

Printer Spacing Chart

Figure 1-17 Record Formats for Card Record and Mailing Labels

As can be seen from the forms above, the programmer will know what the formats of the records to processed are so that he can write his program to conform to the required formats.

The programmer must also be informed of the processing which is to be accomplished within the program which he is to write. Although numerous techniques exist to supply this information, as will be discussed in subsequent chapters, the program narrative would be sufficient for the mailing labels system. The following narrative would be prepared by the systems analyst for use by the programmer.

DATA PROCESSING PROGRAMMING SPECIFICATIONS		
SUBJECT Mailing List System	**DATE** August 2	**PAGE** 1 **OF** 1
TO Programmer	**FROM** John Davis, Systems Analyst	

A program is to be written to prepare mailing labels. The formats of the input card file, the printer spacing chart, and the systems flowchart are included as a part of this narrative. The program to be written should include the following processing:

1. The program should read the input cards and create the printed mailing labels as per the formats illustrated.

2. Each card which is read is to be in ZIP code sequence. Therefore, a sequence check should be included in the program to ensure that the ZIP codes (columns 76-80) are in an ascending sequence. If they are not in ascending sequence, an error message should be written on the computer console and the run cancelled.

3. The program should be written in COBOL.

4. The time allocated for the writing of this program is two days. In addition, two days are allowed for testing and documenting the program. It is estimated, therefore, that that the program should be completed at the end of four days from the beginning of programming.

5. All test data will be prepared by the programmer to test the program. A systems test will be conducted on the fifth day after the beginning of programming.

6. If there are any questions, contact Systems Analyst at Extension 246.

John Davis

Systems Analyst

sjm

Figure 1-18 Program Narrative

As can be seen from Figure 1-18, the details of the program to be written are set forth for the programmer, together with an estimated time of completion. In a simple program such as the mailing labels, this narrative should be sufficient to impart the knowledge the programmer requires in order to write the program.

Once the system has been designed and turned over to the programming staff for the programming effort, the analyst must turn his attention to ensuring that the system will be implemented on time and that the users of the system will be able to use the system when it is implemented.

In the mailing labels system, the systems analyst would check with the programming staff to ensure the program was on schedule. In addition, he would deal with the mailing department, which has the responsibility of attaching the labels to the outgoing mail, to be sure that they were ready for the new labels. He would ensure that the keypunch department understood the format of the records to be punched. Also, since the system is just being started, he would have to ensure that all of the mailing lists for which new cards must be made were available to the keypunch department, and the preparation of the mailing list cards would need to be started.

When the programming is completed, the system must be fully tested prior to implementation.The testing portion is usually done in several phases. In this system, the programmer is to test the program to be sure it prints the labels as he understands they are to be printed. The systems analyst would then give the programmer some systems test data to ensure that the program works correctly.

After the systems testing is completed, the analyst must again return to management and the user to get their final buy-off of the system. He can show them the input, the output, and the steps which are taken in the system. At that time, a final approval by the user and management would allow the analyst to schedule the implementation date.

Prior to implementation, the systems analyst must also have prepared any necessary operations documentation, which could include such things as instructions to the computer operator for running the program, instructions on what to do with the labels after they are printed, and instructions on what to do with the card file after the run is complete. Other documentation would have to go to the user explaining any information which they may need to know to enable the system to operate smoothly.

Case Study - Phase 5 - Systems Implementation and Evaluation

The system will finally be implemented on a given day, and the mailing labels will be prepared by the computer. After a system is implemented, there must be procedures for the analyst to audit the system to be sure that the system is operating properly. There must be feedback from the users to determine what they do not like, if anything, about the system. There should also be procedures for correcting any wrongs within the system as immediately as it is feasible.

After a system is implemented, invariably some user will want more information or will want the information in a different format, etc. Therefore, there must be procedures for making changes to the system. For example, in the mailing list program, it may be desired by management to get a report of all of the persons on the mailing list in alphabetical order. It will be recalled that the labels are printed in ZIP code sequence. Therefore, the analyst must schedule changes to the system in order to accomodate the additional requests.

It should be noted that changes to a system after it is operational are normally more expensive than features which were included in the original system design. Thus, it is one of the major functions of the systems analyst in the design phase of the system to anticipate some of the requests which may be forthcoming from users after the system is implemented and suggest these features at the time of design. In many cases, time and money are saved by the initiative of the systems analyst at the time of design. For example, in the case study if the analyst could have anticipated the need for an alphabetical listing, the procedures for preparing this listing could have been recommended in the original systems design.

The last step in the systems cycle is to conduct a post installation evaluation. This evaluation should be conducted after the system has gone through several processing cycles. During this phase, the analyst should determine if the mailing lists are being prepared as efficiently as projected, if the system handles all possible cases, and if there are any problems or complaints relative to the operation of the system itself.

Case Study - Summary

The previous mailing list system is a very small system. It requires, however, as do all systems, that the basic steps be followed in order to have a successful system. When steps are bypassed and are hurried, there are normally problems with the system. The old adage about ''never have time to do it right the first time, but there is plenty of time to do it right the second time'' is particularly applicable to systems analysis. There is no substitute for designing and implementing the system in the correct manner. In the subsequent chapters, some of the things of which an analyst must be aware will be discussed, together with many of the tools with which the analyst works.

THE SYSTEMS ANALYSIS PROFESSION

As noted earlier, a system is any set or sequence of procedures to be used to reach a given objective. Thus, long before electronic data processing and computers, there were employees within an organization called ''systems analysts'', whose duties included the analyzing and designing of the procedures of a company for the purpose of improving the efficiency of operations. This text, although applicable in a general way to these types of systems analysis jobs, will deal primarily with the techniques, skills, and methods of the data processing systems analyst, that is, the type of systems analyst that is concerned with the implementation of systems on the computer. It is, therefore, important to be aware of the attributes which an individual in the data processing systems analysis field should possess.

In attempting to define what attributes a successful systems analyst must possess, there are many different criteria which may be used. These criteria vary greatly, depending upon the size of the company, the number of employees, and the type of computer system. There are, however, certain guidelines which may be specified.

1 - Educational Background - It is stated by many authorities that the systems analyst should have some type of college degree, preferably in the business or computer science areas. In addition, some companies would prefer that a systems analyst have a master's degree. In actual practice, however, it is found that fewer than 50% of the systems analysts have a college degree, although a much higher percentage have some college or university education. Thus, it should be pointed out that although a college degree is not a prerequisite for employment as a systems analyst, a degree is desirable and often required by large organizations.

2 - Experience - Although a systems analyst must have reasoning capabilities which may be sharpened by a formal education, one of the most critical requirements for a successful systems analyst is experience in the subject field where he will be defining and designing systems; that is, the analyst should be familiar with the terminology, practices, and procedures of the business environment in which he will be working.

For example, if an analyst is to design an accounts payable system, it is desirable that he have some formal courses in accounting on a college level; but it is equally as critical that he realize the function of the accounts payable department within his company and that he be aware of the methods and practices of that particular company or one quite similar to it. Thus, for the analyst to design a successful accounts payable system, he must be familiar with the company's disbursement procedures, that is, writing checks to creditors, and all of the related accounting functions of the department. As will be shown in subsequent chapters, some of this information will be available when a proper investigation is performed; but without a firm awareness of the functions of accounts payable systems, the analyst will have a difficult time coming up with a feasible design for a computerized system.

In addition to experience within a particular area, it is quite desirable that the analyst have a strong general background in all areas of business procedures and applications. It may be, for example, that an accounts payable system will eventually be a part of a general ledger system which keeps track of all moneys spent and received. Therefore, the analyst should have a broad background in accounting and related business functions.

Thus, it will be found in many installations that senior analysts are "well-seasoned" in business functions and have commonly had five to seven years experience in business applications prior to obtaining their current job classification.

In many data processing departments, programmers who have done an outstanding job in programming are "promoted" to systems analysts jobs, which historically have a higher "social" standing and pay more money. It has not, however, been conclusively shown that ex-programmers make better systems analysts. In many cases, making a programmer a systems analyst takes one who is performing very well at one job and places him in another job in which he is not able to perform as well. This is not to say, however, that it is not important for the analyst to have a good knowledge of programming and data processing. In too many cases, an analyst with little or no programming knowledge is unable to design practical systems from the standpoint of programming and implementation. In general, however, it can be said that it is not a prerequisite that an analyst have a background as a programmer.

It should be re-emphasized, however, that the analyst should have a broad background in business and data processing, and this general knowledge should include a thorough familiarity with computer hardare and software capabilities.

3 - Personal Attributes - As with any given job, there are always certain personal attributes which are desirable for the systems analyst. A systems analyst must be many things to many people.

A good portion of the attributes of the successful systems analyst pertain to the fact that he is dealing with many different people at all levels of a company structure. For example, it is not at all uncommon for an analyst to interview a vice-president of a company in the morning and a machine shop foreman in the afternoon. Thus, one of the most important attributes which an analyst can possess is the ability to get along with people and the ability to converse with all levels of individuals within a company. Unlike some jobs within a data processing department, an analyst must be "people oriented." He is daily dealing with people both within and outside the data processing department. In addition, the systems which he designs are to help people in performing their tasks within the company, so he must be aware of "people problems" as well as mechanical and procedural problems in order to design an effective system.

In addition to working with people outside the data processing department, such as high level management, the systems analyst must deal with persons within the data processing department; specifically, programmers, operators, and other analysts who may be on the systems "team" for a given project. Thus, the analyst must contend with all types of problems from within the data processing department and must have the ability to communicate and the diplomacy to alleviate any potential or real problems concerning people.

In addition to dealing with people, of course, the analyst is concerned with designing the procedures which will constitute a system. He must, therefore, have the intelligence and ability to see the system in its overall status as well as being able to isolate and define minute portions of the system. Most competent systems analysts are extremely resourceful and ingenious. They are inquisitive and enjoy exploring alternative solutions to problems while keeping an open mind and remaining flexible until the time comes to make a determination as to the proper way to do a certain job.

After the exploration and fact gathering has been done, and a system established in the mind of the analyst, it remains for him to communicate the mechanics of the system, the advantages of his design, the methods of programming and implementing the system, and other considerations to upper level management, lower level management, programmers, and actual users. Thus, one of the prime requisites for a successful analyst is that he have the ability to communicate clearly and concisely. This communication is both oral and written and, as noted, will be with all levels and skills of personnel. It is contended by many data processors that most problems which develop in data processing systems are there as a result of faulty or misunderstood communications. This goes all the way from the analyst misunderstanding the desires of the president of the company to the computer operator misunderstanding the written directions of the analyst for the processing of the computer run. The ability to communicate is, therefore, one of the necessary ingredients for a successful analyst.

Working Environment and Procedure

As noted, the data processing systems analyst normally works in the systems analysis and design department of the data processing section. Within this department, most systems projects are staffed by more than one analyst. Although this is not true in small installations or for very small projects, it is normally true that there will be a project manager and one or more systems analysts working on the project.

The function of the project manager, who is usually a senior systems analyst, is to oversee the development of the system from the time it is recognized there is a problem to the time the problem is solved, in many cases by implementing a data processing system. The project manager will many times be concerned with scheduling the various phases of the systems investigation and design, and ensuring that the project is on schedule. He may, of course, also be engaged in the actual systems work together with the analysts on his "team"; but normally, the analysts on a systems team will bear the load in terms of the fact gathering, the systems design, and the implementation of the system.

SUMMARY

As can be seen from the previous discussion, the functions performed by a systems analyst are many-faceted. In addition, the types of systems which can be designed and implemented vary over a wide applications area. Regardless of the system being designed, however, the analyst must perform his functions with disciplined, well-defined methods if the system is to be implemented successfully. The following chapters detail many of the considerations which must fit into these methods in order to have successful data processing systems.

STUDENT ACTIVITIES—CHAPTER 1

REVIEW QUESTIONS

1. Define the term "Procedures."

2. Define the term "System."

3. What are the two basic types of business organizations? What are the primary differences between the two?

4. List the basic types of subsystems common in any information processing system of a business.

   1)
   2)
   3)
   4)
   5)
   6)

5. What is the function of an organization chart?

6. Explain the authority relationship in a "line organization."

7. Explain the "staff" function in a Line and Staff organization.

8. List five functions of a Systems and Procedures Department.

1)

2)

3)

4)

5)

9. List the three functions common to most data processing departments.

1)

2)

3)

10. Where does the data processing department commonly fit into the business organizational structure?

11. List and briefly explain the five phases in a systems project.

Phase 1

Phase 2:

Phase 3:

Phase 4:

Phase 5:

DISCUSSION QUESTIONS

1. The typical career path in data processing has been from programmer to systems analyst, that is, after gaining experience, programmers are often ''promoted'' to systems analysts. The following quotations offer two divergent views on this philosophy:

   • *Programmers are ''machine-oriented'' workers and typically like to work alone solving complex logical problems and coding and debugging computer programs...Systems analysts are ''people-oriented'' workers who must learn to communicate and work effectively with people...in my company programmers are never promoted to systems analysts. We take individuals knowledgeable in the industry and train them to be systems analysts...you don't have to be an expert programmer to be a systems analyst...in fact, programmers and systems analysts are different personality types.*

   • *The biggest mistake a company can make is to try and make a systems analyst out of someone knowledgeable in the business but with no technical programming background...I hired one so-called systems analyst who didn't even know how many characters could be recorded in a punched card!...there is no way an analyst can design an effective system unless he is thoroughly familiar with the internal operation of the computer and related peripheral devices... this knowledge can best be obtained by spending a few years as a programmer...even operations experience is helpful.*

   Analyze and comment on each of the above quotations.

2. Discuss the advantages and disadvantages of having the data processing manager report to the chief financial officer of the company, such as the vice-president of finance.

3. Present an argument for and an argument against the following proposition: ''Since the heart of a business is now or will be centered around the data processing operation, members of data processing management will be knowledgeable in all phases of the business and a company should draw exclusively from data processing management to fill vacancies in top-level management, such as presidents and vice-presidents.''

CASE STUDY PROJECT—CHAPTER 1
The Ridgeway Company

INSTRUCTIONS

The Ridgeway Company Case Study is presented throughout the textbook to give the student ''practical'' experience in the analysis and design of a business system. At the conclusion of each chapter a segment of the case study is presented. All assignments in the case study should be completed prior to undertaking the study of the following chapter.

INTRODUCTION

The Ridgeway Company is an organization whose main business is centered around the purchase and development of recreational land. Sales during the past year have been in excess of eleven million dollars. The Ridgeway Company recently acquired a large recreational complex containing both a tennis club and a golf course. Named the Ridgeway Country Club, the facilities include 50 lighted tennis courts, an 18 hole golf course, a ''Pro Shop'' which sells tennis and golfing supplies and related items, a clubhouse containing a restaurant and bar, and other recreational facilities including a swimming pool and exercise room.

Ridgeway Country Club

MANAGEMENT PERSONNEL

Senior level management of the Ridgeway Company includes George Ridgeway, President; Senior Vice-President, Harlin Hill; and three Vice-Presidents who report directly to Mr. Hill. These Vice-Presidents include: William Howell, Vice-President, Finance; Chester Langely, Vice-President, Administration, Research and Development; and Holden Archer, Vice-President, Operations. Arnold Logan is the company's land development consultant and serves in a staff capacity, reporting directly to the President.

STUDENT ASSIGNMENT 1
PREPARING AN
ORGANIZATIONAL CHART OF TOP MANAGEMENT

1. Prepare an organizational chart of the top level management of the Ridgeway Company. NOTE: Retain this chart for future assignments.

DATA PROCESSING PERSONNEL

The Ridgeway Company recently acquired a System/370 Model 125 computer system including a card reader/punch, a high speed printer, and both magnetic tape and magnetic disk storage devices.

The Data Processing Department consists of Lee Ura, Director of Data Processing; a Manager of Systems Analysis and Design; a Manager of Programming; and an Operations Manager.

The Director of Data Processing reports directly to the Vice-President of Finance, William Howell. The Managers of Systems Analysis and Design, Programming, and Operations report to the Director of Data Processing.

STUDENT ASSIGNMENT 2
PREPARING AN ORGANIZATIONAL
CHART OF THE DATA PROCESSING DEPARTMENT

1. Prepare an organizational chart of the Data Processing Department. Use the organizational chart begun in Student Assignment 1. Retain this chart for future assignments.

PHASE I

INITIATION OF THE SYSTEMS PROJECT -
PRELIMINARY INVESTIGATION

INITIATION OF THE SYSTEMS PROJECT - PRELIMINARY INVESTIGATION

CHAPTER 2

INITIATION OF THE SYSTEMS PROJECT PRELIMINARY INVESTIGATION

2

INTRODUCTION

With installation of an electronic data processing system, the data processing manager or the manager of the systems department is faced with the problem of which applications within the business organization are to be converted to computerized processing.

This decision requires a careful understanding and review concerning the objectives, needs and function of the business, the time and costs of the proposed projects, and the computer hardware, software, and personnel available to implement the proposals.

ORIGINATION OF SYSTEMS PROJECTS

Why is there a need for systems studies within a business organization? Historically, the origination of systems projects can be traced to four basic sources. They include:

1. REQUESTS TO SOLVE IMMEDIATE PROBLEMS - Unfortunately, in many installations a great portion of the time spent by systems personnel is in "fire-fighting," a term used in systems work to define that type of systems activity that is devoted to the solving of immediate problems within the organization. Under the fire-fighting approach, the systems analyst is called upon to review problems as they arise within a business organization and recommend any changes necessary to solve the problem, including such action as new controls or the redesign of the system. Although the "fire-fighting" approach tends to solve problems, when this technique becomes the basis of all systems activity, the entire business system becomes a "patchwork" of corrected or improved systems with little consideration to the business organization as an integrated whole.

2. DEPARTMENTAL REQUESTS - Another source of systems projects is through requests from the various departments within the organization. As various departments become aware of the capability of the computer system, numerous requests are likely to be received for computer services. For example, a new sales analysis report may be requested, mailing labels may be needed, or an updated inventory report might be desired. It is expected that the systems group will carefully evaluate all such requests and recommend the course of action to be taken. A definite procedure should be established to provide for requests for a system study or for the preparation of a computerized report. This should take the form of a System Services Request Form so as to render some degree of formality to what otherwise may be a phone call, a memo, or a verbal request. The System Services Request Form provides a starting place for each system proposal.

3. MANAGEMENT DIRECTIVE - Another source of systems projects is a top management directive. This directive may be brought about by new governmental reporting regulations, an additional activity being established within the business, etc. Whatever the nature of the assignment, the systems department in most organizations can expect a portion of its effort to be devoted to carrying out projects initiated through management directives.

4. DESIGN OF NEW SYSTEMS - The fourth source of systems projects is the one which most systems personnel think is best, and this is the systems department's own recommendations for projects which it feels will contribute to the company's overall success. The success of this approach, of course, depends upon the department's knowledge of the operations of the business and its needs. Although this approach may offer a high measure of accomplishment because the system which is developed may be an integrated and efficient system, there have been numerous examples of systems departments designing large scale "management information systems" which have a much higher degree of sophistication, and hence cost, than is necessary for the effective operation of the business. Many of these large scale information systems, after hundreds and even thousands of man hours of effort, have never been successfully implemented. Thus, management has become increasingly critical and wary of large projects originated by systems departments which are designed to solve all of management's problems.

EVALUATION OF SYSTEMS PROJECTS

Within most organizations, the demand for systems projects and data processing services will exceed the capacity of the systems department to carry out the requested activities. Thus, the systems manager is faced with the task of evaluating all requests for services, rejecting some, and selecting and establishing priorities for those requests which appear to offer the greatest benefit to the effective operation of the business.

The ultimate goals of all systems projects in a profit making organization is to reduce costs and increase profits. Therefore, the systems manager must be aware that some requests are not economically feasible and should not be considered for further study. For example, requests for the preparation for a single report, that is, a report that is needed only once, may require extensive input design, preparation, and programming effort. "One-time" reports which are needed immediately can often be more efficiently prepared using manual methods rather than using a computer. Thus, the preparation of such a report may not be economically feasible. When requests of this type are received by the systems manager, he must evaluate the cost and benefits to the company and, perhaps, reject the request for system services at this time and return the request to the user department.

After rejecting those requests for system services that do not appear to be economically feasible at this time, the systems manager is then faced with the task of determining which of the system projects warrant immediate, further study. For example, the systems manager may receive a request to investigate errors occurring in payroll, a request to design a new inventory control system, a request for sales analysis reports and a request for the computerized billing and aging of customer accounts. With limited personnel resources in terms of available systems analysts on the staff, which of the projects would be considered for further study? By what criteria does the systems manager evaluate and decide upon the priorities for the many and varied requests which the department is likely to receive? Under normal circumstances the basic rule is to favor those projects which provide the greatest benefit to the company at the lowest cost in the shortest period of time. However, many factors must influence a decision on the selection and scheduling of systems projects. These questions include:

1. Will the proposed system or changes in the current system reduce costs? Where? When? How? How much?

2. Will the proposed system or changes in the current system result in better information or produce better results? How? Are the results measurable?

3. Is the problem urgent?

4. Will management be receptive to the proposed undertaking?

5. How large is the project and how long will it take to implement? Will the results be lasting?

6. Can the systems effort and conversion costs be justified?

7. Is the current staff capable of carrying out this project? Will additional personnel be required?

8. Is the project feasible with existing hardware?

It is extremely unusual for positive answers to be obtained to all questions. Some proposed systems may not reduce costs but will provide more timely management reports. Other systems may substantially reduce costs but require the purchase or lease of additional hardware. Some systems may be extremely desirable but require several man years of effort. In addition, the decisions which are made concerning a system often involve the weighing of intangible factors, such as improving customer service or acquiring more accurate information. In many cases, it is the intangible factors which will determine whether a system proposal is considered for further study which could lead to the design and implementation of the system.

It is at this point that the systems manager must draw upon his knowledge of the business, data processing, and computer programming, and be able to arrive at a decision with a minimum amount of information; as a decision must be made as to which of the requests for system services are to receive priority and further study.

THE PRELIMINARY INVESTIGATION

After a decision has been made relative to the priority of systems projects, the systems manager will normally direct that a PRELIMINARY INVESTIGATION be undertaken. A preliminary investigation involves one or more analysts investigating the request for system services for the purpose of determining the true nature and scope of the problem and whether a detailed system investigation and analysis of the current system is desirable. For example, if the request for system services concerned a payroll system which was not functioning properly, the analyst would perform a preliminary investigation to determine if the complaints were justified. If the results of the preliminary investigation by the analyst indicated that a problem did exist within the payroll department that could be resolved by a redesign of the current system, the analyst would then submit a report to management recommending a further detailed study and analysis of the payroll system. The **preliminary investigation** is sometimes called the **problem definition** phase of the systems project, for the primary purpose of this phase of the project is to define the true nature of the problem specified in the request for system services.

Objectives of a Preliminary Investigation

The preliminary investigation is conducted when there is the possibility that the problem presented in the systems request can be solved only with some major changes to the current system. It must be noted that the preliminary investigation is not a comprehensive data gathering activity. Rather, the preliminary investigation is aimed at gathering enough information to determine if the information or problems specified in the request for system services warrants a detailed systems investigation and analysis.

The preliminary investigation is not designed to define all of the problems which may exist in a system or to propose an absolute solution to the problems. Instead, the analyst should accomplish the following:

1. Obtain an understanding of the true nature of the problem.

2. Be able to define the size and scope of the proposed systems project.

3. Be able to state the benefits that are likely to occur for the company if problems stated in the request are resolved or if the system is improved.

4. Be able to specify time and money estimates for a detailed investigation and analysis of the current system.

5. Be able to present a report to management indicating the nature of the problem and recommendations relative to the desirability of conducting a full-scale investigation and analysis of the current system.

Understanding the Problem

One of the key factors in the preliminary investigation is an attempt to gain an understanding of the true nature of the problem and the reason for the request for the study. In many cases, the "stated" problem in the request for services may not be the "real" problem. For example, a request for additional hardware, because the current computer system is not "fast enough," may conceivably not be the "true" problem. The true problem may be poor systems design, improper scheduling, or even poor operations and programming. The request for a system for handling complaints may be a problem in quality control in the manufacturing plant.

Thus, the solution to the problem can only take place after the problem has been properly defined. One of the activities to be undertaken in the preliminary investigation is to determine what this problem is. This is normally accomplished through effective interviewing and overall review of existing procedures.

Defining the Scope of the Project

As the true nature of the problem described in the request for services is being determined, it is also necessary to define the scope of the problem. The scope of the systems project is directly affected by whether the objective is to "patch" or improve an existing system or to develop the best possible new system.

The scope of the project is the limit to which the analyst can carry his investigation and a limitation on the solutions which can be imposed. If this is not done, there may be the temptation to investigate and redesign an entire system and related subsystems. In addition, there may be some problems within a system which can be tolerated and which are not economically feasible to solve. If boundaries are not imposed, a project scheduled for completion in several weeks could take months or even years to complete.

The primary basis for defining the boundary is a precise statement of the problem and a statement of the objective of the systems project. For example, the statement of the problem that the "payroll is not being produced accurately" could lead to significantly different approaches relative to the scope of the project than the statement "union dues are not being properly deducted from the Employees' pay." Likewise, a statement that the objective of the proposed systems project is to "redesign the payroll system" is significantly different than a statement that the objective of the project is to "correct the errors occurring when union dues are deducted."

The definition of the problem to be solved and the definition of the scope of the study and objectives of the systems project are critical to the process of designing the proper solutions to the problem. In addition, they must be specified before a project can be submitted to management for approval. It is the job of the analyst during the preliminary investigation to determine the scope and boundaries of the problem and recommend approaches to the solution of the problem which offer the greatest benefit to the company. Too often analysts, during the preliminary investigation, have a tendency to recommend a complete study of the entire system when a less costly approach, in terms of time and effort, could be equally effective in solving the company's problems.

Defining the Costs and Benefits

In the preliminary investigation the analyst should also consider the cost and benefits likely to occur to the company if the problem defined is considered for further study. The ultimate objective of any systems project is the solution of the problem specified in the request for system services; however, if the improvement of the existing or proposed system cannot be cost justified, further study should not be undertaken.

Estimates for Detailed Investigation and Systems Analysis

A further result of the preliminary investigation should be the ability of the analyst conducting the investigation to specify the time and money estimates for a detailed investigation and analysis of the current system. As was noted previously, the preliminary investigation is used to determine the problem and the scope of a proposed system. With these in mind, the analyst should determine the time which it will take to perform an in-depth investigation and to analyze the results of this investigation. The following factors should be considered when making this estimate:

1. What information must be obtained and what is the volume of information which must be gathered and analyzed?

2. What sources of information are to be used and what is the difficulty in gathering and analyzing the information?

3. How many people are to be interviewed and how much time is required for both the analyst and the people to be interviewed?

4. How much time and how many people are required to correlate the information gathered and prepare a report indicating the findings and the alternative solutions to the problem?

Presentation to Management

As a result of these time, personnel, and money estimates, a report will be presented to management which will then make a decision relative to whether the detailed investigation and systems analysis should take place.

CONDUCTING A PRELIMINARY INVESTIGATION

There is a definite series of activities and events which the systems analyst should follow in conducting a preliminary investigation regardless of the size of the systems project. In order to conduct the investigation, the analyst should normally proceed with the following steps:

1. Obtain a letter of notification and authority authorizing the preliminary investigation.

2. Obtain an organizational chart of the areas of the company undergoing study to determine those persons to be interviewed.

3. Conduct interviews to obtain the necessary information.

4. Review current systems documentation to the extent necessary to understand the operations of the current system.

5. Correlate the information obtained so that a presentation can be made to management relative to the feasibility of initiating a full-scale investigation of the current system with the objective of determining alternate plans of action which will result in an improvement in existing methods and procedures.

Letter of Authority

A systems project often results in a significant change from the usual mode of operation within a company and resistance is likely to be encountered in the problem definition phase of the study. In addition, the study of the system will often require the crossing of organizational boundaries within a company and will require the cooperation of not only a variety of department heads within the company, but also a broad spectrum of personnel from various levels within the business.

Therefore, before beginning the study and analysis of the existing system, the analyst should obtain a "letter of authority." A letter of authority is simply written permission from top management to perform the study. The letter of authority should contain a clear statement that a study is taking place, that the study may be crossing organizational lines, and that cooperation is requested of all employees.

This direct evidence of top management's support will do much to lessen the resistance that could occur from those affected and required to participate in the study. It seems to be an inherent characteristic of human nature to oppose "change," and the very nature of the systems project is to bring about change of some type. Thus, the analyst must be clearly aware of the possibility of resistance from those involved in the study when he seeks information to determine the nature of the problem and current operating procedures.

The letter of authority is also important as a means of documenting management's support of the preliminary investigation and can be relied on by the systems department in case any dispute concerning the system study should arise at a later time.

Organizational Charts

In many instances, an analyst assigned to conduct a preliminary study of a system will not know the organizational structure of the departments around which the problem centers, nor in some cases will he know the structure of the top-level management concerned with the problem. Therefore, the analyst should procure organizational charts of any portion of the company which may have an involvement with the problem presented in the request for services. The organizational chart should provide the analyst with detailed knowledge about the personnel and structure of the company so that he may interview selected individuals involved with the system under study.

Conducting Interviews

The primary method of obtaining information in the preliminary investigation is the interview. It is essential that the analyst schedule interviews with those individuals within the scope of the study who can assist in defining the problem and provide information relative to the true nature of the system.

The interview is an important method of fact-finding and should be approached as logically as computer programming. In determining the approach to be taken, it is necessary to remember that the purpose of the interview, and the preliminary investigation in general, is to uncover facts about the currently existing system. Its purpose is not to convince others that a new system is needed. Thus, it is essentially the role of the analyst in these interviews to be a listener. He must attempt to separate facts from opinions while encouraging a free flow of ideas.

There are five basic steps in conducting a meaningful interview. These steps include the following:

1. Determine who to interview.

2. Set objectives for each interview—determine precisely the information needed from each person interviewed.

3. Define specific questions to be asked during the interview.

4. Conduct the interviews as planned. Be sure to make appointments prior to interviewing.

5. Record and evaluate the interview.

When conducting interviews during the preliminary investigation, the analyst must be quite selective in who he interviews. The preliminary investigation is not designed to learn every detail of a given system, so the analyst must not spend time interviewing every person who has anything to do with the system. At the same time, it is important that those persons with the most information be interviewed so that an accurate picture of the current system can be gained.

In general, during the preliminary investigation, the analyst will interview those persons in managerial or supervisory positions. These people should have sufficient knowledge of the system and the problems which have been encountered with it to give the analyst a good idea of the overall operation and problems. If necessary, however, the analyst should not hesitate to interview persons in different levels within the company in order to get a complete picture. When the detailed fact-gathering phase is begun, the analyst will interview persons at all levels of the organization in order to gain all the information concerning the system. During the preliminary investigation, however, interviews will normally be restricted to higher-level personnel.

Resistance to Change

It is at the interviewing stage that the analyst might encounter resistance from company employees. Some people have the misconception that the analyst is an ''efficiency expert'' whose purpose is to change existing procedures or whose prime objective is to ''get people fired'' and reduce personnel.

Resistance to change may take many forms. At one extreme, employees may merely question. At the opposite extreme, reaction may take the form of open opposition, rebellion, refusal to cooperate, and even attempts to thwart the implementation of the proposed new system.

Resistance to change can best be overcome by open communication with those involved relative to the purpose and objective of the study, with emphasis upon the fact that systems and procedures work is aimed at improving human endeavor. In addition to open communication and providing information to employees relative to the study, it is often quite possible to reduce resistance by seeking participation in the planning for the change from employees and supervisors by such activities as soliciting their suggestions for improvement of existing procedures. Personnel specialists point out that resistance to change will decrease to the degree that employees are able to have some ''say'' in the nature and direction of the change.

Reviewing Documentation

Although the interview is an extremely important method of obtaining information concerning a system and the manner in which it operates, there are many occasions when it is necessary for the analyst to further investigate the system by examining documentation which currently exists for the system. This documentation may take the form of written procedures, flowcharts, or other forms of written communication. The analyst should pursue these investigations until he has enough information to be able to determine the current operations and problems of the system and present the information to management.

Presentation of Findings

After the preliminary investigation has been completed, the results must be presented to those persons who can decide whether a detailed systems investigation is to take place. In many organizations, the presentations are made directly to management, which may consist of high-level management such as vice-presidents and departmental management.

The Systems Review Committee

In other organizations, the committee method of evaluating preliminary investigations is used. By using a committee to evaluate findings, rather than relying upon the judgment of a single individual such as a vice-president, there is the obvious advantage of drawing upon the knowledge and talent of a wide range of individuals with specialized talents within the business organization. For example, a typical evaluation committee could be composed of the Systems Manager, the Data Processing Manager, a Senior Analyst, a Vice-President of the area in which the system would have application, and one or more line managers who would be implementing and using the system.

On the basis of the information presented as a result of the preliminary investigation, the members of the committee express their views on the project and a decision is made relative to whether the full-scale investigation of the problems and related system will take place. Although the committee organization has the advantage of utilizing the talents and knowledge of various members within an organization, it should be noted that the responsibility for successful implementation of the system rests with the systems department. Therefore, those from the systems department who are members of the committee must ensure that all members of the committee are aware of any difficulties which may be encountered with a new system so that the systems department will not have the impossible task of attempting to implement a system which will not work.

When the presentation is made, it is imperative that the analysts present the findings and their recommendations in a manner which is straight-forward and clear to those listening to the presentation. Therefore, it is quite important that the analyst making the presentation plan the approach to be used.

The important point of the presentation to keep in mind is that it is to be used to communicate the findings of the preliminary investigation to the listeners. Communication is the interchange of information so as to produce a mental picture in the mind of the receiver which duplicates the picture in the mind of the sender. Therefore, when planning the presentation, the analyst must always bear in mind that his main task is to project his ideas and thoughts to the person to whom the presentation is being made.

The primary element in preparing for the presentation is defining the objectives of the presentation. As has been noted, the objectives of presenting the results of the preliminary investigation are to inform the listener of what is currently taking place within the system, to direct their attention to the problems which have been occurring within the system and why they have been occurring, and to present a recommendation relative to whether a full-scale investigation should take place.

After the results of the preliminary investigation have been presented, the management personnel or the committee who heard the presentation will make the decision on whether further work will be done on the proposed system.

CASE STUDY - JAMES TOOL COMPANY

In order to illustrate the techniques which are utilized by a company to initiate a systems study and to design and implement a solution, a case study of the James Tool Company will be utilized. In order to solve this problem, the techniques for systems analysis and design will be presented in this and subsequent chapters.

Case Study - The Facts

The James Tool Company was organized in 1964 by Howard James because he felt that there was a need in the tooling business for a company which could respond to the needs of smaller machine shops and manufacturing plants for tooling and dies which the shops could not afford to produce themselves.

James started with six employees and incorporated within the state of California. Because of clever marketing, together with producing a timely product at a very competitive price, James Tool Company has grown to 456 employees. These employees consist of five officers of the company, office and clerical help, and foremen and workers within the shops which produce the tools.

An organization chart of the top-level management is illustrated below.

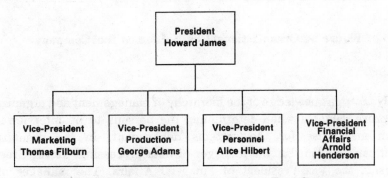

Figure 2-1 Organization Chart - James Tool Company

As can be seen from Figure 2-1, the four vice-presidents report to the president of the company, Howard James. Each of the vice-presidents have organizations below them. A complete layout of the high-level management positions within James Tool Company is illustrated in Figure 2-2.

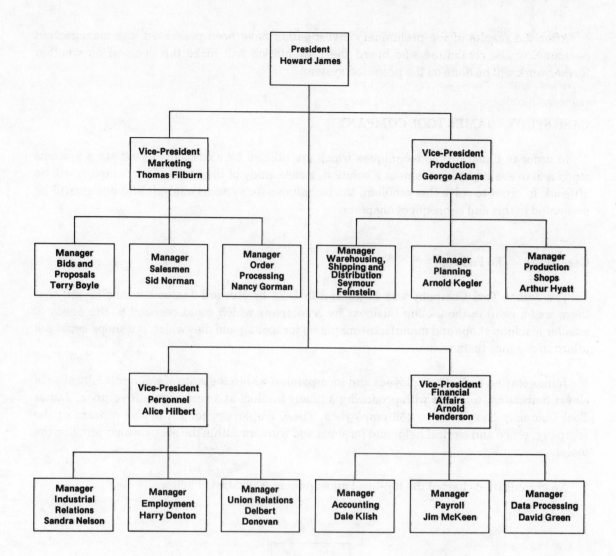

Figure 2-2 Organizational Chart of James Tool Company

As previously noted, a knowledge of the hierarchy of management and organization within a company may prove valuable when determining the persons to be interviewed concerning problems which may arise for the systems department. In the organizational chart in Figure 2-2, it can be seen that the manager of data processing, David Green, reports to Arnold Henderson, the Vice-President of Financial Affairs. The manager of the payroll department, Jim McKeen, also reports to Henderson.

Case Study - Structure of the Data Processing Department

The organization of the data processing department consists of the data processing manager, David Green, a systems and programming manager, and an operations manager. The organizational structure is illustrated below.

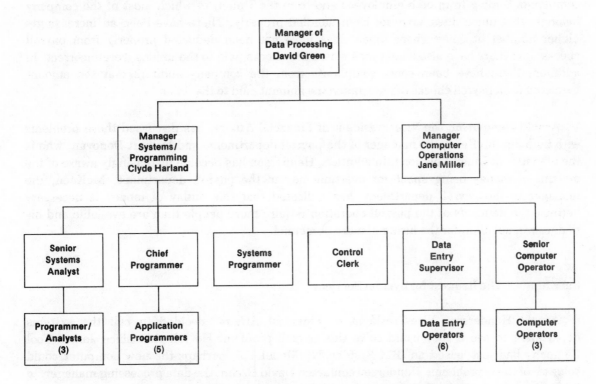

Figure 2-3 Organization of Data Processing Department

As can be seen from Figure 2-3, the senior systems analyst, the chief programmer, and the systems programmer report to Clyde Harland, the manager of Systems and Programming. The primary function of the senior systems analyst is to design systems for the data processing department, while the programmer/analysts who work for him spend their time both designing systems and writing application programs. The application programmer's primary work is in writing business application programs which are required to implement the various business systems. The operations manager and her people are primarily concerned with running the computer and performing all of the input, output, and data handling procedures which must be processed in the data processing department. All of these people must work together in order for the department to function smoothly, and all are of concern to the systems analyst when designing a business data processing system.

Case Study - The Payroll System

The payroll system of the James Tool Company was designed by Jim McKeen, the current manager of the payroll department, in 1968, when there were 42 people employed by the company. The system has functioned reasonably well since then, but lately there have been complaints coming from both employees and from the Union, to which most of the company belongs, that union dues have not been handled properly. There have been an increasingly higher number of times where union dues have not been deducted properly from payroll checks and there have also been times where the checks paid to the unions were incorrect. In addition, there have been some complaints from the company auditors that the amount deducted from payroll checks did not match the amount paid to the union.

Arnold Henderson, the Vice-President of Financial Affairs, has discussed these problems with both Jim McKeen, the manager of the payroll department, and Delbert Donovan, who is the manager of Union Relations. In addition, Henderson has become increasingly aware of the amount of money being spent for overtime pay in the payroll department. McKeen, the manager of the payroll department, has indicated that this outlay of money is necessary because the demands of the payroll operation require more people than are available and his budget will not allow for the hiring of more personnel.

Case Study - The Request for System Services

Arnold Henderson, Vice-President of Financial Affairs, decided to call the systems department to see if they could solve this payroll problem. He felt that since James Tool Company had just leased an IBM System/370 Model 125, perhaps the new computer could solve all of these problems. Henderson contacted David Green, the data processing manager, to inform him that he had called the systems department.

Upon the request of the systems and programming manager, Clyde Harland, the "Request for System Services" illustrated in Figure 2-4 was completed by Arnold Henderson and forwarded to the systems department for action.

It can be seen that Henderson has specified that there is some type of problem with the payroll system and he is requesting assistance. In no way has he pinpointed the cause of the problem nor has he proposed a solution. This is a typical type of request which the systems department can expect from those outside the department who know of a problem but do not know either the cause or the solution to the problem.

REQUEST FOR SYSTEM SERVICES		
DATE September 10	**PAGE** 1 OF 1	
FROM Arnold Henderson **TITLE** Vice-President **DEPT.** Financial Affairs	**TO** Clyde Harland **TITLE** Manager **DEPT.** Systems and Programming	
SUBJECT Payroll System		

Brief Statement of Problem:

       There have been some problems in our payroll system. In particular, an extraordinary amount of overtime has been paid to Payroll Department employees. In addition, some problems have been encountered in the deduction of union dues from employees' checks, and the unions have been complaining to me. I have spoken with Jim McKeen, manager of the Payroll Department, and he says the only problem is lack of personnel--he wants two more people. I think, however, that there may be more to it than that, so I would like you to look into the situation. Please contact me at your earliest convenience.

A. Henderson

A. Henderson

AH:sjm

To be Completed by Systems Department

DATE:

ACTION:

Figure 2-4 Example of Request for Systems Service

Case Study - Is there a Problem?

Upon receiving the Request for System Services from Arnold Henderson, the Vice-President of Financial Affairs, the manager of the systems department, Clyde Harland, must take steps to determine if a preliminary investigation is required. As noted previously, the manager of the systems department will normally review the Requests for System Services and reject those requests which either would be impossible to solve based upon present or foreseeable capabilities of the data processing department or do not present a problem which would merit the expense of a preliminary investigation.

When Harland received the Request for Services from Henderson, he contacted David Green, the manager of the data processing department, and they decided that it may well be that the systems department could find some of the problems in the payroll system, especially since it was somewhat antiquated. Therefore, they advised Henderson, the Vice-President of Financial Affairs, that they would recommend a preliminary investigation of the payroll system. The memo which Harland sent to Henderson to this effect is illustrated in Figure 2-5.

M E M O R A N D U M

DATE:	September 14
TO:	Arnold Henderson, Vice-President, Financial Affairs
FROM:	Clyde Harland, Manager, Systems and Programming
SUBJECT:	Preliminary Investigation, Payroll System

From the information you have supplied to us, it would appear that there is a good probability that our department would be able to assist you in recommending solutions to some of the problems which have been occurring within the payroll department concerning the union dues and, also, the problems with overtime.

Therefore, I suggest a preliminary investigation be undertaken by our department. I would like a letter of authority from you to all department heads authorizing this investigation. Our department can begin the study two days after the letter of authority is distributed to department management.

Clyde Harland

Clyde Harland

CH:sjm

Figure 2-5 Memorandum from Systems Department

Upon receipt of this memorandum, Henderson called Harland and informed him that he would distribute a letter of authority for the authorization of a preliminary investigation for the payroll system. The letter of authority would be sent to the department managers on September 15.

Case Study - Letter of Authority

Clyde Harland, manager of the systems department, assigned two analysts, Don Mard and Howard Coswell, to the task of conducting the preliminary investigation of the payroll system. Therefore, before the letter of authority was written, Harland informed Arnold Henderson, the Vice-President of Financial Affairs, of his choice for the analysts. The following letter was then distributed by Henderson.

M E M O R A N D U M

DATE: September 15
TO: Company Personnel
FROM: Arnold Henderson, Vice-President, Financial Affairs
SUBJECT: Payroll System Investigation, Letter of Authority

At my direction, Mr. Howard Coswell and Mr. Don Mard of the Systems Department are conducting an investigation into the payroll system currently in operation. This investigation is to determine where improvements can be made so that some of the problems which we have been experiencing can be rectified.

These gentlemen may be contacting you or people who work for you in order to arrange for interviews, to observe current procedures, or for other fact-gathering reasons. Your complete cooperation is requested so that an accurate accumulation of data will take place.

Thank you.

A Henderson

A. Henderson

AH:sjm

Figure 2-6 Example of Memorandum from Vice-President

Note that the letter of authority is not at all detailed but merely gives high-level management approval for the preliminary investigation. Through this letter, the persons to be interviewed and their bosses are aware that if either of the analysts call for interviews or other reasons, there is complete approval. This will many times overcome any problems in terms of establishing interviews or other investigation procedures.

Case Study - Organizational Charts

To begin the preliminary investigation, Don Mard and Howard Coswell requested from Arnold Henderson, Vice-President of Financial Affairs, an organizational chart of the payroll department. Arnold Henderson suggested that the analysts contact Jim McKeen, manager of the payroll department.

Upon contacting Jim McKeen of the payroll department the analysts were informed that the personnel department had job descriptions for all his employees and that they should also have formal organizational charts of the company. When contacting the personnel department, the analysts were informed that personnel did not have any organizational charts, but they did have job descriptions for the employees in the payroll department. The manager of the personnel department, Alice Hilbert, suggested that the analysts contact Arnold Henderson, Vice-President of Financial Affairs—she was sure his office had organizational charts of those departments under his control!

From the above sequence of events, it can be seen that the process of gathering information may not be an easy task. In many organizations, standards and documentation, if available, are difficult to obtain.

From the information received from personnel and from additional contact from Arnold Henderson the following organizational chart was constructed.

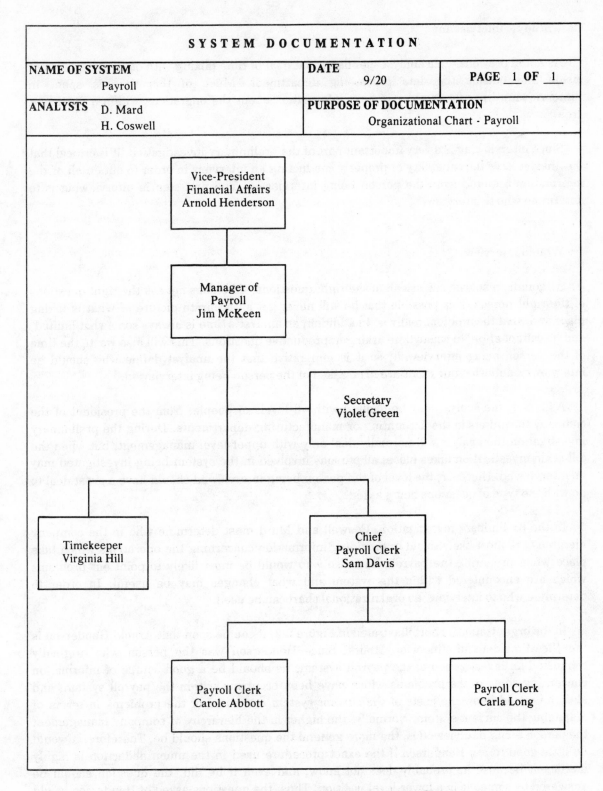

SYSTEM DOCUMENTATION		
NAME OF SYSTEM Payroll	**DATE** 9/20	**PAGE** _1_ **OF** _1_
ANALYSTS D. Mard H. Coswell	**PURPOSE OF DOCUMENTATION** Organizational Chart - Payroll	

Vice-President
Financial Affairs
Arnold Henderson

Manager of
Payroll
Jim McKeen

Secretary
Violet Green

Timekeeper
Virginia Hill

Chief
Payroll Clerk
Sam Davis

Payroll Clerk
Carole Abbott

Payroll Clerk
Carla Long

Figure 2-7 Example of Payroll Organizational Chart

Case Study - Interviewing

As noted previously, an analyst spends a great deal of time talking with other persons, both inside and outside the data processing department. Much of this time is spent in "interviewing" people in order to gain information about the operations within a particular area.

Since interviewing is a very important part of the preliminary investigation, it is critical that the analyst have the capability of properly conducting an interview in order to obtain all of the information required from the person being interviewed. The first step in interviewing is to determine who to interview!

Who to Interview

Although an analyst can ask all of the right questions, if he does not ask the right questions of the right person, it is possible that he will never get an accurate picture of what is taking place and what the problem really is. In addition, an analyst's time is always somewhat limited, and he cannot afford to waste time asking unproductive questions. This will also waste the time of the person being interviewed, so it is imperative that the analyst define who should be interviewed and what questions are to be asked of the person being interviewed.

As noted, the analyst may be dealing with all levels of people, from the president of the company to workers in the accounting or manufacturing departments. During the preliminary investigation the analyst will normally deal only with upper-level management, but when the full-scale investigation takes place, all persons involved in the system being investigated may be interviewed. However, the level of the person being interviewed should have a great deal to do with the type of questions being asked.

In the preliminary investigation, Coswell and Mard must determine who in the company hierarchy are most likely to aid in providing information concerning the operations which take place when preparing the payroll, and also who would be most likely to point out problems which are encountered within the system and what changes may be useful. In order to determine who to interview, an organizational chart can be used.

In the organizational chart illustrated in Figure 2-2, it can be seen that Arnold Henderson is the Vice-President of Financial Affairs. Since Henderson was the person who originally requested an investigation of the payroll system, he should be a good source of information concerning some of the problems which have been occurring within the payroll system; and also, he would know the costs of the current system and some of the problems in terms of managing the current system. Normally, the higher in the hierarchy of company management the person being interviewed is, the more general the questions should be. Therefore, it would do little good to ask Henderson if the exact procedure used in the union deduction is highly inefficient because he probably does not know; and even if he did, the question should be answered by someone in a lower-level position. Thus, the questions asked of Henderson would likely be more general in nature.

Another person who quite likely will have valuable input to the systems investigation is Jim McKeen, the manager of the payroll department. Since McKeen has the responsibility of managing all of the payroll department and therefore is in daily contact with the problems and procedures which are found within the department he should be able to provide valuable information concerning the reasons for certain procedures. It is quite definite that McKeen should be interviewed, perhaps several times, depending upon the information gathered from other sources.

Since one of the major complaints which have been lodged against the existing payroll system is that the union deductions have not been handled properly, another important manager to interview is Delbert Donovan, the manager of Union Relations (see Figure 2-2). Donovan should be familiar with both the problems which have been occurring and also with the company's relations with the various unions. He, perhaps, could suggest alternatives which would be acceptable to the unions in order to correct the present situation.

There will be other persons who should be interviewed also. Some may be on a lower level than upper-management, but during the preliminary investigation, the analysts will normally deal only with management. When the full-scale investigation takes place, the analysts will speak with timekeepers, payroll clerks, and others who work within the system in order to determine details concerning the current operation of the system. When the analysts move from the vice-presidents and managers to the ''line'' people, their questions can become more detailed concerning specific operations which are followed and where bottlenecks, problems, and inefficiencies may exist.

From the above discussion, it can be seen that it is quite important that the analyst choose the proper people to interview at the proper time. Posing good questions to persons who cannot answer the questions will do him no good at all.

Objective of the Interview

Prior to the interview itself, it is extremely important that the analyst determine what knowledge he wants to obtain from the person being interviewed. What knowledge is required will, of course, depend upon the role of the person being interviewed. Different information will be obtained from the vice-president of the company than from a payroll clerk.

The information to be obtained will also vary, depending upon what stage the investigation process is in. For example, if the investigation is just beginning, and this is the first interview with the person, different aspects of the subject matter may be discussed than if the investigation is nearly complete.

It is the responsibility of the analyst to determine what subjects will be discussed and how the questions will be asked. When the investigation begins, the analyst will normally want to obtain the facts which indicate why there has been a complaint. Therefore, he will ask questions which satisfy the problems of WHO indicated there was a problem, WHAT prompted the complaint and the request for a systems analysis of the situation, WHEN the complaint was registered, HOW OFTEN the complaint has been received, WHERE the complaint was received from, and WHY is there a problem or complaint.

After the investigation has progressed, the analyst has the further responsibility of determining what the current procedures are. It should be noted that during the preliminary investigation it is not the function of the interviews to uncover every fact concerning a system. The function of the preliminary investigation is to determine the nature of the problem, define the scope of the proposed system, and be able to specify time and money estimates for a full-scale investigation and analysis of the current system. Therefore, the interviews will not be conducted on an in-depth basis such as would take place in the full-scale investigation.

In the sample problem, it is known by Coswell and Mard that they are going to interview Arnold Henderson, Jim McKeen, and Delbert Donovan. As noted, others will also be interviewed; but for the case study, the interviews with Arnold Henderson and Jim McKeen will be discussed.

Henderson, the Vice-President of Financial Affairs, was the person who originally contacted Clyde Harland, the manager of the systems department, to request the services of the systems department. Since it can be assumed that he had good reasons, the investigators should determine what these reasons were. Thus, for Mr. Henderson, the set of objectives illustrated in Figure 2-8 were drawn up by Coswell and Mard.

Note from the list prepared for Henderson that the basic questions of WHO, WHAT, WHEN, HOW OFTEN, WHERE, and WHY are to be asked. It will be noted also that the subjects to be discussed are of a general nature not involving specific details concerning the operation of the payroll department or any particulars concerning the payroll system. As has been noted, when dealing with top-level management, such as a vice-president, company policy and general statements concerning the problem are to be discussed rather than specific information concerning the operation of a system. These specific questions are better asked of someone who is more involved in the day-to-day operation of the system.

S Y S T E M D O C U M E N T A T I O N		
NAME OF SYSTEM Payroll	**DATE** 9/23	**PAGE** _1_ **OF** _1_
ANALYSTS D. Mard H. Coswell	**PURPOSE OF DOCUMENTATION** Objectives of Henderson Interview	

OBJECTIVES FOR HENDERSON INTERVIEW:

1. Determine the budget for the Payroll Department and if there are any problems meeting the budget.

2. Determine the source of complaints concerning the payroll system. Have they been from Personnel; from the union, or are they complaints which Henderson has?

3. What are the complaints which he has heard and what are his own complaints about the payroll?

4. When did the complaints begin and was there any significant event that transpired right before the complaints began?

5. How often have there been complaints?

6. Are the personnel in the Payroll Department satisfactory with Henderson or are there personnel problems which may be contributing to the other problems?

7. Who would be the best people to talk to to gain more information?

8. Why does he think a computerized payroll would solve the problems as he sees them?

9. How much money is he willing to spend on a new payroll system?

Figure 2-8 Memo Detailing Objectives of Henderson Interview

For the interview with Jim McKeen, the manager of the payroll department, a similar list of objectives would be established by the systems analysts. Again, these objectives are established so that the analysts can detail what information they wish to know about a system and from what sources this information can be obtained. For the interview with McKeen, however, the objectives would differ somewhat from those for the interview with Henderson. Since McKeen is a good deal closer to the day-to-day operation, and in fact is in charge of the entire payroll operation, he would be able to give a more detailed accounting of the actual operations. The list of objectives for the McKeen interview are illustrated in Figure 2-9.

As with the specification of objectives for the Henderson interview, the objectives specified in the McKeen interview are used to outline those subjects which the analysts believe will aid them in determining the methods used in the payroll procedure and also where any problems may be found. Note that the questions to be asked of McKeen are more detailed concerning the actual operation of the payroll department. In addition, since it is known that McKeen wrote the current payroll procedure, he can be asked questions concerning its proposed function when it was written and where it stands now. Note also that questions concerning personnel, overtime, and such operational questions as these can be asked of him.

SYSTEM DOCUMENTATION		
NAME OF SYSTEM Payroll	**DATE** 9/23	**PAGE** _1_ **OF** _1_
ANALYSTS D. Mard H. Coswell	**PURPOSE OF DOCUMENTATION** Objectives of McKeen Interview	

OBJECTIVES FOR McKEEN INTERVIEW:

1. How many people work in the Payroll Department and what is the breakdown of their functions?

2. When was the current payroll system written, and at that time what was the projected maximum number of persons to be using the system?

3. What are the sources of any complaints he has heard?

4. When and how often have the complaints been heard?

5. Get a description of the system currently being used and determine how it differs from the procedure described in the Procedures Manual.

6. Get detailed breakdown of personnel and their functions within the department.

7. Get detailed breakdown of hours worked by function within the department.

8. What does McKeen think is the problem, and does he have a proposed solution?

9. Get his assurance of complete cooperation concerning the systems study, including making available any documents and people necessary.

Figure 2-9 Objectives for McKeen Interview

The listing of objectives to be learned from Delbert Donovan, the union coordinator, and any other persons to be interviewed, would be prepared in a similar manner to those previously illustrated. It is mandatory that the analyst prepare these objectives so that there is direction in the interview, and so that he can determine when he has the information which is required. Without these objectives, the interview itself may wander and the analyst, if he has not prepared the course which he wants to follow in his information gathering process, will not be able to determine when he has enough information to define the current process and to isolate the problems within the current system.

Case Study - Conducting the Interview

Once the analyst has determined the objectives of the interviews which he is to conduct and has determined who to interview, he must prepare for the interview. This preparation is quite important, because the probability of gaining the required information by merely sitting down to ''chat'' is quite small.

One of the keys to successful interviews is to have complete cooperation of all parties involved. This includes the person to be interviewed, the supervisor or manager of the person to be interviewed, and top-level management of the company. Complete cooperation is many times dependent upon these persons having some knowledge of the purpose of the interview. Therefore, the letter of authority, which should normally be prepared before the investigation is under way, must be distributed to all parties who will be interviewed or whose people may be interviewed. As discussed previously, this letter will reduce much of the reluctance on the part of these persons in cooperating with the analyst doing the interviewing.

The systems analyst, when preparing for interviews, should always follow accepted methods for establishing business conferences; that is, he should always make an appointment on a given day at a given time with the person to be interviewed. In addition, it is normally a good idea to call ahead about an hour before the interview to be sure that the person to be interviewed will be available. It should always be remembered by the analyst that the interview is not the primary job of the person being interviewed. Therefore, the analyst must be aware that other business may arise for the interviewee which will cause a postponement of the interview. If this occurs, another appointment should be made.

An interview should never be scheduled without the knowledge of the manager of those persons to be interviewed. Most persons in authority in a department of a company like their people to be involved in their assignments within the department, and interruptions for the interviews may not be the type of thing appreciated by managers unless they are aware of them. Thus, in addition to any memo which is prepared by top management, it would normally be a good idea for the analyst to send a memo to a department manager indicating what appointments he has for the week with people within the department. In this way, a manager will be aware that some of his people are going to be tied up with interviewing and he will not view the analyst as an intruder. Again, in order to perform an adequate investigation of problem areas, it is absolutely mandatory that there be complete cooperation; and anything which the analyst can do to ensure this cooperation will lead to a more complete and authoritative investigation.

As noted previously, when preparing for the interview, the types of questions to be asked will depend upon who is being interviewed. If, however, the objectives to be obtained have been well defined, as discussed previously, the analyst should be able to structure his questions without too much difficulty. Whenever an interview is to be conducted, the analyst should have a prepared list of questions or topics to be covered prior to beginning the interview. To be sure, there may be some subjects of value which are raised during an interview which were not a part of the planned questions; but without a prepared list of questions, there is very little chance of satisfying the objectives of the interview.

When the list of topics to be discussed has been prepared, it is normally a good idea to supply the person to be interviewed with these topics several days in advance. In this way, the person is aware of what topics will be discussed, and, if necessary, can do some personal research in order to be prepared for the interview. By supplying this information, the analyst can many times save another interview because the interviewee will be prepared at the first meeting.

An example of the list of questions and topics which were sent to Jim McKeen by Don Mard is illustrated in Figure 2-10.

M E M O R A N D U M

DATE: September 24
TO: Jim McKeen, Manager, Payroll
FROM: Don Mard
SUBJECT: Interview for Payroll Investigation

As per our telephone conversation, I am planning on our appointment on October 3 at 2:00 p.m. If this is not convenient, please contact me.

The following is a list of questions and topics which I would like to cover when we meet. There may be additional ones between now and then, but I definitely want to talk about the following:

1. How many people work in the payroll Department?

2. What are the various functions of the people in the Payroll Department and how many people work in each area?

3. When was the current payroll system written?

4. What was the projected number of employees which the payroll system was designed to accommodate?

5. What are the sources of any complaints which you have heard?

6. When are the complaints heard? Are they recent? Are there any particular times during the week or month when they are more predominant?

7. How often are the complaints heard?

8. How many hours are worked in each of the areas of the department?

9. What is the ratio of regular hours to overtime hours in the Payroll Department?

10. What do you feel is the problem causing the complaints?

11. Do you have a proposed solution to the problem?

12. What documents would be helpful to study in terms of seeing the overall operation of the department?

13. What people would be the most productive to interview concerning their individual functions within the department?

I think that these questions can form a basis for our first interview. If you have anything else which may prove useful in this investigation, I would appreciate it if you could make it available. I look forward to talking with you on the 3rd.

D. Mard

D. Mard

DM:sjm

Figure 2-10 Example of Memo Detailing Questions and Topics

Note from the memorandum in Figure 2-10 that the analyst has given the person to be interviewed a list of questions to be answered in the interview. If required, McKeen could check the hours of overtime recently worked or any other area in which he felt he did not have the answers immediately available. In this way, McKeen can be properly prepared to answer Mard's questions.

In addition to preparing for the interview in terms of people, the analyst should also prepare in terms of materials which may be required for reference in the interview. Although this condition may not arise in the preliminary study too often, it will certainly arise in the full-scale investigation. Such things as documents, forms, and so on, which may be used daily in the interviewee's job and about which the analyst has questions, should be gathered so that they are available for the interview. In the interview with McKeen, there are no documents of this type required; but when the analyst interviews the payroll clerks in the full-scale investigation, it would be advantageous to have copies of payroll registers or payroll checks or any other documents which can be explained by the person being interviewed. In addition, the analyst should include these documents with the memorandum of questions to the person to be interviewed so that there are no surprises in the interview.

Case Study - The Interview

Once all the preliminaries have taken place, the interview will be conducted. The first important decision to be made is where the interview is to take place. It is quite important that the interview be conducted with a very minimum of interruptions so that the person being interviewed can concentrate completely on the questions and subject matter, and also so that it takes the least amount of time for both the analyst and the interviewee.

Thus, if the interview is to take place in the office of the person being interviewed, it is important for the analyst to tactfully suggest that all calls and interruptions be held until the conclusion of the interview. In most cases, if the set of questions and other materials have been made available to the interviewee, it will be relatively easy to estimate the time the interview will take. If this is not possible, in many instances it will be prudent to set a time limit so that the interviewee is not taken away from his primary duties for too long a period.

It must always be remembered by the analyst that even though the interviewee may eventually benefit from a new system, at the time of the interview the analyst is benefitting, so all consideration must be given to the time of the person being interviewed.

If it appears that it will be difficult to have a private interview in the office of the person to be interviewed, then perhaps alternate locations should be suggested. Such places as conference rooms or vacant offices may be used so that the person being interviewed can be away from his daily activities and have the opportunity to totally concentrate on the material being discussed.

Regardless of the location of the interview, there are certain items which the analyst must attempt to ensure happening in order to have the interview be successful. Perhaps the most important element to establish, especially in the first interview, is a rapport between the analyst and the person being interviewed. It is difficult to define exactly the type of rapport which should be established, but is is important that a confidence be established between the analyst and the person being interviewed so that any question which may be asked will be answered in a candid manner. Thus, the analyst should establish at the outset the purpose of the preliminary investigation and where the interviewee can benefit from this study. Too often, there is a feeling on the part of persons being interviewed that the study is merely a time study attempting to show that they are inefficient or that the computer is going to take their job away. To be sure, this is a possibility, but the analyst must assure the person being interviewed that the system being studied and its likely successor, if there is a new system, will merely aid him in his job and that there is no intention of replacing people with the computer. In fact, computer systems will normally create jobs which were non-existent previously, and this fact may be made known to the person being interviewed.

Above all, the analyst must put the person being interviewed at ease so that all questions are answered completely and candidly. It will do the analyst no good whatsoever if the answers to his questions are evasive or if the answers are not complete. The purpose, once again, of these interviews is to merely find out the facts concerning a particular application or system. There are, however, on the part of the person being interviewed, apprehensions which must be broken down by the analyst in the first part of the interview so, as mentioned, there is complete candidness on the part of the interviewee.

Another area in which an analyst many times fails is the ''art'' of listening. Since the analyst is asking questions, his main function is to listen to the answers. In too many cases, however, he does not listen properly to the answers which are given to him. There is an adage which goes, ''Learn to Listen, Listen to Learn'' and this should be the motto of every analyst who is conducting an interview.

Thus, the analyst, after asking a particular question, should allow the person time to ponder the question and to formulate an answer. It is a proven fact that the average maximum pause time between one person speaking and another person speaking is 3-5 seconds; that is, after this period of time, one or the other expects some answer. The analyst, when conducting interviews, must allow more time than this for the interviewee to answer questions because he may be thinking about the answer. Needless to say, this type of activity requires practice, and analysts should constantly practice their interviewing techniques so as to get the maximum information from an interview. The interview is probably the primary source of information for the analyst when he is attempting to isolate problems within a system and the properly conducted interview, therefore, is critical to his information gathering process.

In order to illustrate a "typical" interview, the interview between Don Mard and Jim McKeen which took place October 3 is recounted below.

Jim: Hi, Don, come on in. I'll be through with the phone in a minute. Have a seat.

Don: Jim, did you have a chance to review the memorandum I sent you concerning some of the topics which I would like to talk to you about?

Jim: Yes I did, and I think I have most of the answers which you are interested in. Excuse me, Don, let me get that phone.

-After phone conversation-

Don: Is there any chance we could go to the conference room so that we can get away from the phone for a few minutes. I think this will go a lot faster if we're not interrupted.

Jim: That sounds like an excellent idea. Let me tell my secretary where I'll be in case of an emergency, but I'll tell her not to disturb us in the mean time.

-In the conference room-

Don: Yes, this is better. Do you have any questions before I start asking some? I thought perhaps there might be something which I could clear up concerning what we're going to be doing in this preliminary investigation of the payroll system.

Jim: No, I don't think so. I think I understand that Henderson asked you guys to find out what is going on within the payroll system and I'll be as much help as I can.

Don: Good. We might as well go down the list of questions which I sent to you last week. It seems like a good place to start. How many people work in the payroll department?

Jim: Including myself and my secretary, there are 6 people in the department. The 4 other persons are all involved in the preparation and production of the company payroll. I've prepared a listing of the persons in the department and their jobs. I thought this might make it easier for you in determining what each person did.

Don: Thank you, Jim, that's great. I think I'll be able to use this in determining the roles of the various persons in the department. Can you tell me a little bit about the current payroll system that's being used. I believe you designed and wrote the procedure, didn't you?

Jim: Yes, I designed it in 1968. At the time it was designed, it was entirely adequate and we felt that it would be adequate for up to 500 persons in the company. It seems, however, that there have been some problems. The biggest problem which I hear about is from Henderson concerning the overtime which we have to pay each week to get the payroll out. The only trouble is that he will not give me authorization to hire anyone else. We particularly need someone in the check preparation area, because all of these checks are prepared from the registers, and it's just a very time consuming job. Those payroll clerks put in between 50 and 55 hours a week, and there's just not any way to make it go any faster. The addition of one or two people would, in my opinion, solve the overtime problem and it would be a whole lot cheaper than redesigning the payroll system and putting it on the computer.

Don: Do you have any idea why he won't give you authorization for one or two more people?

Jim: Yeah, he says that we're over our budget right now, which we are, but that budget was figured before we added that new machine shop. Now, there is no way we can get the payroll out on time without this overtime.

Don: What about the complaints I've heard about the union deduction problems? Do you have any knowledge about those ?

Jim: Yes, apparently when some employees were promoted and required to join the union after a while, their union dues were not deducted immediately. The union has been pretty upset over it and so have the employees because they had to pay a lump sum when the error was detected. You'd think they would let us know when they join the union.

Don: May I see the form sent from Personnel to your department when an employee joins the union?

Jim: There's no specified form, just an interoffice memorandum or a phone call from Personnel.

Don: I've heard that it has occurred several times. Do you see any problems with the employees themselves?

Jim: Not at all. It's the manager of Union Affairs. He's supposed to call us when an employee joins the union. I run a pretty tight ship and I don't think you'll find any problems with my people. There's no need to worry about them because I check them out closely, and so don't worry about that. I don't let anything get by me if I can help it. So just don't worry about the people or how I run them in this department. I understood you only wanted to find out about the procedure, anyway.

Don: Yes, that's right. Is there any particular peak period when the overtime becomes heavier?

Jim: No, I don't think so, although sometimes it becomes a little heavier during certain vacation times because two or three checks must be made out for one employee, or one of our people may be on vacation. Otherwise, it is pretty consistent.

Don: Can you give me a breakdown of the hours worked in each section and by each employee? I ask this so we can see if there is a concentration of effort in one area and a slack in another area.

Jim: Well, I'll have my secretary prepare a listing for you, but I can guarantee you that I wouldn't let things get fouled up like that. I know what I'm doing in Payroll—I've been in it for 16 years and there is no way I would let there be slack areas and peak areas, believe me.

Don: Other than overtime and union dues then, you don't see any real problems in the payroll system, is that correct?

Jim: Yeah, that's right. Give me two more girls and get that manager of Union Affairs straightened out and there won't be any problems at all.

Don: Jim, can you tell me who might be the most helpful in determining some of the detailed processing which takes place? Perhaps you could mark this sheet you prepared.

Jim: I will, but I think you'll probably be wasting your time. I designed that system and I know it better than anybody. Besides, you'll be taking some of the time away from my people, and they are busy enough being shorthanded like they are. But Henderson seems to think it is a good idea for you to take a look at the system, so I will inform everybody in the department that you will be calling for appointments. I would appreciate it if you let me know who you're going to talk to and when, if you could.

Don: We'll be happy to—we'll send you a memorandum on Monday outlining who we would like to talk to. I don't know of anything else at this time which we could go over. I really appreciate your time, and I would expect that we will be talking several more times before this is all over.

Jim: I'm quite sure of that. If that's it, I'll get back to my office. Talk to you later.

-End of Interview-

At the conclusion of the interview, as illustrated above, it is normally a good practice to synthesize and summarize the interview with the person being interviewed. There is always the chance that statements may be misunderstood and, therefore, the analyst should review his understanding of the facts stated so that there is a meeting of the minds. Thus, Don got Jim's agreement that Jim thought the only problem area was that they were short of people and that the manager of Union Affairs had not properly notified the payroll department of union changes.

Case Study - Post Interview Procedures

After the interview is completed, there are definite steps which should be taken to ensure that the topics discussed in the interview are preserved and will not be forgotten. These primarily involve setting aside time immediately after the interview to record what transpired in the interview.

There are pros and cons about taking notes during an interview, but the accepted views are that note taking should be kept to a minimum during the interview. To be sure, it may be necessary for the analyst to jot down a few things with which to jog his memory, but he should not act as a secretary taking dictation. This is distracting to the person being interviewed, and in general, fails to establish the necessary rapport as discussed previously.

Instead, the analyst should set aside time immediately after the interview to record the facts which were discussed and to evaluate the information which was received in the interview. It is a proven fact that even with a subject which is a life-and-death subject, 50% of what took place in a conversation will be forgotten within 30 minutes. Therefore, it is imperative that the analyst immediately record those facts which have come forth in the interview so that they will not be forgotten.

This written review may be in the form of a narrative describing what took place, or may be in the form of written answers to the questions which were prepared and submitted to the person to be interviewed prior to the interview. The written review which was prepared by Mard after his interview with Jim McKeen is illustrated on the following pages.

SYSTEM DOCUMENTATION		
NAME OF SYSTEM Payroll	DATE 10/3	PAGE _1_ OF _2_
ANALYSTS D. Mard H. Coswell	PURPOSE OF DOCUMENTATION Interview with Jim McKeen	

Interview was the first interview with McKeen. The impression was that he was generally cooperative, but that he seemed to be a little afraid of the possibility of a new system. He defended his use of his people and said we would not find any flaws in the way his people worked for him. In response to questions, the following answers were received:

1. How many people work in the Payroll Department? Six, including McKeen and his secretary.

2. What are the functions of the people in the department and how many people in each area? See the attached listing prepared by McKeen.

3. When was the current payroll system written? 1968.

4. What was the projected number of employees which the payroll system was designed to accommodate? 500 people.

5. What are the sources of complaints? Arnold Henderson concerning the overtime, and to a lesser extent, Delbert Donovan about the union dues deductions.

6. When are the complaints heard, etc? No particular times. It appears Henderson brings up the subject many times with McKeen.

7. How often are the complaints heard? From Henderson, I get the impression they are heard alot. From Donovan, only when the union gets on him.

8. How many hours are worked in each of the areas of the department? The check preparation people work 50-55 hours per week. Others in the department work 40 hours.

9. What is the ratio of regular hours to overtime hours? The people in Payroll are averaging between 10-15 hours per week overtime.

10. What is the problem causing complaints? McKeen feels he just does not have enough people to keep up with the payroll, especially since the new machine shop was opened. Additionally, there is no formal procedure for initiating union dues deductions.

Figure 2-11 Written Report of Interview - Part 1 of 2

SYSTEM DOCUMENTATION		
NAME OF SYSTEM Payroll	**DATE** 10/3	**PAGE** _2_ **OF** _2_
ANALYSTS D. Mard H. Coswell	**PURPOSE OF DOCUMENTATION** Interview with Jim McKeen	

- Page 2 -

11. What is your proposed solution? McKeen says if he had two more girls to prepare the payroll he could do it with no problems, and that there is not a new payroll system required. However, he is over-budget, and Henderson will not approve new personnel for him.

12. What documents would be helpful to study? No particular documents mentioned by McKeen.

13. What people would be best to interview? See the attached sheet which McKeen marked.

When union dues are to be deducted from an employee's pay after he has joined the union, there is only a phone call or interoffice memo sent. There is no form which passes from the Personnel Department to the Payroll Department.

Figure 2-12 Written Report of Interview - Part 2 of 2

The analyst should also send a memo to the person interviewed thanking him for his time both in preparation for the interview and during the interview itself. In addition, this memo should again summarize the facts discussed so that the interviewee has a written copy of the results of the interview and also so that he can correct any misconceptions held by the analyst.

The memo sent to Jim McKeen by Don Mard after the interview is shown below.

M E M O R A N D U M

DATE: October 4
TO: Jim McKeen
FROM: Don Mard
SUBJECT: Interview

    Just a note to say thanks for your time yesterday, and also the time you spent in preparing the listing of the people in your department. It will be of great help to us in terms of defining the various jobs performed within the department.

    Just to summarize, you feel that the only problem with the current payroll system is the fact that it is understaffed, and the addition of several more people would solve both the overtime problem and the problems encountered with the union dues. Outside of that, you feel that the current payroll system is adequate to handle as many as 500 employees.

    If this is incorrect, I would appreciate it if you could contact me as soon as possible so that you can correct any misconceptions which I have. If not, I again would like to thank you for your time and effort in this investigation.

D. Mard
D. Mard

DM:sjm

Figure 2-13 Example of Memo Following Interview

Case Study - Analysis of the Interview

It is usually helpful for the analyst to analyze the interview and the person interviewed, as well as recording the facts which were gained. This is true because many "facts" which may be stated in the interview might possibly be biased by one or more circumstances. For example, it may be that the person interviewed is attempting to protect his "empire" and feels that any information which he gives out is going to destroy this. Therefore, he does not lie when asked questions, but he does not tell complete truths.

In other circumstances, the person interviewed may have very strong feelings about the reasons for the problems, and his insisting on his reasons may distort the facts to the extent that much of the information gained from him is useless. There is also the situation where the person being interviewed wishes to be so helpful that they answer questions when in fact they do not know the answer. Obviously, the analyst must distill the proper answers from those which may not be correct.

It is this type of analysis which must be performed upon the facts which are obtained in an interview. In the case of Jim McKeen, it should be seen that several types of bias may have crept into his answers. First, he designed the currently used system and will naturally feel an inclination to defend it against any suggestions that it may be inadequate. Secondly, since he is the manager of the payroll department, he is somewhat unwilling to say that there may be personnel problems within the department which are contributing to the problems of the department. Lastly, he feels very strongly that he knows the answer to the problems—hire several more people for the payroll department. In this case, it may be a difficult task to have McKeen give totally objective answers because of his biases for his own system and department and the fact that he feels he knows the solution to the problem. It is up to the analyst to take these answers under consideration and then, after all of the facts have been gathered from all other sources, to evaluate all of the answers in light of the complete picture. This is not to say that the entire answer may be to hire two people; only that much more detailed information must be gathered before this conclusion can be reached. It will be noted that some of the attitudes or biases which McKeen had were recorded in the written review of the interview (Figure 2-11 and Figure 2-12) so that they will not be forgotten when the interviews are reviewed after all of the facts have been obtained.

Unsuccessful Interviews

Regardless of the preparation for an interview, there are some interviews which will not be successful. There are many reasons why an interview does not accomplish its stated goal, but one of the primary reasons is that the analyst and the person to be interviewed do not get along with each other. This may be caused by several factors. For example, there may just be a personality conflict, it may be that the person being interviewed is anti-computer and afraid that the analyst is there to put him out of a job, or there may be some internal political problem of which the analyst is not aware which precludes cooperation on the part of the person to being interviewed.

When an interview appears to be not going well it is up to the analyst to terminate the interview with as much tact as possible. It obviously is a waste of two peoples' time to continue with an interview which is not going to produce any results. At an appropriate place, the analyst should indicate that he has all of the information which he needs and terminate the interview. It may be at a later time there will be more cooperation on the part of the person being interviewed, or it may be that the information sought will be found elsewhere. In any event, when an interview is not successful, the analyst must realize this and not attempt to push it any further.

Case Study - Reviewing Documentation and Further Investigation

The interviews with Jim McKeen and others in the company provided the analysts with a great deal of information concerning the operation of the payroll system and the problems which are currently found in the system. When conducting the preliminary investigation, however, it is frequently necessary to investigate other areas as well as conducting the interviews.

After interviewing McKeen and others, Don Mard, the systems analyst, wanted to find out more about the actual sequence of operations which took place within the current system. Therefore, he consulted the current documentation of the system. There he found step-by-step procedures which are followed in the preparation of the payroll. In addition, he noted that the forms which are used by the personnel department when an employee is hired and by the payroll department during the preparation of the payroll were described. As a result of the information in the documentation of the payroll system and the interviews which he conducted, Mard determined that the sequence of events in the payroll is as illustrated in Figure 2-14.

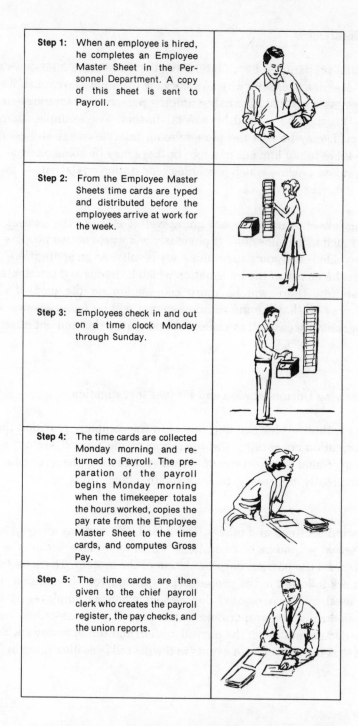

Step 1:	When an employee is hired, he completes an Employee Master Sheet in the Personnel Department. A copy of this sheet is sent to Payroll.
Step 2:	From the Employee Master Sheets time cards are typed and distributed before the employees arrive at work for the week.
Step 3:	Employees check in and out on a time clock Monday through Sunday.
Step 4:	The time cards are collected Monday morning and returned to Payroll. The preparation of the payroll begins Monday morning when the timekeeper totals the hours worked, copies the pay rate from the Employee Master Sheet to the time cards, and computes Gross Pay.
Step 5:	The time cards are then given to the chief payroll clerk who creates the payroll register, the pay checks, and the union reports.

Figure 2-14 Sequence of Events in Payroll Preparation

From the sequence illustrated above it can be seen that a number of forms (Employee Master Sheet, Time Card, Payroll Register, Pay Checks, etc.) are prepared and used in the current payroll system. In the preliminary investigation, the analyst is not concerned with their individual formats and the detailed information which is used with each form. This information will be gathered only after a full-scale investigation is authorized by management. The primary emphasis in the preliminary investigation is to gain an overall understanding of the operation of the system in question.

Case Study - Presentation of Findings

After the systems department has conducted a preliminary investigation, it must report its findings to either a committee of evaluators or to top-level management. The presentation should include what the systems analysts have found concerning the operation of the system in question, the problems which appear to be present in the system, and their recommendations for future action. This report is normally presented in a written form and an oral presentation is also made. The written report generated as a result of the preliminary study of the payroll system is illustrated in Figures 2-15 and 2-16.

MEMORANDUM

DATE: October 7
TO: Arnold Henderson, Vice-President, Financial Affairs
FROM: Clyde Harland, Manager, Systems and Programming
SUBJECT: Findings of Preliminary Investigation of Payroll System

INTRODUCTION

A preliminary investigation of the Payroll System was conducted by the Systems Department as a result of a Request for Services from Arnold Henderson, Vice-President, Financial Affairs. The findings of this preliminary investigation are presented below.

OBJECTIVES OF PRELIMINARY INVESTIGATION

The preliminary investigation was conducted to investigate two major complaints concerning the payroll system. First, it was specified that deductions for union dues from employees' checks were not being performed correctly, especially for employees who joined the union after employment. Second, considerable overtime was required for employees of the payroll department in order to have the payroll prepared on time. The objective of the preliminary investigation was to determine if these problems actually exist and, if so, to recommend a future course of action.

FINDINGS OF THE PRELIMINARY INVESTIGATION

After interviewing those persons directly associated with the payroll system, it is the conclusion of the systems department that the problems specified in the Request for Services do in fact exist. In addition, it would appear that there are a number of other problems which may be associated with the system. In particular, the system is seven years old and was not designed for a number of contingencies which now exist within the company. The method of communication with other departments within the company with the payroll department is not systemized and would appear to be inadequate. The manual methods which are used in the payroll department are not doing the job and it would appear that only a slight increase in the workload will render the system completely inadequate. Costs are rising with the system because of the inefficiencies and it does not appear that there are any adjustments in the system which can keep costs at the present level or reduce costs.

RECOMMENDATIONS

As a result of the preliminary study, the systems department recommends a detailed systems investigation and analysis of the payroll system take place. It appears that there is a need to design a new payroll system. Attempts to correct existing problems, such as errors in the deducting of union dues, would only serve as a temporary solution to the need for a new payroll system.

Figure 2-15 Report - Part 1 of 2

Arnold Henderson   -2-   October 7

The systems department will assign a team of systems analysts the task of determining the detail processing which takes place within the system, the inputs and outputs of the system, and the formats of the data used within the system. In addition, the analysts will analyze the total system in detail in order to present alternative solutions to solve the problems which are present in the current system.

Among the methods used will be a detailed review of the standard procedures of the payroll system, further personal interviews with those who have first-hand knowledge of both the problems encountered in the payroll processing and the operations of the payroll department, personal observation of the procedures within the department, and consultation on various statistics or other material which may be available concerning the operation of the payroll system.

TIME AND MANPOWER REQUIREMENTS

It is estimated that two man-months would be required for the analysts to carry out the investigation recommended above. This would be handled by assigning two analysts for a full-time period of one month.

An additional period of two weeks would be required for the report to be prepared and this would be done by one analyst.

It is estimated that approximately one man-month of time would be required from persons outside the systems department for the purposes of interviewing and supplying information to the investigating analysts. This time would be apportioned on an as-needed basis depending upon when interviews and other meetings could be arranged.

Cost:

The following is an estimate of the costs of performing this full-scale investigation and the subsequent report:

2-1/2 man-months - Systems Analysts	at 1400.00/mo.	$3,500.00
1 man-month - Outside Personnel	at 1200.00/mo.	1,200.00
1 man-month - Secretary	at 650.00/mo.	650.00
		$5,350.00

EXPECTED RESULTS

It is expected that at the conclusion of the detailed systems investigation and analysis, the systems department will have defined in detail the problems which are in the payroll system and will present alternative solutions to the solving of the problems. Recommended solutions will be proposed that will eliminate or substantially reduce overtime and eliminate errors in the deduction of union dues.

G. Harland
C. Harland

Figure 2-16 Report - Part 2 of 2

As can be seen from the report on the preliminary investigation, the objective of the preliminary investigation is first defined. This is necessary so that management or committee personnel receiving the report will all realize why the investigation was initiated.

The findings of the preliminary study are then related. Note that these findings are conclusions of the systems department after their preliminary investigation. Thus, they are opinions and opinions are only as reliable as their source. It is quite important, therefore, that the analysts have the necessary business and data processing experience to be able to express competent decisions and that management has the confidence in the analysts to listen to their opinions. Thus, as the analysts present the Preliminary Investigation report, they must not only put forward their ideas but they must also sell themselves.

The report additionally contains the recommendations of the systems department concerning further action which should take place. It must be noted that the recommendations of the systems department are not always followed and it is up to management to make the final decisions concerning what follow-on action will be taken. In most cases, however, the recommendations of the systems department will weigh heavily in their decision-making.

One of the elements upon which management will make their decision is the cost of the recommendation of the systems department and the benefits to be obtained. Therefore, it is imperative that an estimate of time and costs be included in the report of the preliminary investigation if further action is recommended by the systems department. Management must be made aware of all costs whenever any type of system activity is to take place.

In addition, the expected results of the recommendations by the systems department are specified. Whenever management is to be asked to spend money, it must be told the results which are expected from the expenditure of the money.

Case Study - Result of Preliminary Investigation

After the report was submitted, the manager of the systems department was asked a number of questions by Henderson and other members of management. As a result of this review process, management of James Tool Co. expressed the opinion that a full-scale investigation should take place. Therefore they gave their approval to the next step in the systems project—the **Detailed Systems Investigation and Analysis.**

SUMMARY

The formal steps in the Initiation of the Systems Project and Preliminary Investigation are an important aspect in the overall process of systems analysis and design. In far too many cases, businesses fail to properly define the steps in this phase. As a result, the true problem is not determined and the scope and boundary of a system project is not defined, leading to projects being late, projects exceeding costs, and projects failing to remedy existing problems.

In summary, the basic steps in the first phase of the Systems Project include:

1. Establish a formal procedure for reviewing system requests.

2. Establish criteria and methods for the acceptance of systems projects.

3. Make a determined effort to understand the true nature of the problem.

4. Define the scope and boundary of the study.

5. Present to management the recommendation of the systems department. If the recommendation is for further study of the system, it should include:

    a. The investigation method.
    b. The time and manpower requirements.
    c. The cost of study.
    d. The expected results.

6. Obtain management approval to begin the detailed investigation and analysis phase of the system project.

Any systems project, no matter how simple in nature, will normally entail a significant expenditure of company resources; therefore, in order to insure maximum results, great care should be taken before undertaking the detailed systems analysis and designing phase of a systems project.

STUDENT ACTIVITIES—CHAPTER 2

1. List and briefly discuss the four basic reasons for the origination of systems projects.

   1)

   2)

   3)

   4)

2. What are the ultimate goals of all systems projects in a profit making organization?

3. List the factors which influence a decision on the selection of systems projects.

   1)

   2)

   3)

   4)

   5)

   6)

   7)

   8)

4. What is the purpose of the Preliminary Investigation?

5. What is another name for the Preliminary Investigation phase of the Systems Project?

6. List what should be accomplished during the Preliminary Investigation.

    1)

    2)

    3)

    4)

    5)

7. What factors should be considered when estimating the time and cost requirements for a detailed investigation?

    1)

    2)

    3)

    4)

8. List the steps which should be undertaken in a preliminary study.

    1)

    2)

    3)

    4)

    5)

9. List the five basic steps in conducting a meaningful interview.

   1)

   2)

   3)

   4)

   5)

10. During the Preliminary Investigation who will the analyst normally interview?

11. What is the function of the Systems Review Committee?

12. Who has the final responsibility for approval for a detailed investigation of a systems project?

DISCUSSION QUESTIONS

1. It was stated in the text that one source of new systems projects is a "Management Directive." You are the systems manager and you receive a "management directive" from the Vice-President of Marketing to prepare 200 mailing labels on the computer for a "one-time" mailing of advertising literature. As systems manager you know that the mailing labels can be prepared more efficiently by merely assigning a typist to prepare the mailing labels. How would you handle this problem?

2. Some authorities in systems analysis say "the only way to interview is to tell the interviewee that you are going to take detailed notes, have the notes typed up, and then have the notes approved by the interviewee. If you don't take notes, most of what you have heard will be forgotten or misinterpreted at a later date."

   Other authorities say "Interviews should be confidential...in systems analysis you are normally dealing with 'problem situations'...most of the information which will be beneficial to you as the analyst will be 'off the record.' Such statements as 'quality control is incompetent' or 'the time clerks are careless' would never be made if you tell the interviewee that whatever he says will be recorded."

   Which position do you support and why?

3. What would you do as the Senior Systems Analyst if a Vice-President of your company came to you and **insisted** that a system be computerized but at the conclusion of the preliminary investigation you knew that the system could not be cost-justified?

CASE STUDY PROJECT—CHAPTER 2
The Ridgeway Company

INTRODUCTION

The Ridgeway Country Club was opened one month after acquisition by the Ridgeway Company. Holden Archer, Vice-President, Operations, was charged with the responsibility for the development of this new project. Within a short time membership grew to over 1900 members. There are 1500 full members of the club and 400 social members who are permitted use of the clubhouse and swimming pool only.

Although the Ridgeway Country Club has been a financial success since its acquisition and membership has been higher than expected after three months of operation, George Ridgeway, President of the Ridgeway Company, recently received the following letter in the mail.

December 10, 1976

Mr. George Ridgeway
President
Ridgeway Company
3030 Deuce Drive
Oceanview, CA 92641

Dear Mr. Ridgeway:

During the three months in which I have been a member of your club, I have very much enjoyed the use of the fine facilities. However, the monthly statements which I have received have been incorrect in each of the last three months. During the first month I was overcharged by $25.00, the second month undercharged by $42.00, and when I received my last monthly statement I was overcharged by $250.00!

During the past three months it has been necessary, each month, to take a copy of the receipts I have received from the Pro Shop, restaurant, and bar and sit down with your bookkeeper and reconcile my account. I have been told that "things" should get better when the system gets worked out.

If "things" don't get worked out soon I shall find it necessary to request a refund on my membership fee.

Sincerely,

Seymour Long

Seymour Long

REQUEST FOR SYSTEM SERVICES

Because of the critical nature of the problem, Mr. Ridgeway contacted William Howell, Vice-President, Finance, and asked him to investigate the billing problem. Mr. Howell in turn contacted Lee Ura, the data processing manager, for assistance. Mr. Howell was requested to complete a REQUEST FOR SYSTEM SERVICES form. The form received from Mr. Howell is illustrated below.

<table>
<tr><td colspan="4" align="center">R E Q U E S T   F O R
S Y S T E M   S E R V I C E S</td></tr>
<tr><td>DATE</td><td>December 11, 1976</td><td>PAGE</td><td>1 OF 1</td></tr>
<tr><td>FROM
TITLE
DEPT.</td><td>William Howell
Vice-President
Finance</td><td>TO
TITLE
DEPT.</td><td>Lee Ura
Director of Data Processing
Data Processing</td></tr>
<tr><td colspan="4">SUBJECT   Billing</td></tr>
<tr><td colspan="4">Brief Statement
of Problem:

    I recently received a call from George Ridgeway relative to a letter he received from one of the members of the Ridgeway Country Club complaining about inaccurate billing. This apparently has been a continuing problem for the members.

    In talking with the Accounting Department it was indicated that additional billing clerks are needed to properly service the 1,900 members of the club.

    It seems to me that the billing for the newly acquired Country Club would be an ideal application for our computer system. If you have any questions please let me know.

 W. Howell </td></tr>
<tr><td colspan="4" align="center">To be Completed by Systems Department</td></tr>
<tr><td colspan="4">DATE:

ACTION:

</td></tr>
</table>

Upon receiving the Request For System Services from Mr. Howell, Lee Ura, Data Processing Manager, reviewed the request and sent the following memorandum to William Howell, Vice-President, Finance, recommending that a preliminary investigation be undertaken.

MEMORANDUM

DATE: December 12, 1976
TO: William Howell, Vice-President, Finance
FROM: Lee Ura, Director of Data Processing
SUBJECT: Preliminary Investigation, Billing System

In reviewing your Request For System Services it would appear that our department would be able to assist you in recommending some solutions to the billing problem which appear to exist in the newly acquired Country Club.

Therefore, I would suggest that a preliminary investigation be undertaken by our department. Prior to undertaking this study I would appreciate the distribution of a Letter of Authority to management and department heads of the Country Club advising them of this investigation. Jerry Schock, Senior Analyst, is available to begin work on the project immediately.

Lee Ura

sjm

LETTER OF AUTHORITY

Lee Ura then assigned Jerry Schock, Senior Systems Analyst, to the project. Before beginning the Preliminary Investigation, a Letter of Authority was drafted for distribution throughout the company.

MEMORANDUM

DATE: December 13, 1976
TO: Management Personnel, Ridgeway Country Club
FROM: William Howell, Vice-President, Finance
SUBJECT: Letter of Authority

At my direction, Jerry Schock, Senior Systems Analyst for the Ridgeway Company, will be conducting an investigation of the billing system of the Ridgeway Country Club. This study is being made to determine if improvements can be made to the current system.

Mr. Schock may be contacting you or the people working for you in the next few weeks to arrange for interviews or observe current operations relative to the club's billing system.

Your cooperation will be appreciated.

W. Howell

sjm

FACT GATHERING

In preparing for the preliminary investigation the analyst obtained the following information relative to the organization of the Ridgeway Company. Holden Archer, Vice-President, Operations, is the senior management level person in charge of the Ridgeway Country Club. Reporting directly to Holden Archer is Gunther Swartz, General Manager of the entire club. Working directly under Gunther Swartz is Arnold Holey, Manager of the Golf Facilities; Rodney Scott, Manager of the Tennis Facilities; Christopher Connely, Manager of the Pro Shop; and Pierre La Franc, Manager of the Clubhouse, Restaurant, and Bar.

The accounting function is handled by Percy Carbols, Chief Accountant. Two bookkeeping clerks work in the accounting department. The accounting department is responsible for the preparation of the billing for the members of the club. Percy Carbols reports directly to Mr. Howell, Vice-President, Finance.

The manager of the Golf Facilities, Arnold Holey, has five instructors working for him giving golf lessons.

Rodney Scott, manager of the Tennis Facilities, has seven tennis teachers under his direction.

Christopher Connely, manager of the Pro Shop, has five part-time clerks employed throughout the week.

Pierre La Franc, manager of the Clubhouse, Restaurant and Bar has two cooks, two bartenders, ten waitresses and five bus boys which he supervises.

STUDENT ASSIGNMENT 1
PREPARING AN ORGANIZATIONAL
CHART OF THE RIDGEWAY COUNTRY CLUB

1. Prepare an organizational chart of the Accounting Department and other members of the staff of the Ridgeway Country Club and combine this organizational chart with the chart prepared in Chapter 1.

STUDENT ASSIGNMENT 2
CONDUCTING THE PRELIMINARY INVESTIGATION

1. Analyze the organizational chart. List below the individuals you would interview when conducting the preliminary investigation.

    1)
    2)
    3)
    4)

2. Prepare a list of objectives for the interviews with each of the individuals named above.

3. Prepare a list of specific questions to be presented to each of the individuals to be interviewed.

4. Conduct interviews (contact instructor). Prepare a written report of the interviews.

5. Prepare a memorandum thanking those interviewed for their cooperation.

STUDENT ASSIGNMENT 3
PRESENTATION OF FINDINGS

1. Prepare a report summarizing the findings of the Preliminary Investigation. The report should include:

    a) An Introductory Statement.

    b) The objectives of the Preliminary Investigation.

    c) The findings of the Preliminary Investigation.

    d) Recommendations.

    e) Time and Manpower requirements, if the detailed investigation is to take place.

    f) Expected results of a detailed investigation.

PHASE II

DETAILED SYSTEMS INVESTIGATION AND ANALYSIS

DETAILED SYSTEMS INVESTIGATION AND ANALYSIS

CHAPTER 3

DETAILED SYSTEMS INVESTIGATION AND ANALYSIS

3

INTRODUCTION

One definition of a systems analyst is that he is one who is skilled in the art and science of problem solving. Inherent in this definition is the assumption that there is a problem to be solved. In the Preliminary Investigation the basic job of the analyst is to determine if a problem does, in fact, exist. In the Detailed Systems Investigation and Analysis the objective of the analyst is to determine "What takes place in the current system?" and then to determine WHY a problem exists. The determination of what takes place is the "fact-gathering" stage of the systems project and is vital to the success of the project.

In the DETAILED SYSTEMS INVESTIGATION, the emphasis is on WHAT is taking place, that is, what procedures are used, what documents are used, how many people are involved in a particular operation, how many transactions are processed, what information is generated and used within the system which is under study, and other such questions. In the ANALYSIS of this information, there are two basic questions which must occupy the mind of the analyst—WHY are these activities being performed in the manner found in the detailed system investigation, and WHERE can improvements and changes be made to eliminate the problems which have been found within the system.

Management has long used a simple three-step approach to decision making that is well suited to the task of systems analysis. These steps include:

1. Get the Facts
2. Analyze the Facts
3. Make a Decision

By applying these simple rules to the systems project, the analyst can be assured of a scientific approach to the solution of the problem within the study.

THE DETAILED SYSTEMS INVESTIGATION

In the detailed systems investigation the analyst will use many of the techniques utilized in the preliminary investigation to gather the facts. The prime difference lies in the amount and depth of the investigation that is taking place. The basic fact-gathering techniques utilized in the detailed systems investigation include:

1. The Interview
2. The Questionnaire
3. The Detailed Review and Analysis of Current Documentation and Operating Forms
4. Personal Observation of Procedures

During and following the fact-gathering process the analyst must develop a critical and questioning approach to each of the procedures within the system, for the desired outcome of this phase of the systems project is for the analyst to be able to submit recommendations to management of alternative solutions to the problems found within the current system.

SYSTEMS ANALYSIS

After the analyst has gathered together as much information as possible about the system and the current operating procedures have been documented, the analyst must then begin the critical task of determining what is wrong with current operating procedures. Although many difficulties will be revealed during the fact-gathering steps, a systematic approach to analysis should be undertaken. The basic tools of analysis consist of obtaining the answers to the questions of WHO? WHAT? WHERE? WHEN? AND HOW? Within each of these questions the analyst should ask WHY? Included should be answers to the following questions relative to the system under study.

1. WHO - Who performs each of the procedures within the system? WHY? The analyst should question why—is the correct person performing the activity? Should the job duties be assigned to someone else?

2. WHAT - What is being done, what procedures are being followed. WHY? Often procedures have been followed for many years and no one knows why. The analyst should question why a procedure is being done at all. Is it necessary?

3. WHERE - Where are operations being performed? WHY? Where should they be performed? Could they be performed more efficiently elsewhere?

4. WHEN - When is a procedure performed? WHY? Why is it being performed at this time? Is this the best time?

5. HOW - How is a procedure performed? WHY? Why is it performed in that manner? Could it be performed better, cheaper, more efficiently in some other manner?

Note that in each step of the analysis the predominant question is WHY? By developing an objective, questioning, and inquiring approach to analysis the analyst is more likely to succeed in developing the most efficient design that will contribute to the company's objectives.

The first step in determining the Whys of the current system is to adequately define and document the current system. This requires that the analyst understand the inputs, outputs, flows, files, and restrictions and controls which are currently a part of the system. Only after this complete understanding will the analyst be able to isolate the features of the system which are causing the problems and therefore be able to propose solutions.

The following chart summarizes the important questions which the analyst must ask during the Detailed Systems Investigation, the System Analysis and the System Design phases of the systems project. The Systems Design Phase will be discussed in detail in subsequent chapters.

DETAILED SYSTEM INVESTIGATION	ANALYSIS	SYSTEMS DESIGN
What—is done? Where—is it done? When—is it done? Who—does it? How—is it done? Should—it be done at all?	Why—is it done? Why—is it done there? Why—is it done at this time? Why—does this person do it? Why—is it done this way? Why—should (or should not) it be done?	What—should be done? Where—should it be done? When—should it be done? Who—should do it? How—should it be done? Should—it be changed (or eliminated)?

Figure 3-1 Questions in Systems Project

When conducting the systems analysis phase of the systems project, the analyst must learn to be objective in his analysis of the present system and should not presuppose that there is no existing method for coping with the current problem as stated in the original proposal. The existing system may not be operating properly, but this does not always indicate that a major redesign effort is necessary. The analyst should not only look for inadequacies of the present system, including the lack of adequate controls, problems of policy and procedures, and organizational problems, but also should develop an awareness of the good features of the system as well.

STEPS IN THE DETAILED SYSTEMS INVESTIGATION AND ANALYSIS

The task of analyzing a problem large or important enough to warrant a change in the system is often a complex and lengthy job, and often affects the job duties of individuals associated with the system for many years to come; therefore, this "fact-finding" phase of the systems project must be carefully planned. There are a number of well-defined steps which must be taken to assure the success of the project. These steps include:

1. Review the organizational structure of supervisory and line personnel associated with the system under review.

2. Conduct interviews with supervisory and line personnel and/or obtain information on the operating level by use of questionnaires.

3. Obtain copies of operating documents used in all procedures throughout the system.

4. Conduct detailed observation of all procedures in the system.

5. Document and record all procedures.

6. Analyze the existing procedures within the system for both good and bad features.

7. Make a presentation to management recommending alternative solutions to the problems uncovered in the current system.

REVIEW THE ORGANIZATIONAL STRUCTURE

During the preliminary investigation the analyst should have obtained and documented the organizational structure of all levels of personnel associated with the procedures under review within the system. At this time, however, the analyst should review this organizational structure in order to determine which supervisory or line personnel to interview in the detailed investigation. If large numbers of employees are involved, the analyst may want to determine "key" personnel or senior employees for interviewing. With other individuals, the analyst may utilize a questionnaire in order to elicit information pertaining to the system from them.

THE INTERVIEW

As previously discussed in Chapter 2, the interview is one of the most valuable fact-gathering techniques available for use by the analyst. In the preliminary investigation, the basic purpose of the interview was to discover if the problem existed as specified in the request for system services; and interviews at that stage were conducted with management-level personnel. In the Detailed Systems Investigation it is necessary to uncover specific facts and learn of the detailed operating procedures that are taking place within the system; therefore, the interviews at this phase of the systems project must be conducted on a lower level. In the detailed investigation, interviews should be arranged with lower-level supervisory personnel, foremen, and line personnel such as key production workers or clerical personnel—that is, those persons who are actually doing the day-to-day work.

It should be noted that even though the systems analyst is interviewing on a much lower level in terms of the structure of the company, the same procedures and care should be followed in conducting the interview as previously discussed; that is, the objectives of the interview should be clearly established, specific questions should be prepared in advance, appointments should be made with the person to be interviewed, and the results of the interview should be recorded and formally documented.

QUESTIONNAIRES

In large systems projects where it is not possible to interview all individuals associated with the procedures, the questionnaire is sometimes a valuable tool of the systems analyst. The questionnaire is merely a form containing a number of questions seeking facts or opinions from a large segment of employees that may be able to contribute to the success of the study. The questionnaire is most useful in obtaining answers to brief questions such as to whom do you report?, Who reports to you?, etc. Questionnaires normally seek to obtain information such as job duties, workloads, reports received, volume of transactions handled, type of job duties, difficulties, and perhaps opinions of how the job could be better or more efficiently performed.

Questionnaires should be brief and easy to complete. One advantage of the questionnaire, when used extensively, is that it tends to provide all employees with input and suggestions for the improvement of their job duties. This sometimes can be an important human relations factor, particularly when it is known or "rumored" that a change in method of operation or the system is to take place. Another advantage is that there is an unlimited range of topics which can be covered and it is much more economical in terms of time and related costs when compared to the personal interview.

Disadvantages include the difficulty in writing a questionnaire that can be easily understood and not misinterpreted. A questionnaire reportedly used by one company stated "HOW MANY PEOPLE DO YOU HAVE WORKING FOR YOU - (Broken down by sex)." The respondent replied "None that I know of broken down by sex but several have a drinking problem."

Other deficiencies of the questionnaire are that the questionnaire may not truly uncover specific problems existing with a procedure, replies may be inadequate, and people may be unwilling to put controversial comments in writing. Many employees may be willing to comment or talk to the analyst personally about operating problems, but may not be willing to go on record in writing.

Specific disadvantages include the following:

1. People object to answering numerous, time consuming, tedious questionnaires.

2. People may not understand the purpose of the questionnaire and it may create undue and unnecessary apprehension among the workers.

3. It is difficult to design questionnaires which will obtain the exact information desired.

4. Clerical expense in preparing a questionnaire may be prohibitive.

5. Problems of interpreting questions and subsequent interpreting of answers may lead to inaccuracies.

In drawing up a questionnaire the following rules should be applied:

1. Make the questionnaire as brief as possible and as easy as possible to answer.

2. Arrange questions in a logical order.

3. Phrase questions to avoid misunderstandings.

4. Phrase questions to avoid giving clues to expected answers.

5. Avoid questions which require long narrative type answers. If this type of information is needed schedule an interview.

6. Do not ask questions that might appear threatening to a person's job if an answer is given.

7. Determine carefully what questions are required to obtain the information desired.

In summary, the questionnaire should be used in the detailed systems investigation when it is desirable to obtain information from a large number of people on a variety of topics related to the system.

OBTAIN COPIES OF OPERATING DOCUMENTS

During the preliminary investigation the analyst will normally review the system documentation to obtain an overview of the operating procedures and documents that are supposed to be utilized in the system. As pointed out, in many organizations the system documentation may not be current, forms may no longer be utilized, and procedures may have been modified. It is important, therefore, that the analyst obtain copies of the actual forms and operating documents that are currently being used in the system. The analyst should obtain blank copies of these forms, but in addition, should obtain samples of completed forms that have actually been utilized in the procedures during a processing cycle. Samples of documents can normally be obtained when interviewing individuals associated with each procedure. These completed forms then provide the basis for the analysis of the existing procedures within the system.

OBSERVING CURRENT OPERATING PROCEDURES

After the analyst has reviewed the current documentation, obtained copies of operating documents, and interviewed key personnel in an attempt to gather information relative to the detailed operation of the system and its problems, it is then necessary to actually observe current operating practices. This is often a very revealing part of the detailed systems investigation, for the analyst is likely to discover that neither the system documentation nor the statements from those interviewed truly reflect the system in actual operation.

Personal observation has a number of special values, including:

1. Personal observation affords the only positive means of measuring the dependability of statements made in the interview, or of determining if current procedures as specified in the documentation manuals are being followed.

2. Personal observation can correct "hazy" understandings or erroneous impressions obtained through interviewing.

3. Personal observation can protect sources of "off the record" information. By observing operations and basing subsequent actions and recommendations on what is seen rather than what is heard, the analyst can protect employees' confidential information.

4. Recommendations usually receive better acceptance when it can be shown that they are based upon personal observation of actual operations.

5. Personal observation helps the analyst acquire the "know-how" needed when he is asked to assist or supervise the testing or installation of the changes which have been recommended.

6. As observations are made the analyst can become better acquainted with the operating personnel who may be implementing any new systems which are recommended and subsequently designed.

Methods of Examining Operations

To gain first-hand knowledge of the procedures that are taking place within a system, it will be necessary for the analyst to personally examine each procedure, including each form, report, record, and facility involved in the system. This is the only way in which it can be assured that the analyst understands the inner workings of the present system.

Review Transactions

The analyst should examine, in proper sequence, each processing step within the procedure. During exploration the analyst should:

1. Ask sufficient pertinent questions to ensure that he thoroughly understands the present operations of each procedure, including the methods for handling exceptions or difficult cases, or problems which arise within the procedure which are not covered by standard operating procedures. In many procedures there are exceptional conditions which arise that are handled by using "common sense." These exceptions should be noted. For example, what happens in a payroll system if an employee loses a time card? What is the procedure if an employee checks in five minutes late but works 10 minutes overtime? Often the rules for cases such as this are not written down or formalized; however, when designing a new system the analyst should attempt to document as many exceptions as possible.

2. Learn the frequency of each transaction. Are there numerous additions and deletions? Learn the frequency of changes to master files. Such information is significant during the design phase of a study in determining the type of file which should be designed.

3. Observe the beginning steps in a procedure, the completed output from the procedure, and unfinished or intermediate steps in the procedure during a processing cycle.

4. Examine each pertinent form, record, and report with regard to the actual value of its contents. What purpose does each item of information serve?

5. Consider the work of each person associated with the system, keeping in mind the following questions: What information does he receive from other people? What information does he generate on his job? What tools does he use in the procedure? Is there alternative equipment which might be used? To whom does he pass on information? Does the recipient actually require all of it?

6. Examine the overall personnel and related information associated with the system. What data are required in the processing? In what volume? What processing is now given to the data? What items, in what volume, could be processed using the computer? What changes can be made to simplify or improve present procedures without EDP?

7. Consult those receiving current reports to determine if the reports they receive are complete, timely, accurate, and in the most useful form. Inquire as to what information received can be eliminated or improved upon, and what information not presently received would be helpful.

Forms

As the analyst is tracing a sample transaction through the system, copies of forms used in each procedure should also be obtained. This analysis should include:

1. Samples of blank forms, and samples of forms that have been completed.

2. The origin of each item on the form, that is, who records the information on the form. Does the information originate from the employee, such as the social security number, or from another source such as the deductions amount for union dues.

3. Frequency and volume of usage of all forms, including the number of copies.

4. Distribution of copies.

5. End use of the copies at the point of distribution.

6. End use of each item of information on the form.

RECORDING THE FACTS

The importance of adequate records of interviews, facts, ideas, and observations cannot be overemphasized. Where facts are voluminous, it is difficult at times to appreciate their significance; and ideas and understandings or comments on the present system may be easily forgotten as the analyst moves from procedure to procedure through a system. Therefore a basic rule is: WRITE IT DOWN. The following principles should be followed in documenting the detailed investigation:

1. Record all information as it is obtained.

2. Use the simplest recording manner possible consistent with completeness.

3. Record your findings in such a way that they can be understood by someone who is not a member of the systems staff.

4. Arrange your recordings so that related information can be easily brought together and coordinated.

Although many analysts use specialized forms for documenting a system, such as special forms for interviews, special forms for summarizing document contents, etc., one of the most effective methods is a simple narrative form in which a statement is merely made of what is taking place, problems that are apparent, and suggestions for improvement.

SUMMARY OF FACT-GATHERING PHASE

In the fact-gathering phase of the systems project, the analyst must develop a critical and questioning attitude toward every aspect of the current operation. Being "critical" and "questioning" means being critical of the "system," not the personnel within the system. If procedures are not being followed, if forms are poorly designed, or if there is a lack of documentation, the analyst must guard against making personal criticism of the individuals assigned to the task of carrying out the procedures within the system. It must be remembered that the analyst must gain the confidence of those employees in the current system and not antagonize them; for these are the same employees who must implement any new systems and procedures which may be developed by the analyst.

There have been numerous studies which have indicated that new systems have failed because the line employees did not cooperate when the new system was installed. Thus, the analyst must constantly be aware that he is not only dealing with system, procedures, and methods of doing work, but with people as well.

ANALYSIS OF FACTS

As was noted previously, once the facts have been obtained it is necessary to analyze the entire system with the question WHY? utmost in mind. In order to analyze and interpret the facts which have been collected, however, it is necessary to arrange them in significant relationships, verify them, and fill in the gaps that may not be currently documented. One of the basic tools used by the analyst in this step is the flowchart.

There are numerous types of flowcharts used by systems personnel, including process flowcharts which illustrate the flow of paper through a system, work simplification-type flowcharts which are designed to measure and record specific details about operating procedures, etc. One of the most commonly used flowcharts used by the data processing systems analyst is the SYSTEMS FLOWCHART.

The systems flowchart is merely a general pictorial representation of the procedures which take place within the system. The systems flowchart includes a pictorial representation of all input and output operations and associated processing, including the creation of all files. It shows the flow of data through all parts of the data processing system; hence, the systems flowchart emphasizes the media involved and the work stations through which this media passes. The following symbols are commonly used in systems flowcharting.

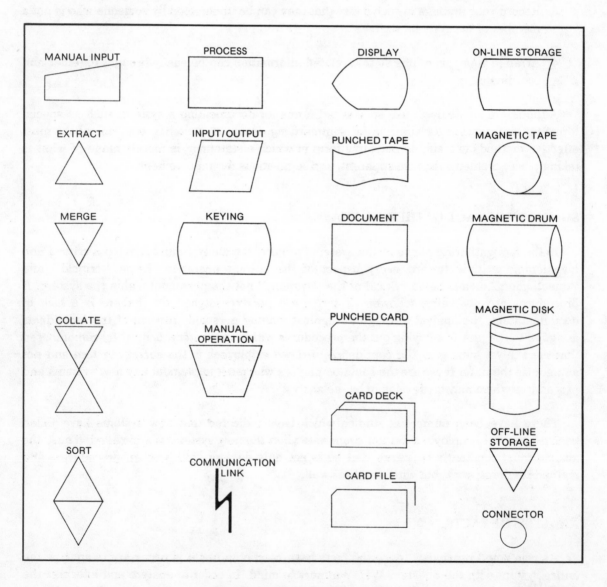

Figure 3-2 Flowchart Symbols

The following is an example of a systems flowchart.

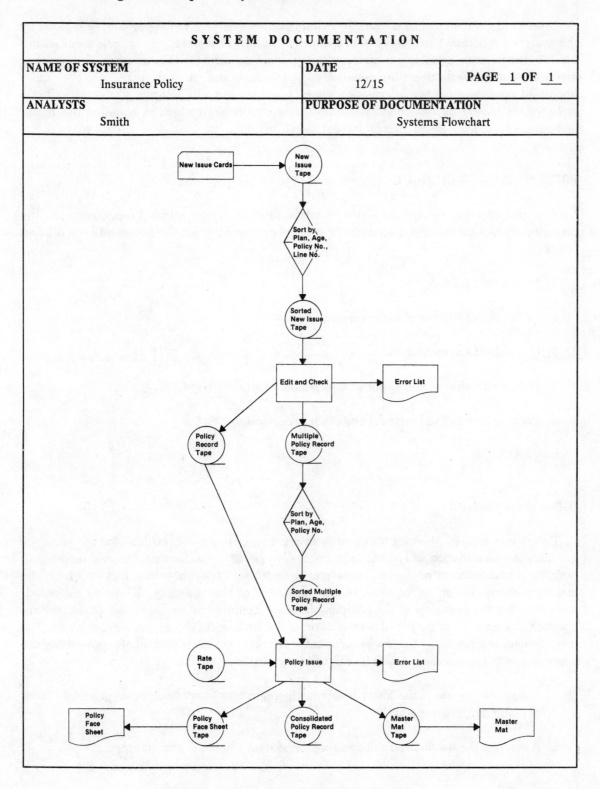

Figure 3-3 Example of Systems Flowchart

PRESENTATION OF FINDINGS

Once the system has been documented and possible solutions to the problem developed by the analysts, it becomes necessary to present the findings of the study to management and to indicate the possible solutions as seen by the systems department. The presentation which is made to management after the detailed fact-gathering and analysis takes place is quite important because, from the information given, management will determine whether or not a new system should be designed. Therefore, it is worthwhile to look at some of the better techniques for making management presentations.

PRESENTATION TECHNIQUES

The presentation by the analysts consists of two major areas: Preparation for the presentation and the presentation itself. The steps in preparing for the presentation are listed below:

1. Define the audience.

2. Define the objectives of the presentation.

3. Organize the presentation.

4. Define all technical and application-oriented terms to be used.

5. Prepare audio-visual material and other presentation aids.

6. <u>PRACTICE</u>

Define the Audience

Before any detailed planning for the presentation takes place, it is critical that the audience to whom the presentation will be made is defined. In general, the presentation will be made to persons in management positions. These management positions, however, can vary from the manager of the payroll department to the president of the company. It should be noted, however, that the president of the company or other higher-level management personnel are not normally involved in the initial presentation of the findings of the systems department. The management is more likely to be a department manager or a vice-president. By presenting the analysis first to management at this level, the analyst can accomplish two purposes:

1. It does not appear to the lower-level management that they are being bypassed in favor of going right "to the top."

2. It involves the people who will be using the system when it is implemented.

By accomplishing these purposes, the analysts can many times ensure cooperation by the persons who will be implementing and using the system. As has been noted previously, any effort which the analysts can make to reduce the possible negative reaction to the change of a new system will be beneficial. Therefore, including the lower-level management in decisions concerning the design and implementation of a new system is quite important.

Define the Objectives

The objective of any communication is to produce a mental picture in the mind of the receiver which duplicates the picture in the mind of the sender. Therefore, it is quite important that the analyst define the "picture" which he wishes to impart to the management to whom the presentation is given.

In general, the objectives of a management presentation concerning a detailed systems investigation and analysis are:

1. To apprise management of the status of the current system.

2. To express the findings of the systems department relative to the problems which have been found to exist within the current system.

3. To explain alternative solutions which have been determined by the systems department to be feasible.

4. To detail costs and implementation times for the alternative solutions.

It is quite important that these general objectives then be broken down into specific items which are to be presented. Without this definition of objectives, it is likely the analyst will not be able to plan a presentation which will adequately inform management of the results of the detailed systems investigation and analysis.

Organize the Presentation

Once the objectives of the presentation have been determined, the analyst must then set about to organize the presentation so that the objectives will be met. There are two major areas which must be determined when the presentation is organized:

1. In what sequence should the material be presented?

2. On which subjects should the major emphasis be placed?

The sequence in which the presentation is made can be quite critical because if the material is presented in such an illogical and unorganized manner that the persons listening to the presentation are confused and not sure about what was said, then all of the good work put into the investigation and analysis of the system will have been wasted.

In addition, it is quite important that emphasis be placed upon the proper subjects. In general, a presentation of the results of a detailed systems investigation and analysis should place emphasis on the problems encountered, the solutions proposed, and the costs of the solutions proposed. In most cases, the management personnel will not be concerned with the details of the system operation. If any of them are interested, they can consult the documentation which was prepared by the analyst and questions can be answered outside the meeting. Instead, they will be interested in what is wrong, how it can be fixed, and how much it will cost to fix it.

Define Terms

When a data processing systems analyst makes a presentation to management, it is likely that there are some terms, concerning both data processing and the system in question, which will not be familiar to the audience. It is important, therefore, that the analyst determine those areas where the persons hearing the presentation are not knowledgeable and to prepare material, either written or spoken, which will enable the participants to understand the terminology used in the presentation.

Prepare Auxiliary Materials

It has been shown that 75 percent of everything which is learned is learned through seeing. Only 13 percent is obtained through hearing and the remaining 12 percent is learned through the other senses. Therefore, it is quite important that the use of visual-type displays be planned for and used in a presentation.

There are a number of different types of audio-visual materials which can be used in a presentation. Some of the more useful ones are:

1. Chalkboard
2. Flip Charts
3. Overhead Projectors
4. Slides
5. Film Strips
6. Motion Pictures
7. Video-tape

These materials, used either separately or together, can add greatly to the effectiveness of a presentation. It must be noted, however, that this type of material must be prepared before the presentation is made. Therefore, a very important part of the planning process is to determine what use should be made of audio-visual materials and then to prepare them for presentation.

Practice

After all of the preparation discussed above has been completed, the analyst will have his presentation ready. One important element is missing, however, and this is PRACTICE. The analyst should never go into a presentation without thoroughly reviewing all of the material which he is to present and at least go over the presentation in his mind. In most cases, it is beneficial to actually go through several "dry runs" to be sure that all of the material which will be necessary is ready and that the timing of the presentation is correct. It is vitally important that the analyst be able to keep a smooth, enlightened presentation going and it is likely that this will not be possible unless he has "rehearsed" several times so that he is completely familiar with the presentation.

THE PRESENTATION

After the preparation has been completed, the time will come to make the presentation to management. In order to have a successful presentation, there are several elements which must be present, including the following:

1. The analyst must "sell" himself and his reliability.

2. The analyst must control the presentation.

3. The analyst must have the ability to answer any questions in terms understandable to the audience.

4. The analyst must use good speaking techniques.

Selling Reliability

In order to be successful in a presentation, the analyst must sell himself and the reliability which he represents as well as the material which has been accumulated from the detailed systems investigation. In some cases, a brilliant presentation will not convince management that the system should be designed and implemented because they are not sold on the individual who gave the presentation and who will be working on the project. As a corollary to this, it is also true that many systems which are not truly deserving of being implemented have been because of the "sales job" done by the analyst at a meeting with management.

Therefore, the analyst must give a presentation which shows confidence about the subject and the suggestions made by the systems department. He should avoid any conflicts with members of management who may attend the meeting while at the same time facing criticisms with the straight-forward approach of presenting the view of the systems department. This aspect of selling himself and his reliability should always be in the analyst's mind while he is making his presentation.

Control of the Presentation

The analyst must have control of the presentation during the time it is given. Obviously, the analyst will have the floor and preside over the meeting, but in having control of the meeting the analyst will be directing attentions to those things which he feels are important and will not let the meeting wander. In addition, if handled properly, the audience will feel that he has completely answered all of their questions and that he has convinced them that he is correct. If the analyst loses control of the meeting, then it is likely that no one will be convinced when the meeting is over.

One of the more important aspects of having control over a meeting is in identifying with the audience and having them identify with the analyst. It must always be on the analyst's mind that he cannot take a superior attitude toward the audience simply because he is more familiar with the subject matter. This can be accomplished by adequate preparation so that the material presented will be right for the audience, but can also be done by such things as having eye-contact with the audience so that they feel that the analyst is talking directly to them. Humor, when done properly, can have a very good effect on an audience. In general, the safest humor is for the analyst to poke fun at himself. This gives the audience the feeling that the analyst is an ''all right'' guy and they feel more at ease with him.

Again, keeping control of the presentation and keeping the audience attentive and receptive is mandatory for a good presentation.

Answering Questions

Inevitably there will be questions in a presentation and sometimes the questions can be quite difficult, especially from those who are pessimistic about the work of the systems department or from those whose jobs would be most affected by a new system. It is necessary that the analyst listen very intently to the question being asked and then respond with a straight-forward answer which answers the question.

Too often analysts will respond with an answer which does not answer the question of the audience or which, although technically correct, does not put it on a level which can be understood by the person asking the question. This, of course, will not only leave the question unanswered but will also cause the audience to feel that the analyst is not relating to them and can be one means of quickly losing control of the presentation. Therefore, answering questions asked from the audience can be very important.

Good Speaking Techniques

All of the previous elements of a good presentation are part of good speaking technique. There are others, however, which should be kept in mind by the analyst when making the presentation. Most good speakers concentrate on using techniques such as good voice and accent, making no distracting movements, and a relaxed approach in order to be as effective as possible. Another important element is the pace at which the presentation is given. If the pace is too fast for the audience, they will be lost and will lose interest; on the other hand, if the pace is too slow, the concentration of the audience will be lost and the presentation will not be effective.

It is also quite important that the analyst not read his speech. There is very little which will turn off an audience faster than the speaker reading his speech. This should be avoided at all costs.

All of the techniques and elements of speech-giving which have been discussed can, in most cases, be overcome by proper preparation. Thus, it is imperative that the analyst spend a great deal of time in preparation and practice so that the presentation will be effective.

SUMMARY

Once the detailed investigation has been completed, the analyst must analyze the data which has been gathered to determine exactly what is taking place in the system and to isolate and define any areas which may be causing problems within the system. The systems and the problem areas in particular must then be documented for presentation to management.

The presentation to management is a critical step in the process since without management approval no solutions to the problems can be developed by the systems department. Assuming that approval is received from management, the analysts will begin design of a new system. Subsequent chapters will deal with the designing and implementing of a new system.

CASE STUDY - JAMES TOOL COMPANY

Phase 1 of the systems project, the Initiation of the Systems Project and the Preliminary Investigation, revealed that the James Tool Company was having difficulty in the payroll department because union dues were often not deducted properly, the dues were not being forwarded to the union accurately, and there was an excessive amount of overtime being paid to those in the payroll department. Upon the request of Arnold Henderson, Vice-President of Financial Affairs, Clyde Harland, manager of the systems department, made a preliminary study of the problem. Harland subsequently recommended that a detailed study of the existing payroll system be undertaken. Thus, the second phase of the systems project begins—the Detailed Systems Investigation and Analysis.

Case Study - Organization Chart

On the basis of the job descriptions from Personnel and a list of the employees in the payroll department, a payroll organizational chart was constructed by the analysts Mard and Coswell in the preliminary investigation (Chapter 2). As one of the first steps in the detailed investigation, Mard and Coswell also prepared an organizational chart of the personnel department, and of the Employment section, in particular. These charts are illustrated in Figures 3-4 and 3-5.

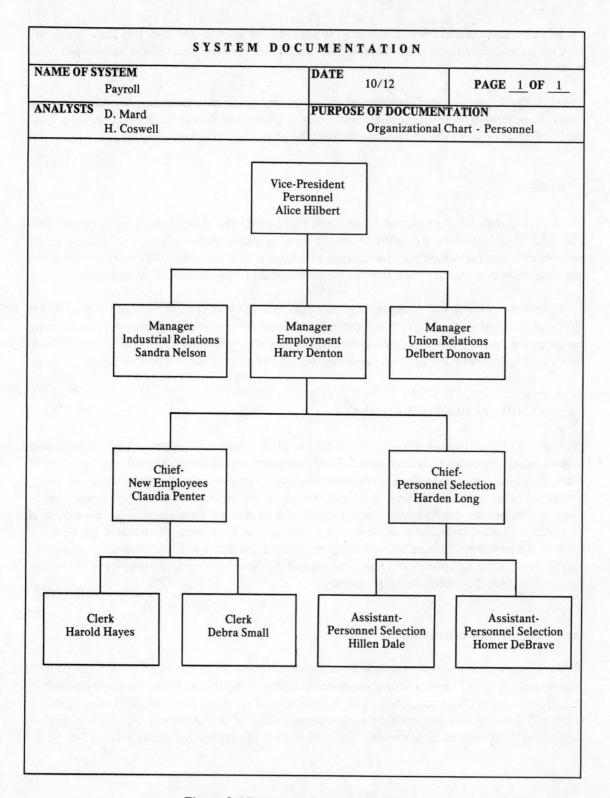

Figure 3-4 Personnel Organization Chart

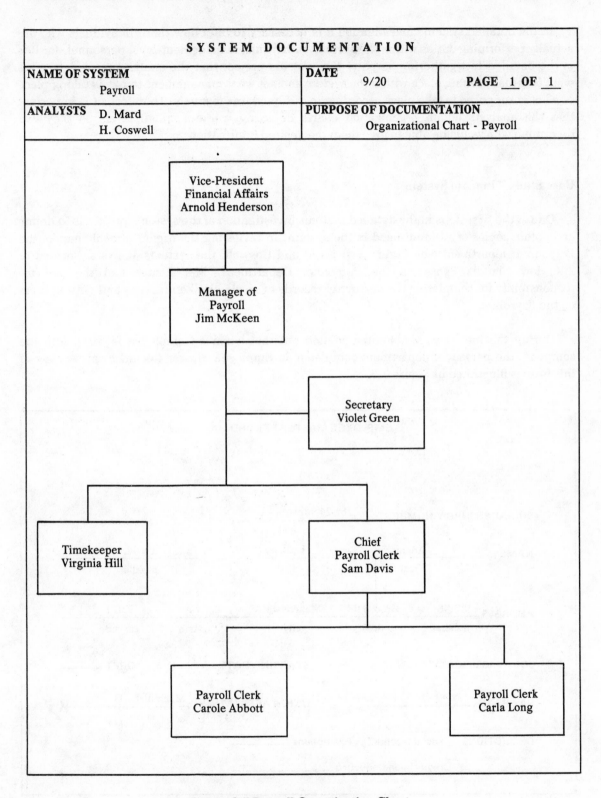

Figure 3-5 Payroll Organization Chart

In the detailed systems investigation it is necessary to interview those individuals who are actually performing duties within the system, rather than management-level personnel, for it is in the detailed fact-gathering stage of the systems project that the analysts must determine what is actually taking place within the system and not what management thinks is taking place or what is supposed to take place according to the standards manuals. Therefore, it is important that the analysts obtain organization charts depicting workers on all levels so they can determine who would be able to give them the desired information.

Case Study - Input to System

One of the first steps in the detailed systems investigation of the systems project is to define and obtain copies of all input used in the system. In reviewing the organizational chart of the Personnel Department (see Figure 3-4) Mard and Coswell, the systems analysts, decided to interview Claudia Penter of the Personnel Department. Ms. Penter had the primary responsibility for completing the personnel records of newly hired employees and getting them on the payroll.

During the interview Ms. Penter pointed out that when an employee is hired into the company, the personnel department completes an Employee Master Record form. A copy of this form is illustrated in Figure 3-6.

EMPLOYEE MASTER RECORD

SOCIAL SECURITY NUMBER _____ 261-24-9021 _____

NAME _____ Adams _____ Anthony _____ David _____
                       last   **first**   **middle**

ADDRESS ____ 7148 _____ Groove Rd. ___ Manchester, _____ CA _____ 92101 _____
            number   **street**   **city**   **state**   **zip code**

JOB CLASSIFICATION ___ 3F ____   STARTING DATE __ 1/21 ___   DEPT. __ 01 __

PAY RATE ___ $4.40/Hr. _____   UNION AFFILIATION __ Machinists (1) _____

DEDUCTIONS   Federal Income Tax Exemptions ___ 3 _____

                 Credit Union Deduction _____

Figure 3-6 Example of Employee Master Record Form

When the person is hired, the personnel department fills out the form listing the essential information about the employee, including the social security number, name, address, job classification, etc. This form is then sent to the payroll department. When the form is received in the payroll department, it is filed in alphabetical order by last name.

The analysts noted that although the form contained the essential information about the employee needed by the payroll department, there was no place on the form to record changes in employee status, such as a change in pay rate, a change in union classification, etc. In addition, there was no unique identifying employee number which could be used to identify the department in which the employee worked or which could be used for other identifying purposes.

When questioned about the procedures which are followed when there are changes in the status of an employee, such as a change of pay, Ms. Penter informed the analysts that another form was used. This form is the Deduction and Pay Rate Change Form and is illustrated in Figure 3-7.

```
┌──────────────────────────────────────────────────────────────────────┐
│                                                                        │
│               DEDUCTION AND PAY RATE CHANGE FORM                       │
│                                                                        │
│                                                                        │
│                                                                        │
│   SOCIAL SECURITY NUMBER _____ - _____ - _____          │
│                                                                        │
│   NAME _____       │
│             last              first               middle              │
│                                                                        │
│   DEDUCTION CHANGE  ☐   Income Tax Exemption   Old _____  New _____│
│                                                                        │
│                         Credit Union Deduction  Old $_____  New $_____ │
│                                                                        │
│   PAY RATE CHANGE   ☐   Old $_____   New $_____                   │
│                                                                        │
│   EFFECTIVE DATE        _____ / _____ / _____                       │
│                                                                        │
└──────────────────────────────────────────────────────────────────────┘
```

Figure 3-7 Deduction and Pay Rate Change Form

When personnel is notified of a change in status of an employee, such as a change in pay rate, the Deduction and Pay Rate Form is completed and a copy is sent to the payroll department. A copy of this form is then filed with the original Employee Master Record Form.

The analysts noted that there is no place on the Deduction and Pay Rate Change Form for changes relative to union dues. When questioned about the procedure for handling union dues when currently employed individuals join the union, Ms. Penter informed the analyst that the union was supposed to inform the personnel department when an employee became a member of the union. "They're supposed to send an official notification," she confided, "but to save paper work sometimes I am notified by telephone."

Upon conclusion of the interview, Mard and Coswell prepared the documentation illustrated in Figure 3-8

SYSTEM DOCUMENTATION		
NAME OF SYSTEM Payroll	**DATE** October 15	**PAGE** 1 OF 1
ANALYSTS D. Mard H. Coswell	**PURPOSE OF DOCUMENTATION** Interview with Claudia Penter, Personnel Dept.	

Two basic forms are used in the Personnel Department when dealing with the Payroll Department--the Employee Master Record Form and the Deduction and Pay Rate Change Form. When an employee is hired by the company, the Personnel Department fills in the Employee Master Record Form, listing all of the vital information concerning the employee. This form is then sent to the Payroll Department. Whenever a change in the status of the employee takes place, such as an increase in pay or a change in the number of deductions, the Deduction and Pay Rate Change Form is completed and sent to the Payroll Department. This change form is then filed with the Employee Master Record Form.

Several notable deficiencies are apparent in this procedure. There is no formal method for recording information relative to a change in status when a employee joins the union on either the Employee Master Record or the Deduction and Pay Rate Change Form. This undoubtedly is a cause of many of the errors relative to the proper deduction of union dues.

Likewise, the filing of the Deduction and Pay Rate Change Form behind the Employee Master in the Payroll Department could lead to errors in the payroll system, as the Deduction and Pay Rate Change Form provides the only historical record of an employee's current status. A loss or misfiling of the Deduction and Pay Rate Change Form could result in the erroneous preparation of the payroll.

A review of the current documentation of the payroll system also indicates no standardized procedure for the handling of changes in status of the employee relative to employees joining the union.

Figure 3-8 Documentation of Interview

Case Study - Interview - The Payroll Department

After interviewing Claudia Penter of the personnel department, it was decided that the analysts should interview the chief payroll clerk, Sam Davis. During the interview, Davis pointed out that when an employee is hired, a copy of the Employee Master Record Form is forwarded to the payroll department and is filed with the other master sheets by the time clerk in the payroll department. It was then noted that the time clerk in payroll types up a time card each week for all employees in the firm. Thus, each week the time clerk was required to type over 450 time cards containing the employee name, period-ending date, the social security number, and the union code. Figure 3-9 illustrates the time card.

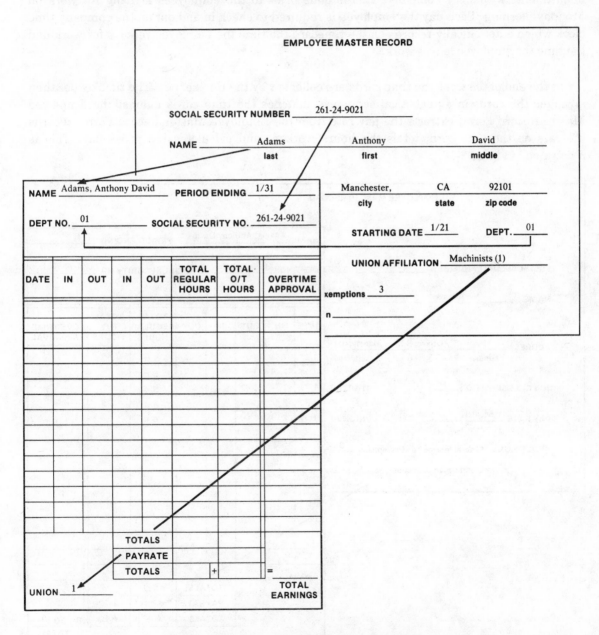

Figure 3-9 Example of Time Card Preparation

The analysts noted that this repetitive task was very time-consuming and subject to errors that could create considerable difficulty in the payroll department. For example, an error in typing the union code or the social security number could result in the erroneous preparation of the payroll!

The analysts were informed by Davis, the chief payroll clerk, that the pay rate was not entered on the time card at this time because an employee's pay rate is deemed confidential.

After the time cards are typed, the cards are manually arranged in alphabetical sequence within department number and distributed to the time clock racks located in each of the departments within the company. This is done prior to the employees arriving for work on Monday morning. Each day the employee is required to check in and out on the company time clock which automatically records on the timecard the time the employee reports for work and the time the employee leaves work.

At the end of the week the time cards are collected by the timekeeper. The timekeeper then arranges the cards in alphabetical sequence, matches the time cards against the Employee Master Record Form, extracts the pay rate from the Employee Master Record Form, records the rate on the time card, totals the hours worked, and calculates the gross pay. This is illustrated in Figure 3-10.

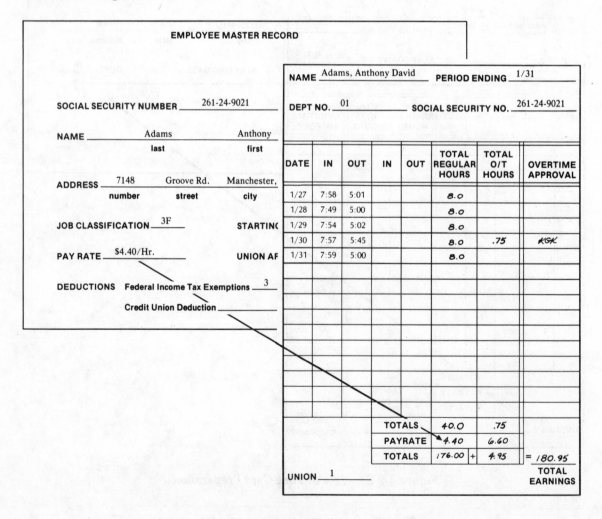

Figure 3-10 Completion of Employee Time Card

By matching the time cards against the Employee Master Records, the payroll department is assured that everyone will be paid on time, for if there is a missing time card it would be detected prior to the distribution of the paycheck on Thursday and an investigation could be undertaken to determine why a time card was not turned in for a given employee.

Likewise, matching the time cards to the Employee Master Records prevents someone from turning in a duplicate time card or a time card for a fictitious employee. This type of activity would be caught because there would be no master record. Thus, the company was protected against fraudulently issued checks, according to the chief payroll clerk.

Case Study - Deducting Union Dues

When asked about the deductions for union dues, Mr. Davis, the chief payroll clerk, stated that this procedure was well documented and the analysts were directed to the company standards and procedures manual.

A detailed analysis of the documentation from the standards and procedures manual revealed the procedure specified in the page from the documentation illustrated in Figure 3-11 was followed.

JAMES TOOL CO.   **PAYROLL**
SYSTEMS AND PROCEDURES   **18**

VII. Union Deductions

Deductions for union membership are taken from the pay of those employees who are members of a union. The union dues are deducted each week from the paychecks and the amount deducted is placed in an account. On the 15th of each month, the total amount deducted from the paychecks is paid to each of the applicable unions. The following is the procedure which should be used to calculate and record the deductions made for each union employee.

A. Unions and Rates

The following unions have employees in James Tool Co. In addition, their monthly dues are given.

Machinists Union	$ 6.00/mo.	$1.50/wk.
Electricians Union	$ 8.00/mo.	$2.00/wk.
Plumbers Union	$ 8.00/mo.	$2.00/wk.
Teamsters Union	$12.00/mo.	$3.00/wk.

B. Deduction Procedure

1. Determine from the Employee Master Sheet for each employee the union affiliation, if any (see Exhibit 1).

2. From the monthly dues in (A) above, determine the weekly deduction which must be deducted from the paycheck.

Figure 3-11 Procedure from Standards and Procedures Manual

The analyst noted that there was no union code stated in the procedure manual, yet there was a union code on the Employee Master Record Form and the Employee Time Cards. When the payroll clerk was questioned, the analysts were informed that one of the payroll clerks suggested using a numeric union code several years ago so that the name of the union would not have to be typed each time on the time card. Everyone thought this was a good idea and so the numeric union code was adopted for use on the time cards.

Case Study - Tax Deductions

The analysts also inquired as to the method used to deduct taxes. The payroll clerk, Sam Davis, explained that tax tables are used. The table illustrated in Figure 3-12 is an example of the tables which are used to determine the deductions made for Federal Income Tax. By determining the wages which are paid and the number of deductions which are to be made, the amount of tax to be withheld is determined from the table. Although the table illustrated in Figure 3-12 is for Federal Income Tax, the tables used for the other taxes are quite similar, and examples of each of these tables would be gathered by the systems analysts.

MARRIED Persons — WEEKLY Payroll Period

And the wages are—		And the number of withholding allowances claimed is—										
At least	But less than	0	1	2	3	4	5	6	7	8	9	10 or more
		The amount of income tax to be withheld shall be—										
$100	$105	$14.10	$11.80	$9.50	$7.20	$4.90	$2.80	$.80	$0	$0	$0	$0
105	110	14.90	12.60	10.30	8.00	5.70	3.50	1.50	0	0	0	0
110	115	15.70	13.40	11.10	8.80	6.50	4.20	2.20	.10	0	0	0
115	120	16.50	14.20	11.90	9.60	7.30	5.00	2.90	.80	0	0	0
120	125	17.30	15.00	12.70	10.40	8.10	5.80	3.60	1.50	0	0	0
125	130	18.10	15.80	13.50	11.20	8.90	6.60	4.30	2.20	.20	0	0
130	135	18.90	16.60	14.30	12.00	9.70	7.40	5.10	2.90	.90	0	0
135	140	19.70	17.40	15.10	12.80	10.50	8.20	5.90	3.60	1.60	0	0
140	145	20.50	18.20	15.90	13.60	11.30	9.00	6.70	4.40	2.30	.30	0
145	150	21.30	19.00	16.70	14.40	12.10	9.80	7.50	5.20	3.00	1.00	0
150	160	22.50	20.20	17.90	15.60	13.30	11.00	8.70	6.40	4.10	2.00	0
160	170	24.10	21.80	19.50	17.20	14.90	12.60	10.30	8.00	5.70	3.40	1.40
170	180	26.00	23.40	21.10	18.80	16.50	14.20	11.90	9.60	7.30	5.00	2.80
180	190	28.00	25.20	22.70	20.40	18.10	15.80	13.50	11.20	8.90	6.60	4.30
190	200	30.00	27.20	24.30	22.00	19.70	17.40	15.10	12.80	10.50	8.20	5.90
200	210	32.00	29.20	26.30	23.60	21.30	19.00	16.70	14.40	12.10	9.80	7.50
210	220	34.40	31.20	28.30	25.40	22.90	20.60	18.30	16.00	13.70	11.40	9.10
220	230	36.80	33.30	30.30	27.40	24.50	22.20	19.90	17.60	15.30	13.00	10.70
230	240	39.20	35.70	32.30	29.40	26.50	23.80	21.50	19.20	16.90	14.60	12.30
240	250	41.60	38.10	34.60	31.40	28.50	25.60	23.10	20.80	18.50	16.20	13.90
250	260	44.00	40.50	37.00	33.60	30.50	27.60	24.70	22.40	20.10	17.80	15.50
260	270	46.40	42.90	39.40	36.00	32.50	29.60	26.70	24.00	21.70	19.40	17.10
270	280	48.80	45.30	41.80	38.40	34.90	31.60	28.70	25.80	23.30	21.00	18.70
280	290	51.20	47.70	44.20	40.80	37.30	33.90	30.70	27.80	25.00	22.60	20.30
290	300	53.60	50.10	46.60	43.20	39.70	36.30	32.80	29.80	27.00	24.20	21.90
300	310	56.00	52.50	49.00	45.60	42.10	38.70	35.20	31.80	29.00	26.10	23.50
310	320	58.40	54.90	51.40	48.00	44.50	41.10	37.60	34.10	31.00	28.10	25.20
320	330	60.80	57.30	53.80	50.40	46.90	43.50	40.00	36.50	33.10	30.10	27.20
330	340	63.60	59.70	56.20	52.80	49.30	45.90	42.40	38.90	35.50	32.10	29.20
340	350	66.40	62.40	58.60	55.20	51.70	48.30	44.80	41.30	37.90	34.40	31.20

Figure 3-12 Example of Tax Tables

Case Study - Summary - Input Fact-Gathering

One of the important reasons for examining the input used in the current system, besides determining what input is required for the system to function, is to look for the duplication of data. Another reason is the possibility that input data is being prepared which is not necessary to the processing within the system. As systems evolve and procedures change, it may be found that old procedures are followed even though they are no longer necessary. Therefore, it is incumbent upon the analyst to make a thorough study of all of the input to the system to determine if it is all necessary.

It is also possible that because of changes a form will not contain enough information for the processing. For example, in the sample payroll system, there is no space provided on the Employee Master Record Form for changes in the pay rate, deduction changes, and changes in union affiliations. Therefore, in modifying the current system or designing a new one, the analysts must be aware that these items must be included on the form.

Another area of concern to the analysts is the error factor associated with the input data. In many systems where problems arise, there is a high proportion of errors which are occurring and which directly contribute to the inefficiencies of the system. It may be, therefore, that the elimination of one part of the input data which may be redundant will reduce the error rate significantly and correct the problems within the system. In the payroll system, the transferring of information from the Employee Master Record Form to the time cards, the matching of the time cards to the Employee Master Record, and the recording of the pay rate on the time card from the Master Record was error prone; therefore, the analysts were aware that this procedure should be changed.

It is also important that the analyst determine the source of the input data and its method of preparation. In some instances, the problems within the system may be directly related to the source of the data. In particular, if there is some time deadline which must be met and the data is not available at the given time, this delay in the subsequent processing may account for many of the problems within the system. A mere change in the source of the input data or the method of preparation may eliminate the problem. There may also be problems with the method of recording the data for use at a later time or in the clearness of the source documents when preparing the input data. In all cases, this type of analysis is necessary in order that the analyst have a complete picture of the system as it currently exists.

Although in most cases this analysis of the input data will not lead directly to a solution, it is important that the analyst look for trends such as those mentioned because they could lead to a solution. It should be noted, however, that it is most difficult to analyze all of the input data without being aware of the output which is required from the system and the method of processing the input data. Therefore, together with the input of the payroll system the analysts must gather detailed facts concerning the output of the system.

Case Study - Output

The same type of fact-gathering must take place for the output currently produced by the system as was done for the input to the system, that is, what output is produced, and in the case of the output, where does it go. In all systems, output is information of some kind. It may be management reports or it may be payroll checks, but the output is information which is utilized in some manner.

It is necessary, therefore, to analyze what information is being produced and whether it is the information which will be most useful to those using it. In many systems, the output being produced is in the form of printed reports. These reports may be prepared by a computer or they may be prepared manually, such as in the sample payroll system, but they are designed to supply information to the user. If a system is not working effectively, one of the problems may be in the information which is being supplied to the user. If the user of the reports is not receiving information which will be useful to him or if the information on the reports is not accurate, then all of the functions of the system are not worthwhile because the end product of the system, information, is not useful or correct. Therefore, the analyst must be careful in analyzing the output from a system to ensure that if the data is not correct or useful, then he has obtained from the users the data which they would find to be most useful.

In order to determine the output which is produced by the system and the use to which it is put, the analysts interviewed Sam Davis, the chief payroll clerk, Jim McKeen, the manager of the payroll department, and others who used the output produced from the system. From these interviews and the systems documentation which was available, they were able to determine the output produced from the payroll system.

Case Study - Payroll Register

The first report which was shown to the analysts is the payroll register. The payroll register is illustrated in Figure 3-13.

JAMES TOOL COMPANY
PAYROLL REGISTER

WEEK ENDING 01 / 31   PAGE 1

EMPLOYEE DATA		EARNINGS			DEDUCTIONS					NET PAY	
Employee Name	Soc. Sec. Number	Regular Earnings	Overtime Earnings	Total Earnings	Federal Tax	FICA	State Tax	Credit Union	Union Dues	Net Amount	Check Number
Adams, Anthony David	261-24-9021	176.00	4.95	180.95	23.30	10.59	3.80		3.00	140.26	02232

Figure 3-13 Example of Payroll Register

Note from the example of the payroll register in Figure 3-13 that each employee is listed together with his earnings, his deductions, and his net pay. In .addition, the number of the check which is written to pay him is recorded on the payroll register. This register is used as a record of payments of the employees by James Tool Co. so that it knows who it paid and how much. In addition, it indicates the amounts which have been deducted each week for taxes, social security, and union dues. These amounts must be used to indicate how much money James Tool Co. must pay to the federal government, the state government, and the unions.

Case Study - Employee Compensation Record

It will be noted that the Payroll Register contains the weekly payroll information for all the employees of the company but does not contain any year-to-date information for each employee. Therefore, the Employee Compensation Record is kept to show the complete payroll history of each employee. An example of the Employee Compensation Record is illustrated in Figure 3-14.

JAMES TOOL COMPANY
EMPLOYEE COMPENSATION RECORD

EMPLOYEE NAME _Adams, Anthony David_   MARITAL STATUS _Married_

EMPLOYEE ADDRESS _7148 Groove Rd., Manchester, Calif. 92101_   NUMBER OF EXEMPTIONS _3_

SOCIAL SECURITY NUMBER _261-24-9021_   DATE EMPLOYED _01/16_

UNION AFFILIATION _Machinists (1)_   DATE TERMINATED _____

| | WEEKLY PAYROLL | | | | | | | | | | YEAR-TO-DATE | | | | | | | | |
| | EARNINGS | | | DEDUCTIONS | | | | | NET PAY | | EARNINGS | | | DEDUCTIONS | | | | | NET PAY |
Week Ending	Regular Earnings	Overtime Earnings	Total Earnings	Federal Tax	FICA	State Tax	Credit Union	Union Dues	Net Amount	Check Number	Regular Earnings	Overtime Earnings	Total Earnings	Federal Tax	FICA	State Tax	Credit Union	Union Dues	Net Amount
1/24	176.00		176.00	21.50	10.30	3.52		3.00	137.68	02015	176.00		176.00	21.50	10.30	3.52		3.00	137.68
1/31	176.00	4.95	180.95	23.30	10.59	3.80		3.00	140.26	02232	352.00	4.95	356.95	44.80	20.89	7.32		6.00	277.94

Figure 3-14 Example of Employee Compensation Record

The information on this report contains the payroll history of company employees. This form is used to prepare the W-2 forms which must be distributed to the employees at the end of the year. The Employee Compensation Record is completed each week by the payroll department.

Case Study - Union Deduction Register

Continuing the investigation the analysts found that a special report was prepared for weekly disbursement to the unions. The Union Deduction Register is illustrated in Figure 3-15.

JAMES TOOL COMPANY
UNION DEDUCTION REGISTER

PERIOD 01 / 16   —   02 / 15   PAGE 1

WEEK NO. 2   UNION Machinists (1)

WEEK ENDING 01 / 31   LOCAL NO. 328

EMPLOYEE NAME	SOC. SEC. NUMBER	TOTAL EARNINGS	UNION DUES THIS WEEK	UNION DUES THIS PERIOD
Adams, Anthony David	261-24-9021	180.95	3.00	6.00

Figure 3-15 Example of Union Deduction Register

The Union Deduction Register is given to the union after the compilation of each week's payroll so that they will know how much union dues have been deducted from the pay of each union employee. Both James Tool Co. and the union use this report to determine the amount of money which will be owed to the union each month. In addition, the union uses the report to compare it to their calculations of the amount of money they should be paid to be sure that they are getting all of the money they are supposed to receive.

Case Study - Overtime Report

Upon further investigation the analysts found that a special overtime report was generated each week and sent to Arnold Henderson, the Vice-President of Financial Affairs. An example of the report is illustrated in Figure 3-16.

JAMES TOOL COMPANY OVERTIME REPORT			
PERIOD ENDING 01 / 31			PAGE 1
EMPLOYEE NAME	DEPARTMENT	HOURS OF OVERTIME WORKED	OVERTIME COMPENSATION
Adams, Anthony David	01	.75	4.95
Barnes, Terry Q.	04	1.80	10.80

Figure 3-16 Overtime Report

The analysts determined that the overtime report had not been part of the original system, but was added when Henderson wanted to know how much overtime was being paid in the company. Although it originally was prepared only upon Henderson's request, eventually someone decided that it should be prepared each week. When the analysts asked Henderson about the report, they found that he only consults it on occasion and that he does not need the report each week. Therefore, it appears that this report could be put on an "as needed" basis rather that being prepared weekly.

Case Study - Payroll Checks

A very important output from the payroll system are the payroll checks which are distributed to the employees. The payroll check is illustrated in Figure 3-17.

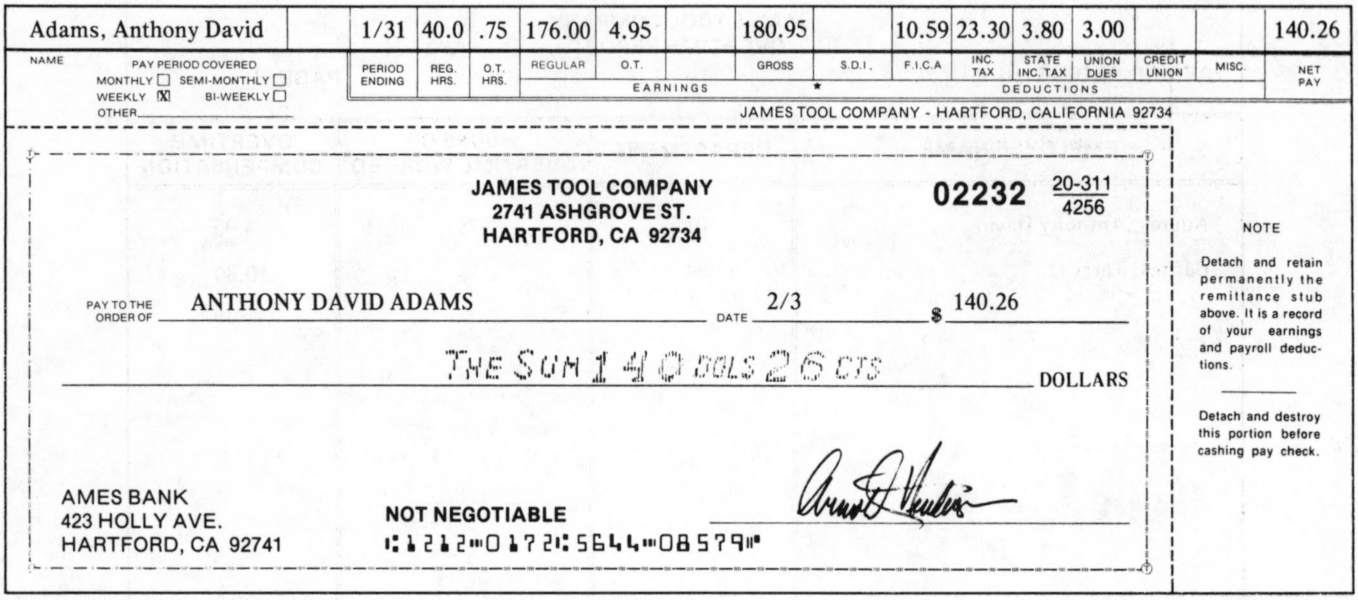

Adams, Anthony David		1/31	40.0	.75	176.00	4.95		180.95		10.59	23.30	3.80	3.00			140.26
NAME	PAY PERIOD COVERED MONTHLY ☐ SEMI-MONTHLY ☐ WEEKLY ☒ BI-WEEKLY ☐ OTHER ☐	PERIOD ENDING	REG. HRS.	O.T. HRS.	REGULAR	O.T. EARNINGS		GROSS	S.D.I. *	F.I.C.A.	INC. TAX	STATE INC. TAX	UNION DUES DEDUCTIONS	CREDIT UNION	MISC.	NET PAY

JAMES TOOL COMPANY - HARTFORD, CALIFORNIA 92734

JAMES TOOL COMPANY
2741 ASHGROVE ST.
HARTFORD, CA 92734

02232   20-311 / 4256

PAY TO THE ORDER OF   **ANTHONY DAVID ADAMS**   DATE   2/3 $ 140.26

THE SUM 1 4 0 DOLS 2 6 CTS _____ DOLLARS

AMES BANK
423 HOLLY AVE.
HARTFORD, CA 92741

NOT NEGOTIABLE

⑆121⑈0172⑆5644⑈08579⑉

NOTE

Detach and retain permanently the remittance stub above. It is a record of your earnings and payroll deductions.

Detach and destroy this portion before cashing pay check.

Figure 3-17 Payroll Checks

In the example above it can be seen that the document consists of the payroll check itself and a "stub" which contains the hours worked, the amount of gross pay, all of the deductions which are taken out of the gross pay, and the net pay which corresponds to the amount which is written on the check. The payroll check is cashed by the employee and he retains the check stub for his own records.

The analysts noted also that each payroll check had to be signed by Arnold Henderson, the Vice-President of Financial Affairs, prior to being distributed to the employees. Therefore, they have to consider this when recommending alternative solutions to problems found in the payroll system.

These reports are the weekly reports which are generated from the payroll system at James Tool Co. As noted, the analysts will analyze these reports to be sure that the information on it is useful and that it is accurate.

Case Study - Distribution of Reports

Another problem which may be found within a system is that correct and accurate output is being produced, but it is not being distributed to the right people. The distribution of the output from a system is extremely critical and it may be that by correcting distribution, many of the positive aspects of a system will be realized and many of the problems with the information coming from the system will be solved.

In the sample payroll problem, copies of the Payroll Register and Union Deduction Register which are generated are delivered to the head of the payroll department, Jim McKeen, to the Vice-President of Financial Affairs, Arnold Henderson, and to a permanent file. The Employee Compensation record is kept in a binder for use in the preparation of the next payroll. The payroll checks, of course, are distributed to the employees. The analysts found that this distribution was sufficient and there appeared to be no problem in this area.

Case Study - Information Generation

It is also a symptom of a system which is not working effectively that too much information is being generated, that is, reports or other information which is not useful to anyone is being generated just because it has always been that way. Thus, one of the tasks of the systems analyst is to be very critical in terms of the output from a system to be sure that only the output required is generated. The production of unused output can be quite costly, and if rising costs are one of the main problems within a system, in some cases a thinning out of unused output can solve the problem.

A related area to the production of unused information is the format of the information produced from the system. There are many systems where virtually every piece of data which is in the system is produced on some type of output report. This means that even though there is not a need to know something, it is still produced on an output report. This problem differs somewhat from the problem of having useless information because, on occasion, the data produced on these reports may be quite useful. However, at certain times, or perhaps even the majority of time, the data is not required and therefore should not be produced by the system.

Thus, the introduction of "on demand reports," that is, reports which are produced only when specially requested, may lessen the burden of producing output from the system and may, therefore, reduce costs or speed up processing in such a way as to allow the current system to function efficiently. In the payroll system at James Tool Company, all of the reports which were generated except the Overtime Report were deemed necessary. The Overtime Report was seldom used on a regular basis and it was therefore decided that the report could be eliminated from the weekly processing. Instead, it should be put on a "demand" basis where it could be generated when required by Henderson.

Case Study - Output Format

In should be realized, of course, that as with the input it is important that the analyst note the format of the data which is contained in the output. This includes the sizes of fields, whether it is alphabetic or numeric data, and other characteristics such as this so that if new output reports must be designed, the analyst will have the information he needs to design them. Another factor which may be important, and which deals with distribution of the reports, is the number of copies which each report requires. Multiple-part paper may be used but one method of saving money either in the same system or in a new system is to cut down in the number of copies which are made. When the new system is designed, however, the analyst must know how many copies are required.

These are some of the elements which must be noted when analyzing the output from a system. It is important to remember that the system must produce that output which is required of it but that producing too much output or useless output may be just as costly and unjustified as producing too little output.

Case Study - Documentation - The System Flowchart

In addition to the input and output used in a current system, the analyst must be concerned with the processing which takes place to process the input and prepare the output. This is true whether the system is a manual system or a system which is processed on a computer. The analyst must, of course, be concerned with the calculations and conditional processing which takes place, but equally as important is an evaluation of the flow of the data from one point to another within the system.

In many systems, there are bottlenecks which occur within the processing of the data in the system. For example, such things as the time cards being received one hour late may cause the entire payroll to be eight hours late. Therefore, it is extremely important for the analyst to note the exact flow of documents and data from the time it enters the system until the time the processing is completed so that any areas which do not function properly can be corrected.

In order to show the flow of data through a system and the processing which takes place within the system, a systems flowchart is normally used. The systems flowchart illustrates, through the use of standardized symbols and descriptions within the symbols, the sequence of processing which takes place as well as the input to the system and the output produced by the system. The system flowcharts in Figure 3-18 and Figure 3-19 illustrate the processing which takes place within the James Tool Co. payroll system.

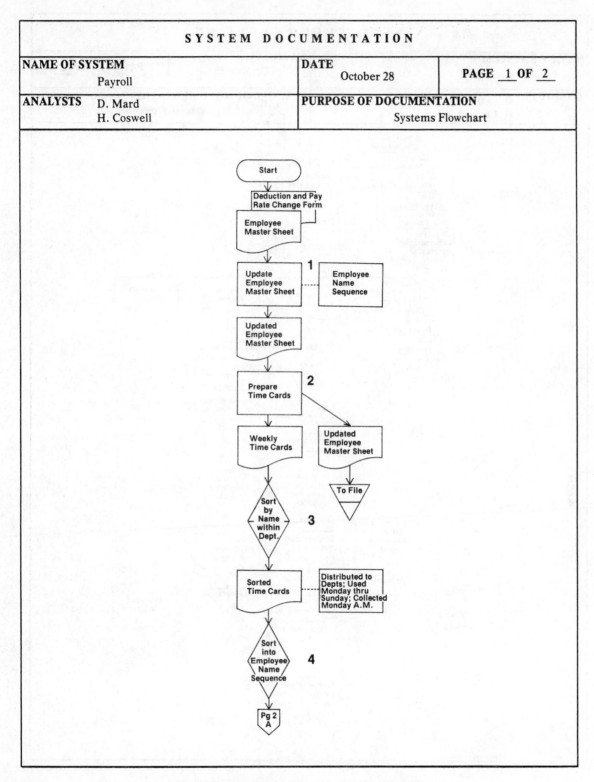

SYSTEM DOCUMENTATION		
NAME OF SYSTEM Payroll	**DATE** October 28	**PAGE** 1 OF 2
ANALYSTS D. Mard H. Coswell	**PURPOSE OF DOCUMENTATION** Systems Flowchart	

Figure 3-18 Systems Flowchart - Part 1 of 2

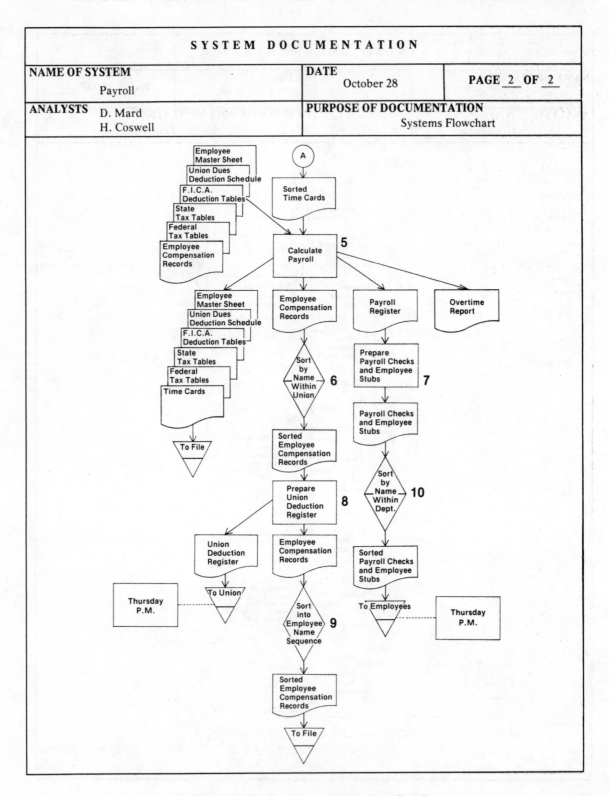

SYSTEM DOCUMENTATION

NAME OF SYSTEM	DATE	PAGE _2_ OF _2_
Payroll	October 28	

ANALYSTS	PURPOSE OF DOCUMENTATION
D. Mard	Systems Flowchart
H. Coswell	

Figure 3-19 Systems Flowchart - Part 2 of 2

The following is an explanation of the processing which takes place within the system:

1. The Employee Master Sheet and the Deduction and Pay Rate Change Form are used as input to the procedure to update and change the Employee Master Sheet. As can be seen from Figure 3-7, the change form contains entries which will change fields on the Employee Master Sheet. It should be noted that the "document" symbol is used for these two forms. Normally, any written or printed document which is used in a systems flowchart is represented by this symbol. Note also the use of the "annotation" symbol, that is, the rectangular process symbol attached to another symbol through the use of a dotted line. The annotation symbol is used for explanation purposes when there is some fact which would be helpful to know in terms of understanding the processing but which is not truly a part of the processing. The example in Figure 3-18 states that the Employee Master Sheet and the Deduction and Pay Rate Change Form are in employee name sequence. This may be helpful in examining the subsequent processing and the resorting of these documents which may occur.

2. After the Employee Master Sheets are updated, they are used to prepare the Employee Time Cards (see Figure 3-9). The information which is recorded on the time cards comes directly from the Employee Master Sheets. After the time cards are prepared, the Employee Master Sheets are placed in the file to await processing in the next week. Note the use of the "off-line storage" symbol to indicate this.

3. The time cards are placed in alphabetical sequence within department number so that they can be distributed to each department to be used by the employees during the week. The reason for placing them in department number sequence is so the timekeeper will be able to distribute the cards easily. The time cards are used during the week by the employees of the company to "clock in" and "clock out" so that the number of hours worked are recorded.

4. When the time cards are collected Monday morning by the timekeeper, they are resorted back into alphabetical sequence so that they can be used in the computation of the payroll. Note that when they are collected by the timekeeper, they will be in department sequence.

5. The payroll is computed. The input to the computation process are the time cards which are sorted in employee name sequence, the employee compensation records, which are also in employee name sequence, the federal tax tables, which are used for computing the amount of federal income tax which must be deducted from the total earnings of the employee, the state tax tables, the social security (FICA) deduction tables, the union dues deduction schedule, and the employee master records. These inputs are used to prepare the Payroll Register as illustrated in Figure 3-13 and to update the Employee Compensation Records as illustrated in Figure 3-14. After the processing is complete, the time cards together with the deduction tables and the employee master records described above are placed in a file from which they will be retrieved the next time the payroll is to be computed.

6. The Employee Compensation Records are then sorted into name sequence within union so that the Union Deduction Register can be made. Note that these records must be in union sequence so that the registers can be prepared properly by union as illustrated in Figure 3-15.

7. After the Payroll Register is prepared, the payroll checks and the associated employee check stubs are prepared from the register. Essentially this is a copying process of transferring the information on the Payroll Register to the checks. After the checks have been prepared, the Payroll Register is filed.

8. The Union Deduction Register is prepared from the Employee Compensation Records. After the Union Deduction Register is completed, it is transported to the various unions for their records.

9. The Employee Compensation Records are resorted back into employee name sequence to be filed for use in the computation of the next payroll. The Union Deduction Register must be completed and sent to the union by Thursday afternoon.

10. After the payroll checks are prepared, they are sorted into employee name sequence within department sequence so that the checks can be batched and distributed to the departments. The checks are then distributed to the employees. Per the union contract, these checks must be distributed to the employees on Thursday afternoon.

As can be seen from the systems flowchart of the payroll system, there is a great deal of handling and sorting of the various documents used in the system. Although this is not necessarily bad, it may indicate that the efficiency of the system may be less than optimal, which leads to problems and errors within the system. In many systems in which problems are found, the processing of the data is haphazard. Thus, the analyst must determine whether there is any lack of efficiency on the part of the staff utilizing the current system. In some applications, a redefinition of the processing within the existing system will be adequate to alleviate any problems. It may even be that a simple training program will allow the persons using the system to use it effectively.

The analyst must also determine whether there are any unique processes within the currently existing system which must be carried forward to a new system. For example, in the payroll system, after the payroll checks are prepared, they must be signed by the Vice-President of Financial Affairs, Arnold Henderson. Regardless of the new system which is designed, there is going to have to be provisions for the signing of the payroll checks by Henderson. Thus, the analyst must search for unique parts of the processing within the existing system and determine if these elements of the old system must be carried forward to the new system. Another example of this are the deadlines of Thursday afternoon for the checks and Thursday afternoon for the Union Deduction Register.

It should be noted that the flow of data throughout the system is the most susceptible to change if a manual system is to be converted to a computer system. In many cases, the input data and the reports and other output data will stay the same because the system is supplying the information which is required and the input data is sufficient to extract the data. Thus, the primary changes would be to the preparation of the input data, the scheduling of the processing, the actual processing which would be on a computer, the scheduling and perhaps the formats of the output.

Case Study - Files

The analyst must make a detailed study of data within the system which is retained for use in the future. Typically, this data is stored in some type of file in such a manner that it can be retrieved when required. Files in typical systems may be used merely to store data for historical purposes, to store data for input to the system at a later time, for providing information on a need-to-know basis, or for other reasons.

Files may also be stored in a number of different ways, varying from file drawers in a file cabinet to disk devices on a computer. Regardless of the function of the file or the manner in which it is stored, the analyst must be aware of all of the files within a system and the role played by each of the files. This is important for two reasons—it must be determined if the file is necessary in any new system which may be designed, and the analyst must determine if the file is being used properly within the current system. As with other portions of the current system, it may be that a realignment in the use of data within a file may solve the problems without a massive redesign job of the whole system.

In many systems, the files contain duplicate data. This may be because the data is accessed through different methods and it was thought that the duplication was necessary. With the access methods available on computer systems, however, it is usually possible to store data in one file and then access this data in whatever sequence or through whatever identification is desired. Therefore, one of the functions of the analyst will be to identify all of the information within the system and to eliminate the data in each of the files which is unneccessarily duplicated.

The analyst must also investigate and define the actual data used within the system files. This data, of course, will be necessary for any redesign of the system, unless it is found that there is unnecessary data within the files. Such attributes as size of fields, sequence in which the data is stored, and other format criteria must be analyzed.

In the sample payroll system, it was found that all of the reports as well as the time cards and other records were stored in normal filing cabinets, each being bound in binders. When there was a need to know any of the information, the reports are merely taken from the file cabinet and read by the person desiring the information. This, of course, is not the most efficient method of making the information available to persons with a need to know and the use of this method will require close scrutiny on the part of the analysts when developing the new payroll system.

Case Study - Restrictions, Requirements, and Controls

In almost every system which is currently in operation there will be certain restrictions and requirements which must be met within the system; there will also be certain controls which are in the system to allow it to function as efficiently as possible. The analyst must be concerned with these because, in all likelihood, the new system will still have the same restrictions, requirements, and in some cases, controls as the old system.

The restrictions and requirements which are inherent within a system can come from either outside forces or from within the company. Typical outside requirements or restrictions would be government regulations, such as tariffs or taxes, which are imposed by either the federal, state, or local governments and which must be accounted for within the system. The analyst would have no alternative except to include these restrictions within any new system which may be designed. Similarly, there may be legal restrictions which must be accounted for within a system even though the system would be more efficient without these restrictions.

Restrictions and requirements which are imposed within a company are normally specified by top-level management. These restrictions will many times be related to costs, that is, management will many times specify that the system merely be capable of processing the data as required, instead of being the most sophisticated system which could be designed for the application. This will usually be a balance between the cost of the system and the service which the system provides. There are others such as the image of the company, the service to outside users or clients, and time limitations which management feels are most important and must be a part of the system even at the cost of other features which the analyst feels would enhance the system.

Controls are those elements of the system which are intended to ensure that the system is functioning properly. Controls will vary with the type of system which is being studied, but certain controls such as verifying social security numbers or other identification numbers, checking totals on reports to ensure that they balance, checking value limits on certain documents like payroll checks to ensure that the values are not too great or too small, are common types of controls. The use and design of codes will be discussed later when input and output design are discussed. At the investigation stage, it is merely important that the analyst be aware of all of the controls which are a part of the system.

In the payroll system for James Tool Co., it will be noted that there are few controls. It was mentioned by the chief payroll clerk that the time cards were matched against the employee master records to ensure that time cards were returned for each employee and that no more than one time card was received for each employee. This was the only control which was pointed out to the analysts. There were no controls, for example, to ensure that the proper pay rate was copied from the Employee Master Record to the time card; nothing was done to ensure that limits were not exceeded in the amount of pay for certain classes of employees; the analysts did not find any balancing reports which were generated or used to ensure that the total payroll amount was correct. These types of controls should be present in any system, but particularly in a system such as payroll where company money is being disbursed. Thus, it is easy to see why errors could occur in the processing of the union dues and it is the job of the analysts to determine what corrections should be made in the new system.

Case Study - Documentation and Presentation

The purpose of the investigation and analysis is to determine what processing takes place within the system and to determine where the problems are within the system. During the analysis of the information gathered, many of these problems may come to light. In order to make any improvements or to redesign the system for a computer application, however, there must be management approval. This approval will normally come only after the current system has been documented and a presentation made to management indicating what currently takes place, where the problems are, and the recommendations of the systems department in solving the problems.

Thus, before the design of a new system can begin, the analysts must present the documentation of the old system to management and suggest possible solutions. In many cases, in addition to presenting the old system, the members of the systems department who present the system to management will have to cost-justify any new system which may be proposed. Therefore, it is necessary to examine the types of documentation which should be prepared for management.

Case Study - Documentation

As was illustrated previously in Figure 3-18 and Figure 3-19, the systems flowchart is one of the primary methods of documenting a system. It illustrates the input and output which are used in the system as well as the sequence of processing and the types of procesing which take place. In any type of documentation of a system, a systems flowchart is mandatory.

In addition to a systems flowchart, however, it is normally necessary to include other documentation which will explain the processing which takes place. Part of this documentation, of course, is the description of the various inputs and outputs used and created in the system. In some cases, the drawing or picture of the input and output, such as illustrated previously, will suffice. In other cases, however, accompanying explanations will be required, such as the definition of the data which will be contained within a field or a file, whether the data is alphabetic or numeric, and similar descriptive narrative. In most manual systems, such as the payroll system for the James Tool Co., this type of documentation is not as important as it is when the application is run on the computer.

Another part of the documentation which is normally necessary to describe the present system is a narration which describes the processing which takes place. For example, in Figure 3-18 and Figure 3-19, it can be seen that although the systems flowchart depicts the sequence in which processing takes place, there is no way to tell, for example, what the steps are when the time cards are prepared. An example of the narration which may accompany the systems flowchart is illustrated in Figure 3-20.

SYSTEM DOCUMENTATION		
NAME OF SYSTEM Payroll	**DATE** October 28	**PAGE** _1_ **OF** __
ANALYSTS D. Mard H. Coswell	**PURPOSE OF DOCUMENTATION** Systems Flowchart Narration	

II. Time Cards

The Employee Time Cards are prepared beginning Thursday morning and must be completed by Friday afternoon. They are prepared by filling in the Name, Social Security Number, and Union Code on the time card from the information contained on the Employee Master Record Form. The period ending date is then filled in on the time card. The period ending date is always seven days from the Sunday which ends the current period. The name is always specified Last Name First.

Figure 3-20 Example of Narrative in System Documentation

Note from the example in Figure 3-20 that although there is not a detailed and complete description of what exactly must take place in the entire procedure, there is enough information in the narrative to describe what will take place and anyone who reads the narrative together with looking at the system flowchart would be able to determine what is taking place. This is the function of the documentation which is developed for the presentation of the current system to management—they can look at the documentation and see what is happening now and how these processes are contributing to the problems which exist within the system.

In addition to the documentation which must be developed for a presentation to management, it is also necessary that there be documentation developed which the analyst can use in the design phase. Thus, such things in the flow of the system as number of documents used, the amount of time to process each document, the timing, that is, when certain documents are available to the processing department, the number of hours or days used in performing certain processes, and other detailed information concerning the system will be important to the analyst who designs a new system because without them, he will not be able to include all of the necessary elements. This information is normally recorded in some type of tabular form or in a narrative form for reference by the analyst.

Once the documentation of the currently existing system is complete, the analysts should explain to management their analysis of the problems within the system and also propose solutions to the problems. In some cases, this may be as simple as suggesting new personnel or more personnel or a change in personnel, or requiring that the standardized operation of the department be followed more closely. It is more likely, however, that a more involved solution will be required. This is especially true if the system to be revised or corrected is a manual system, such as the payroll system for James Tool Co.

The analysis and proposals are normally contained in a report prepared for management which forms the basis for the management presentation. The report prepared by Coswell and Mard for presentation to the management of James Tool Company is shown in Figures 3-21 through 3-25.

MEMORANDUM

DATE: November 11
TO: Arnold Henderson, Vice-President, Financial Affairs
FROM: Clyde Harland, Manager, Systems and Programming
SUBJECT: Detailed Investigation and Analysis of Payroll System

A detailed systems investigation and analysis of the payroll system was conducted by the Systems Department as a result of the approval given the preliminary investigation report. The findings of the investigation are presented below.

Objectives of Detailed Investigation and Analysis

The study was conducted to investigate two major complaints concerning the payroll system. First, it was specified that deductions for union dues from employees' checks were not being performed correctly, especially for employees who joined the union after employment. Second, considerable overtime was required for employees of the Payroll Department in order to have the payroll prepared on time. The objective of this study was to determine where the problems existed and to develop alternative methods of solving the problems.

Findings of Detailed Investigation and Analysis

The following problems appear to exist within the payroll system:

1. Lack of Controls and Checking Features: The current system has virtually no controls or other features which tend to verify the data being used in the system or the calculations which are performed within the system. This could lead to many errors in processing which would go undetected until reports and paychecks are delivered.

2. Excessive Forms Handling: The forms used within the system are sorted and otherwise handled by a number of people and this leads to errors in omission and duplication.

3. Duplication of Data: Much of the data within the system is duplicated. Thus, when changes must be made to data, there are numerous areas where the data must be updated. This is time-consuming and error prone.

4. Lack of Formalized Procedures: Formalized procedures do not exist to handle all situations. Name changes are handled informally. No provision is made on the Employee Master Record Form for recording pay rate changes or union changes; deductions for union dues are obtained from the standards manual which is not current.

Figure 3-21 Detailed Systems Investigation and Analysis Report - Part 1 of 5

Arnold Henderson   -2-   November 11

5. Poor Operating Procedures: The pay rate is transferred each week from the Employee Master Record Form to the time card. This is an unreliable and error-prone procedure.

6. Communication: The method of communication between the Payroll Department and other departments within the company is not systematized and is not adequate.

7. Costs: Costs are rising with the system because of the inefficiencies and it does not appear that there are any adjustments in the system which can keep costs at or below the present level.

Alternative Solutions

There are several alternatives which can be pursued in order to solve the immediate problems of overtime and incorrect union deductions.

OPTION I

Action: Increase by two persons the number of people in the Payroll Department and design forms which would allow for a formal method of communication between the Payroll Department and other departments.

Result: This would decrease or eliminate the overtime and should also solve the difficulties of union deductions which are a result of the lack of adequate communications.

Estimated Costs:
      A. Two payroll clerks   14,400.00/Year
      B. Design of Forms   2,500.00
            Time of Systems Analyst
            Printing of Forms
            Distribution of Forms
            Instruction on Use of Forms

Estimated Savings and Benefits:
      A. Overtime for Payroll Employees   (12,000.00/Year)
      B. Intangibles: More effective communication, better union relationships, happier employees with less overtime.

Advantages: Can be quickly implemented, costs very little.
Disadvantages: Provides only a temporary solution, may not be effective for more than a few months.

Figure 3-22 Detailed Systems Investigation and Analysis Report - Part 2 of 5

Arnold Henderson -3- November 11

OPTION II

Action: Redesign the payroll system for manual processing.

Result: A more responsive system which would reflect the current needs of the company relative to payroll processing.

Estimated Costs:
A. System Design 8,000.00
B. Preparation of Forms, etc. 1,000.00
   required for system
C. Implementation of System 2,000.00

Estimated Savings and Benefits:
A. Elimination of Overtime 12,000.00
B. Intangibles: More effective payroll system, better controls over payroll.

Advantages: Can be implemented within two months, cost is reasonable.
Disadvantages: Estimated life of a manual system is one year, based upon the expected growth of the company, does not provide as much information as may be desired.

OPTION III

Action: Redesign payroll system for implementation on electronic data processing equipment - IBM System/370-125.

Result: An up-to-date, responsive, controlled payroll system which would satisfy needs both currently and in the foreseeable future (3-5 years).

Estimated Costs:
A. Systems Analysis and Design 8,000.00
B. Programming and Implementation 16,000.00
C. Training, Forms Cost and Maintenance (2 yrs) 6,000.00
D. Computer Costs (2 yrs) 14,000.00

Estimated Savings and Benefits:
A. Elimination of Overtime 12,000.00
B. Transfer of one employee from Payroll Dept. 7,200.00
C. Intangibles: More effective payroll system, better controls over payroll, more responsive system relative to management information, ability to place system on-line in the future.

Advantages: Estimated five year life of system, utilizes latest techniques for payroll processing, once implemented should run trouble-free.
Disadvantages: Implementation time (estimated 8 months), cost, possible reaction of employees to computerized system.

Figure 3-23 Detailed Systems Investigation and Analysis Report - Part 3 of 5

Arnold Henderson   -4-   November 11

OPTION IV

Action: Purchase payroll system from outside vendor for implementation on
electronic data processing equipment - IBM System/370-125.

Result: An up-to-date, responsive, controlled payroll system which would satisfy
needs both currently and in the foreseeable future (3-5 years).

Estimated Costs:
A. Purchase of Package   8,000.00 - 30,000.00
B. Modification and Implementation   5,000.00
C. Training, Forms Cost and Maintenance (2 yrs)   6,000.00
D. Computer Costs (2 yrs)   14,000.00

Estimated Savings and Benefits:
A. Elimination of Overtime   12,000.00
B. Transfer of one employee from Payroll Dept.   7,200.00
C. Intangibles: More effective payroll system, better controls over
payroll, more responsive system relative to management infor-
mation, ability to place system on-line in the future.

Advantages: Estimated five year life of system, utilizes latest techniques for
payroll processing, once implemented should run trouble-free,
system is already debugged and proven, minimum amount of
time and trouble for in-house systems and programming staff,
relative fast implementation time.

Disadvantages: Cost, possible reaction of employees to computerized system,
possible difficulties in dealing with outside vendor, possible reliability and maintenance problems.

Recommendations

Although all of the alternatives have their advantages and disadvantages, the
Systems Department recommends that Option III be selected. This specifies that the
payroll system be redesigned by the Systems Department to be processed on the
System/370-125. There are several reasons for this recommendation:

1. Two analysts in the Systems Department have extensive experience in designing
and working with payroll systems. This means there should be minimum
problems both in designing and implementing the payroll system.

2. The payroll system as it now stands will be inadequate in a short time if hiring
progresses as planned and the quick fixes as specified in Option I would be
obsolete in a short time.

Figure 3-24 Detailed Systems Investigation and Analysis Report - Part 4 of 5

Arnold Henderson   -5-   November 11

3. The manual system also would be inadequate within an estimated one year because of the expected increase in personnel and at that time a computerized payroll system would have to be developed.

4. Computerized payroll packages from outside vendors do not appear to be able to handle the multiplicity of unions which are found in our company. Therefore, modifications to the packaged system would have to be made and it has been our experience that modifications of this type can cause unforeseeable problems in the system.

5. The computerized system developed by our department would be designed with an eye toward eventual use of on-line systems, data base applications, and other "state of the art" methods of processing which certainly would not be available with a manual system and likely would not be available with payroll systems from outside vendors.

6. The computerized payroll system developed by the Systems Department would be more cost-effective, in our opinion, than any of the other approaches.

For these reasons, we recommend that the payroll system be redesigned by our in-house Systems Department for implementation on the System/370-125.

Clyde Harland

Clyde Harland
Manager, Systems and Programming

CH:sjm

Figure 3-25 Detailed Systems Investigation and Analysis Report - Part 5 of 5

As can be seen from the Detailed Systems Investigation and Analysis Report, the objective of the study is first defined. This is necessary so that the management or committee personnel receiving the report will realize why the study was initiated.

The findings of the study are then related. Note that these findings are conclusions of the systems department after their investigation. Thus, they are opinions and opinions are only as reliable as the persons expressing them. It is quite important, therefore, that the analysts have the necessary business and data processing experience to be able to express competent opinions and that management have the confidence in the analysts to listen to their opinions. Thus, as the analysts present the report, they must not only sell their ideas but they must sell themselves.

The alternative solutions which the analysts present in the report are also a result of their expertise in the area of data processing systems analysis and design. These alternatives are a result of their analysis of the information they gathered during the detailed investigation and their best judgment concerning the solutions to the problems found.

As can be seen, there are four options. Whenever the results of the study are presented, management must be given alternatives. If they are not, it is possible that the one solution presented to them will not be acceptable and, therefore, no solution will be developed. Thus, as noted previously, one of the results of the detailed investigation and analysis must be alternative solutions to the problems which have been found.

The report also includes the recommendations of the systems department and their reasons for the recommendations. Although alternatives are presented, the systems department will normally, because of their expertise concerning the problems and the system in general, have recommendations concerning the best method for solving the problems. It must be noted that the recommendations of the systems department are not always followed and it is up to management to make the final decisions concerning what follow-on action will be taken. In most cases, however, the recommendations of the systems department will weigh heavily in their decision-making.

Case Study - Result of Detailed Systems Investigation and Analysis

After the report was submitted, the manager of the systems department was asked a number of questions by Henderson and other members of management. As a result of this review process, management of James Tool Co. expressed the opinion that the best course of action was that suggested by the systems department, that is, designing and implementing the system by the in-house systems department. Therefore, they gave approval to the next step in the system project—the design of a new payroll system for implementation on the computer.

STUDENT ACTIVITIES—CHAPTER 3

1. List the basic fact-gathering techniques utilized in the detailed systems investigation.

    1)

    2)

    3)

    4)

2. List and briefly explain the steps in the detailed systems investigation and analysis phase of the systems project.

    1)

    2)

    3)

    4)

    5)

    6)

    7)

3. What level of personnel in an organization are normally interviewed during the detailed systems investigation? Why?

4. Briefly discuss the advantages of the use of interviews as a fact-gathering technique in the detailed systems investigation.

5. List the advantages and disadvantages of the use of questionnaires as a fact-gathering technique in the detailed systems investigation.

Advantages

1)

2)

3)

4)

5)

Disadvantages

1)

2)

3)

4)

5)

6)

7)

6. List the advantages of personal observation as a fact-gathering technique.

1)

2)

3)

4)

5)

6)

7. What is the purpose of the systems flowchart?

8. List the steps in preparing for the presentation of the findings of the detailed systems investigation.

1)

2)

3)

4)

5)

6)

9. List the elements of a good presentation.

1)

2)

3)

4)

DISCUSSION QUESTIONS

1. Some authorities feel that after the detailed systems investigation and analysis, the job of the systems analyst is to ''sell'' to management the system which the analyst feels offers the greatest benefit to the company.

   Other authorities argue that the job of systems analysts is not to ''sell'' anything but to objectively present the facts uncovered in the detailed investigation, present alternative solutions to the problem and let management assume all responsibility for deciding upon the solution to the problem. After all, they argue, management gets paid for decision making — systems analysts don't.

   Which position do you support?

2. In the recommendation to management, upon conclusion of the preliminary investigation, the senior systems analyst estimated the time required to perform a detailed investigation of the payroll system to be 4-1/2 man months at an estimated cost of $5,350.00.

   Shortly after entering the detailed systems investigation and analysis, it became apparent to the systems analyst that the payroll system was much more complex than anticipated and that the detailed investigation would require at least 10 man months of effort.

   What action would you take as systems manager?

CASE STUDY PROJECT—CHAPTER 3
The Ridgeway Company

INTRODUCTION

Upon review of the recommendation from the systems analyst and as a result of the preliminary investigation, approval has been given to conduct a "detailed systems investigation and analysis" of the billing system of the Ridgeway Country Club.

STUDENT ASSIGNMENT 1
THE DETAILED SYSTEMS INVESTIGATION—INTERVIEWING

1. Prepare a list of the steps to be taken in the detailed systems investigation and analysis.

   1)

   2)

   3)

   4)

   5)

   6)

   7)

   8)

   9)

   10)

2. Review the Organizational Chart previously prepared for the Ridgeway Country Club. List the individuals that you would interview during the detailed investigation.

   1)

   2)

   3)

   4)

   5)

   6)

   7)

   8)

   9)

   10)

3. Prepare a list of objectives for the interviews to be conducted for each of the individuals to be interviewed.

4. Prepare a list of specific questions to be asked of each of the individuals interviewed.

5. Conduct the interviews (contact instructor).

6. Prepare a written report of the information gained from each of the interviews.

7. Prepare a questionnaire for the members of the Ridgeway Country Club to determine if they have any suggestions relative to what they would like to see in the billing system.

8. Distribute the questionnaire and when they are returned, prepare a summation of the results of the questionnaire (contact instructor).

OPERATING DOCUMENTS

As a result of the interviews conducted by the systems analyst it was determined that three basic forms are used in preparing the billing for the members of the Ridgeway Country Club. One Sales form is used for recording tennis and golf lessons and sales in the Pro Shop, a special form is used for purchases made in the restaurant and bar, and a statement is sent to each member at the end of the month.

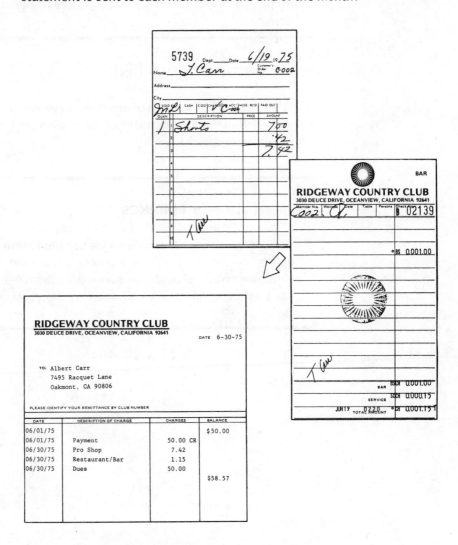

At the end of the month the charge sales slips are arranged in sequence by member number and the billing statement is then typed. The statement only is mailed to the customer. Copies of the sales slips are not mailed with the statement.

STUDENT ASSIGNMENT 2
DETAILED SYSTEMS
INVESTIGATION — SYSTEMS DOCUMENTATION

1. Prepare a Systems Flowchart to document the billing system of the Ridgeway Country Club.

2. Prepare a narrative of the billing system explaining the systems flowchart.

PRESENTATION OF FINDINGS

After reviewing the problems you have been asked to make a presentation to the management of the Ridgeway Company relative to your recommendations.

STUDENT ASSIGNMENT 3
PRESENTATION OF FINDINGS

1. Prepare your recommendations for the improvement of the existing billing system or the development of a new billing system to eliminate errors, provide members with itemized bills, and provide sales information relative to sales in the restaurant, bar and pro shop.

PHASE III
SYSTEMS DESIGN

DESIGN OF SYSTEMS OUTPUT

CHAPTER 4

DESIGN OF
SYSTEMS OUTPUT 4

INTRODUCTION

It should be apparent that in the development of any systems project the activities undertaken by the systems analyst cannot be "compartmentalized" to specific phases; instead, activities in each of the phases often overlap and repeat themselves in the activities of another phase. For example, in the first phase of the systems project, the Initiation of the Systems Project and Preliminary Investigation, it was necessary to define the problem. This initial problem definition normally requires some preliminary data gathering, interviewing, and analysis. In the second phase of the study, the Detailed Systems Investigation and Analysis Phase, a further, more detailed set of activities takes place requiring the gathering of data, further in depth interviewing, and a more detailed analysis of existing operations. Likewise in the next phase of the systems project, that of SYSTEMS DESIGN, there will be a continued and even more detailed analysis of existing procedures with attention directed toward the redesign and improvements in the existing procedures of the system. In many cases, the analyst will conduct further interviews and perform other fact-gathering activities in an effort to design that system which will be most beneficial to the company.

The phases in a systems project have often been described as consisting of an "iterative" process, that is, a process in which the basic steps are repeated and refined in each of the phases until a practical and effective solution is implemented. The importance of the use of the "phase approach" to a systems project is to provide a general set of activities and events in which to organize the study of the system in question.

PURPOSE OF SYSTEMS DESIGN PHASE

After gathering information and analyzing the existing system, the analyst is ready to move into the next phase of the study, that of SYSTEMS DESIGN. The design of a new system is a creative act of devising in part, or in full, new methods and procedures for processing data. Whereas analysis involves fact-finding and data gathering, systems design calls upon the analyst's imagination, creativeness, and awareness of what can be done to improve existing operations. The knowledge gained in the data gathering and systems analysis phase of the study should have revealed the problems and difficulties inherent in the current system. The analyst must now consider how the system can be improved, the demands and needs of management and line personnel for informational output, and the availability of hardware to provide an effective solution to the problem.

After management has reviewed the presentation made by the systems analysts relative to the problems within the current system and has given its approval for the design of the proposed system, the analysts must begin to design that system which will correct the problems inherent within the old system and at the same time, if needed, provide new information which will be of assistance to management. Fundamental to practical systems design is a knowledge of the latest developments in computer hardware and programming systems, a knowledge of business systems and procedures, a knowledge of how problems in the application area under study have been solved elsewhere, and a knowledge of how to perform the systems design process efficiently.

STEPS IN SYSTEMS DESIGN

The Systems Design Phase of a systems project requires the analyst to perform the following activities:

1. Design the System Output
2. Design the System Input
3. Design the Files and Processing Methods
4. Present the System Design to Management for Approval

Design System Output

The first step in the systems design process is to determine what is to be produced from the system, that is, what output is to be created by the system which will be used for informational purposes. It is difficult, if not impossible, to design the input to the system or to define the processing which is to occur within a system if the desired output from the system is not first defined. Therefore, the analyst must first be concerned with the important considerations of output design, both from the standpoint of what information is to be produced from a system and also in what format the information is to be made available.

Design System Input

After the type and format of the output has been determined the analyst must then direct his attention to the design of the system input. This step requires a consideration of the devices and input media to be used for inputting data to the system, a further analysis of the data required as input to the system, an analysis of available source documents and a redesign of source documents if necessary, and the design of the input records.

Design of Files and Processing

After the input and output have been designed it is then necessary to consider the types of files and processing which will be required to produce the given output from the given input while at the same time allowing for the most efficient operation of the system. For example, the analyst must consider if the files are to be stored on magnetic tape or magnetic disk or some other storage media, he must determine if all of the programming is to be done by the programming department or if some outside software routines, such as utility programs, will be used, and other such considerations.

The analyst must also consider the design of the files in terms of access methods, storage methods, and usage. Such factors as the number and frequency of updates and changes to a file and whether there is a need for inquiry or direct access to the data in the files must be determined and then the files must be designed with these requirements in mind.

System Design Presentation to Management

At the conclusion of the systems design phase it is essential that a presentation be made to management describing the proposed design, the design benefits, the cost reductions, and the implementation costs associated with the proposed system. Approval of the system design must be obtained from all levels at this time. Any required changes to the system or recommendations for changes should be considered, for when the systems project enters the next phase, that of Systems Development, changes become extremely costly and time-consuming.

DESIGNING THE SYSTEM OUTPUT

During the Detailed Systems Investigation and Analysis the analyst should have gained a thorough understanding of the type of output that is being produced in the existing system; therefore, during the systems design phase the analyst must review the current output to determine if the reports being produced are meeting the needs of management in providing the necessary information for directing and controlling the activities of the business. When designing the new system the analyst must also determine if changes, either in the form of new reports or in the design of existing reports, are necessary, and if the elimination of some of the existing reports or copies of the reports will contribute to the efficiency of the system being designed.

The following steps should be undertaken when designing the system output:

1. Review the types of output media that may be useful in the new system.

2. Define the output requirements for the new system.

3. Define the specific content of the output reports and other information that must be produced.

4. Design the output documents.

5. Consider the methods and techniques for the disposition and handling of the output.

REVIEW TYPES OF OUTPUT AVAILABLE

It is essential that the systems analyst be aware of the latest developments in computer hardware and peripheral devices that are available when undertaking the design of a new system. With the ability of modern computer hardware to produce a variety of outputs at a high rate of speed it is mandatory that the analyst review the latest technology and developments relative to the forms of output that may be useful in the system under study.

TYPES OF SPECIALIZED OUTPUT

Output from a computer commonly includes the printed report, punched cards, punched paper tape, magnetic tape, and magnetic disk. In the Output Design step of the systems project the analyst is primarily concerned with the design of that type of output that will be of value to the USER, that is, that type of output that will provide information to management and others within the organization that will assist in directing and controlling the activities of the business.

Historically, the printed report has been the most important and widely used form of output for users, but with new and more sophisticated input/output capabilities available on computer systems today, the analyst should review and consider other forms of output in addition to printed reports.

Some of the types of output that may be obtained from computer systems, in addition to the printed report, include:

1. Computer Output Microfilm
2. Cathode Ray Tube (CRT) Devices
3. Plotters
4. Audio Response Units

COM - COMPUTER OUTPUT MICROFILM

In large business organizations, enormous amounts of information in the form of historical records and operational business reports are generated daily. Not only is the storage of large volumes of business information expensive, but this "paper work" explosion has taxed the printing capability of even our fastest computers.

In the past there has been a great imbalance between the output speed capabilities of computers in providing user information and the speed at which that information can be processed within the computer. Some firms have found that microfilm technology offers a significant breakthrough as a means of producing useful computer output rapidly. Microfilm technology reduces images, such as a page from a printed report, from 20 to 90 times. By taking output from the computer on microfilm, not only is it possible to reduce the physical size of the volumes of reports produced in many businesses but the output speed is 10 to 30 times faster than the typical line printer printing at 600 to 1,100 lines per minute.

With computer output microfilm, a page of output from a computer is stored as a very small image on a roll, strip, or sheet of film. One method commonly used is to store output reports on a sheet of film called MICROFICHE. Microfiche range in size from 3 x 7 inches to 4 x 6 inches, with each sheet of film commonly holding from 90 pages to over 260 pages. Figure 4-1 illustrates a microfiche.

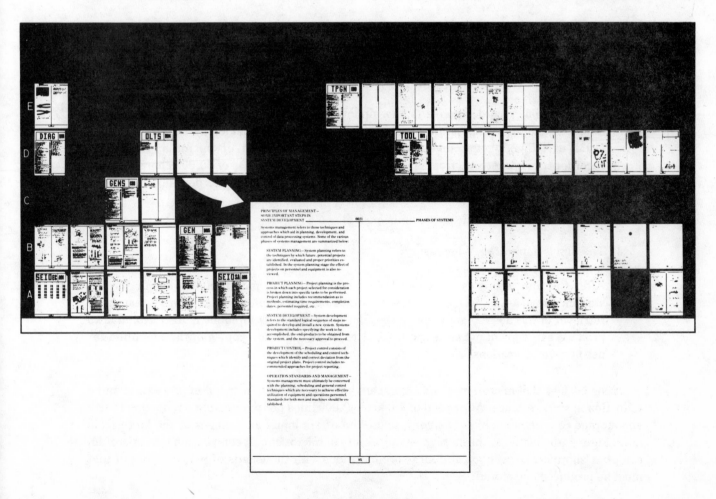

Figure 4-1 Microfiche

Components of a COM System

The basic components of a COM system include: (1) a computer system, normally with magnetic tape output; (2) a microfilm recorder; (3) a film processor; (4) a film duplicator; (5) a microfilm viewing or inquiry station. A machine to produce hard copy when required is also available. These components are illustrated in Figure 4-2.

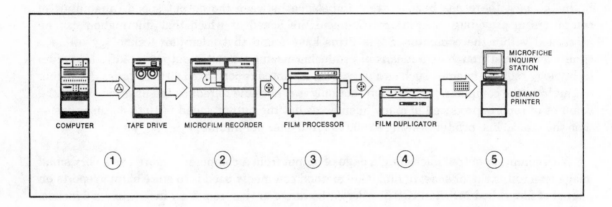

Figure 4-2 Operation of COM

The basic operation of a COM System is explained in the following steps.

1. The input is fed into a computer and processed and output is produced on magnetic tape. The magnetic tape is read directly into a COM converter.

2. The microfilm recorder converts the alphabetic or numeric characters stored on the magnetic tape into characters which are displayed on a cathode ray tube (CRT). The COM unit then photographs the images onto some form of microfilm. Microfilm recorders can accept data from magnetic tape at a rate in excess of 30,000 characters per second, and can produce from 7,000 to 40,000 computer printed lines per minute.

3. The film is processed.

4. The film is duplicated, if necessary.

5. After the film is processed it can then be viewed at a microfilm viewing or inquiry station. Provision is normally made for indexing the resulting microfilm so that rapid retrieval of the desired segment of the film is possible. Equipment is also available to produce hardcopy of pages as required or to make duplicate copies of the microfilm for use at several locations.

It can be seen that microfilm offers significant advantages both in speed of processing and a reduction in storage space. At a speed of 7,000 to 40,000 lines per minute and a reduction in size and storage requirements by 98 percent, microfilm offers many advantages to the large-scale user. Figure 4-3 illustrates the storage saving ability of microfiche. It can be seen that microfilm can offer significant advantages to large organizations with thousands of pages of output that must be produced and stored.

Figure 4-3 Microfilm Storage

Disadvantages include the initial investment for additional COM hardware, including the COM converter, film processor, etc., and the volume of work required to justify conversion to COM. It is estimated that 80,000 to 100,000 pages of output per month are required to justify the consideration of COM.

CRT - CATHODE RAY TUBE OUTPUT

With increasingly greater frequency computer terminals with cathode ray tubes are being used as a form of output. In this type of output device the output from a computer system is displayed on the face of a cathode ray tube. See Figure 4-4.

Figure 4-4 CRT Terminal

Data displayed on the CRT can include numbers, letters of the alphabet, and in some systems pictures in the form of line drawings. Typical systems can display 24 lines at one time, 80 characters per line, for a maximum display of 1920 characters.

This type of output is commonly used for inquiry types of applications where a permanent hard copy is not required. For example CRT output is commonly used in hotel reservation systems, air line reservations systems, and banks and savings and loan institutions. In these application areas an inquiry is commonly made to determine the status of an account, if seats are available, etc. As there is no need for a printed copy, once the information is displayed and reported the next request can be serviced. Data can be displayed on the CRT much faster than it can be displayed on a typewriter or printer and the CRT is an effective form of output for specialized applications.

CRT terminals also offer the ability to enter data into the computer system. The keyboard which is normally a part of the unit can be used to enter alphabetic and numeric data into the system which can then be further processed. In addition, some terminals offer the ability to enter and change data through the use of a ''light pen.'' By using the light pen, the operator can add, delete or rearrange data which appears on the screen of the CRT. This data can then be further processed by the computer system to which the CRT is connected.

PLOTTERS

Although commonly associated with scientific and mathematical applications, plotters offer a form of output suitable for some business applications. Charts and graphs are common tools of many business executives, as they offer an effective way of summarizing business activity. The computer with a plotter used as an output device can be effective in business applications which require the production of charts and graphs.

Plotters may produce pictorial representations including any desired combinations of letters of the alphabet, numbers, special characters and lines with unlimited scale factors. Some machines can produce plots over 100 feet in length. See Figure 4-5.

Figure 4-5 Plotter

Although a variety of plotting devices are available, a common system operates by moving a pen type mechanism on the surface of a sheet of paper. Each movement of the pen occurs at increments of 1/100 inch under control of the computer. Plotting occurs at a speed of up to 16 inches per second.

Because of the specialized nature of the output, plotters have not gained wide acceptance in business oriented applications; nevertheless, in application areas which require frequent graphing or charting, the plotter has proven to be an effective form of output.

AUDIO RESPONSE UNITS

A highly specialized form of output is the audio response unit. The audio response unit provides for an on-line, computer controlled inquiry response system in which the digital output of the computer is converted to speech. Audio response systems are used effectively in those applications where the output can be restricted to short messages, not sufficiently extensive to require visual display. For example, some banks have implemented audio response systems for inquiry into the balances of the checking and savings accounts of their customers. When an inquiry is made to the computer system, the balance is extracted and the output from the computer is converted from digital information to a vocal response such as "CUSTOMER 111119 BALANCE 12525." Audio response systems provide for vocabularies up to 300 words selected by the user.

The specialized nature of audio response systems, the limited vocabulary, and the cost of the systems have limited the use of this type of output to large industries with specialized applications.

COMMON OUTPUT MEDIUMS - THE PRINTED REPORT

The most generally used output from most computer systems is the printed report which is produced on the high-speed printer. An example of a high-speed printer is the IBM 1403 illustrated below.

Figure 4-6 Example of High-Speed Printer

It is important to note several of the characteristics of high-speed printers because the analyst may need to consider them as he is designing the printed output to be produced from a system and also when considering the disposition of the printed output.

The reports prepared on high-speed printers are printed at speeds of up to 3000 lines per minute on continuous forms, that is, forms which are connected from page to page so that they can be run through the printer at a high rate of speed. Printers are available that can print up to 144 characters on a single line. Character sets are available which include numbers, letters of the alphabet and a variety of special characters. It is also possible to obtain printer character sets that allow the printing of both upper and lower case letters of the alphabet. Because of the relative low cost of printers, their flexibility (such as the ease of producing multiple copies through use of carbon paper), and because of the wide acceptance of the printed report in businesses, the high-speed printer will undoubtedly be the most widely used form of output for many years to come.

COMMON OUTPUT MEDIUMS - THE PUNCHED CARD

One of the most useful forms of output is the punched card. Punched cards do not provide a means of disseminating information as readily as the printed report, but they do provide a method of creating documents from the computer system which can be useful in the operation of the system. For example, when the telephone company bills customers, they may send a punched card to the customer which is to be returned with payment. This punched card may then be used as input back into the system to indicate that the phone bill was paid. Figure 4-7 illustrates a typical application in which a punched card which was output from a computer system also serves as an input document.

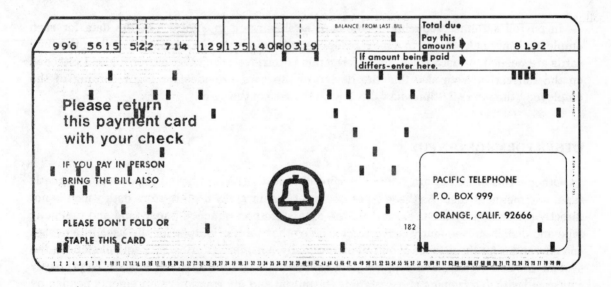

Figure 4-7 The Punched Card as Output and Input

When a punched card is used in this manner, it is sometimes called a "turnaround document" because it is output from the system and then turns around and becomes input to the system. Cards are typically used as turnaround documents by utility companies, credit card companies, and in billing and ordering-type applications.

Another type of turnaround document which is commonly found is a time card used in a payroll system. An example of this type of card is illustrated in Figure 4-8.

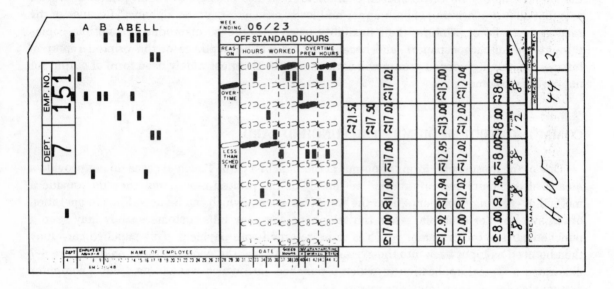

Figure 4-8 Example of Time Card

In payroll systems producing a time card as illustrated in Figure 4-8, the data for each employee is normally stored in a master file, usually on magnetic tape or disk. When the time cards are needed for each week, the information is retrieved from the master file and punched on the timecards. Note also that the design of the card provides for the recording of the employee "time-in and "time-out" directly on the face of the card.

OTHER FORMS OF OUTPUT

Other forms of output are commonly considered to include punched paper tape, magnetic tape, and magnetic disk. As these types of output are normally used to store data which is not directly referenced by the user, that is, the data stored on punched paper tape and magnetic tape and disk is subsequently converted to some other form of output for the user, the design considerations for these types of output are normally considered during subsequent steps of the system project. Again it should be pointed out that at this step the analyst is primarily concerned with determining and designing the output and information requirements needed by individual users within the business organization.

DEFINING OUTPUT REQUIREMENTS

In defining the output requirements of a new system, the analyst is concerned with determining what information is needed as output from the system to meet the needs of the business in the application area under study. As the present system may be producing much of the required information, the analyst can obtain a good idea of the output needed by reviewing the documents obtained in the detailed systems investigation and analysis phase of the study. However, it is important to note that in determining the output requirements for the new system, the analyst should not merely redesign the old output reports for production on the computer. Instead he must give consideration to the current needs of management and the current needs of operating personnel. In addition, the analyst should be concerned with new information that can be produced which will be of value to management, and conversely, which reports could be eliminated or made more efficient utilizing the capabilities of the available or proposed hardware. Far too often in the past, manual systems have been converted directly to the computer resulting in nothing more than a very fast ''manual system'' rather than a system that effectively utilizes the capabilities of the computer.

It should be noted, however, that the analyst must have a clear understanding of his role in defining output requirements. IT IS NOT THE ANALYST'S RESPONSIBILITY to **establish** the informational needs of the proposed system—this responsibility belongs to management. Nevertheless from the information obtained in the detailed systems investigation the analyst may formulate strong opinions relative to the needs and improvement of the existing reporting system. His thinking should be conveyed to management but any decision must always be made with the knowledge of management and the consideration of user personnel. Too often analysts approach the output design step of the system project with the intent of dictating to user personnel the type of report which the analyst ''feels'' they should have, when the proper approach is to ask management and the users to determine the reports and form of reports which are needed for the effective operation of the business.

In defining the output requirements, the analyst should question each item of output which is produced in the current system or proposed for the new system to ensure that it is necessary to meet the information needs of the company. The approach used should involve the asking of the following questions to those individuals who have requested output from the system:

1. Is the output report required?

2. Is all of the information on the report essential? If not, which parts of the report may be deleted?

3. Is all or any part of the information on the report contained in any other report?

4. How often should the report be produced?

5. How many copies of the report are necessary?

When defining the informational requirements of a system, the analyst often finds that management and operating personnel will insist that every bit of information used in the current system is needed and that each copy of a report is essential and must be carried over to the new system. This desire to retain the same information is apparently the "resistance to change" syndrome that so often prevails during the system project. A technique reportedly used by some analysts (but not necessarily recommended) to determine the value of management type reports is to deliberately NOT distribute a given report during a reporting cycle. If no complaints are received, it is obvious that the report has little value. In his review of several companies it was reported by one consultant that over 50 percent of management informational-type reports were not missed when he withheld distribution of reports during a reporting cycle!

It should again be pointed out that although the analyst does not have the authority to arbitrarily establish the information needs of the business, the analyst does have the authority to question the need for any informational requirement. Because of the analyst's unique and detailed knowledge of the system, and because of his knowledge of the capabilities of data processing equipment, the analyst is in a unique position to be able to recommend the production of new reports, the improvement of old reporting procedures, and to specify methods for improving the efficiency of existing operations. Thus, although the analyst does not have the authority to dictate the informational needs of the company, he does have the responsibility for the design of a data processing system that is responsive to the needs of the company in producing timely information at the most economical cost.

DEFINING THE CONTENT OF THE REPORTS

After defining the overall output data required within the system the analyst must direct his attention to the specific content of each report. The logical starting place for the design of new reports is to review the utility of the old reports existing in the current system.

To determine the format and specific content of each report to be designed for the new system, the analyst must consult with management and the operating personnel who will be using the system. During this consultation the parties should decide the formats of the data which is to be contained on the reports and which data should be together on the same report. Such elements as sizes of fields and the type of data to be contained in the fields should be determined at this time. This is an important step in the design phase for if the analyst can obtain the cooperation and assistance of user personnel relative to designing the content and format of reports, the task of "selling" the overall design of the new system will be greatly simplified when the final presentation is made to management.

Based upon the discussions with management and operating personnel as well as the review of the established output requirements, the analyst will then draft a report analysis sheet. This analysis sheet should contain the basic information relative to the content of the report, listing the fields required, the sequence of fields, the maximum size of the totals that may occur, etc. An example of a report analysis sheet is contained in Figure 4-9.

```
+-------------------------------------------------------------------+
|                    S Y S T E M   D O C U M E N T A T I O N         |
+------------------------------------+------------------------------+
| NAME OF SYSTEM                     | DATE                         |
|          Payroll                   |      4/25      PAGE  1 OF 1  |
+------------------------------------+------------------------------+
| ANALYSTS                           | PURPOSE OF DOCUMENTATION     |
|          Smith                     |    REPORT ANALYSIS Form      |
|                                    |   Departmental Payroll Report|
+------------------------------------+------------------------------+
```

FIELD	DESCRIPTION	NUMBER OF CHARACTERS	COMMENTS
Department No.	Numeric	2	
Name	Alphabetic	20	
Hours	Numeric	2	Zero suppress
Rate	Numeric	3	Edit with decimal point
Pay	Numeric	5	Edit with decimal point

GENERAL COMMENTS

Report should include computer generated report and column headings. The current date and page number should be printed on each page.

Figure 4-9 Report Analysis Form

DESIGNING THE OUTPUT REPORTS

After the analyst (in consultation with management and other user personnel) has determined the output required for the new system and has further defined the specific content of each report, he is now faced with the task of designing the exact format of the report as it is to be prepared using the computer.

It is important that the analyst have a thorough understanding of the capabilities of a computer system in producing output reports, including the number of characters that can be printed on one line, the special characters available on the printer, the ability of the computer to accumulate totals and print headings and constant information, and the ability of the computer to produce summary reports and exception reports.

As has been noted previously, the primary purpose of a data processing system is to produce information. The information may be concerning a payroll system, or it may be concerning any number of other business, engineering, or scientific topics. Regardless of the application area, the format of the information which is produced can directly determine how effectively that information can be used. It is important, therefore, for the analyst to understand the various formats that a computer generated report can take, in order to determine the form that will be most useful to those receiving reports.

Detailed Printed Reports

A general type of report that is commonly produced is called the "detailed printed" report. In a detailed printed report one line of information is printed for each record processed. All of the fields in the record do not have to be printed, nor do the fields need to be in the same sequence as on the original input record, but in a detailed printed report some information will be printed from each record. Figure 4-10 illustrates a detailed printed report.

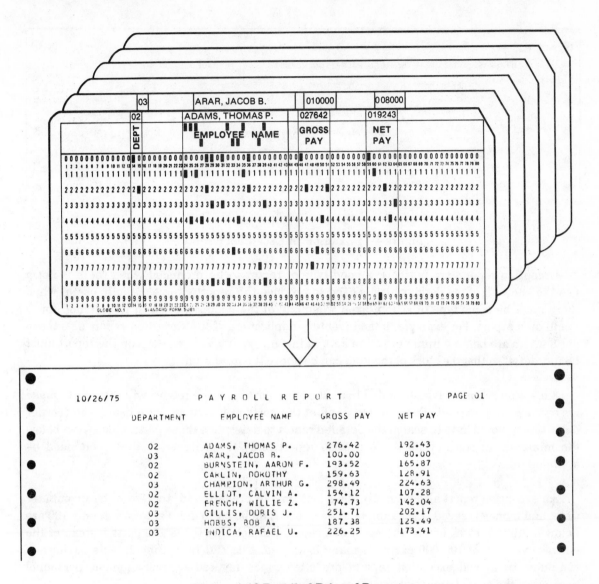

Figure 4-10 Detailed Printed Report

Note in the illustration that, in addition to each detail line printed from the input cards, report and column headings are generated by the computer and that a page number and date are also printed. The spacing of the column headings and spacing between each of the fields is designed by the systems analyst.

Exception Report

Another type of report which is commonly produced from a data processing system is the Exception Report. An exception report is one which produces data which satisfies a specific condition which is deemed worthy of reporting. This condition may be one which indicates an other than normal condition, such as a part in an inventory which is in low supply, or it may merely be a condition which someone finds important enough to note. For example, an exception report may be produced listing all persons who receive over $200.00 per week in gross pay. This report is illustrated in Figure 4-11. Note from the report illustrated that only those employees with gross pay greater than $200.00 are listed.

```
 10/26/75              PAYROLL EXCEPTION REPORT            PAGE 01

             DEPARTMENT      EMPLOYEE NAME       GROSS PAY

                  02         ADAMS, THOMAS P.       276.42

                  03         CHAMPION, ARTHUR G.    298.49

                  03         GILLIS, DORIS J.       251.71

                  03         INDICA, RAFAEL U.      226.25
```

Figure 4-11 Example of Exception Report

Exception reports are normally a good type of report to be produced when there is not a need to know all of the detail which is used to produce the report. It is especially useful when there must be some type of action taken when one of the conditions which is searched for is found on a report. For example, in an inventory application, if the exception report lists those parts which are below a minimum to be kept in inventory, any part which is on the report must be reordered so that the stock of the item can be above the minimum.

Exception reports have the added benefit of not burdening the reader with having to make his own exception report. For example, if all of the parts in inventory were listed, the reorder department would have to search the detailed report to determine those parts which were below the minimum in stock. With the exception report, only those items on the report need be attended to.

An exception report also normally takes less time to prepare and, therefore, less computer time, and money is saved. For example, if there are 100,000 parts in inventory, but only 100 are below minimum stock levels, the exception report will list only those 100 parts instead of the entire inventory of 100,000 parts. In many instances, this can be a considerable savings of computer time, and exception reports are often much favored by management personnel because of the advantages they have over detailed reports.

Reports with Totals

Regardless of the type of report which is to be generated in the system, there are several formats in which the report can be produced. The first is the detailed printed report as illustrated in Figure 4-10, where the individual fields on the input record are merely listed. With most business applications, there is a need to take "sub-totals" when there is a change in selected control fields and final totals after all the records have been processed. These totals are taken for the purpose of control and for the purpose of balancing the computer generatred reports to the original accounting records of the company.

When utilizing the computer in the preparation of reports, final totals can easily be accumulated on any number of fields. In addition to final totals, it is possible to take sub-totals or "minor totals," as they are commonly called in data processing. Minor totals may be taken when there is a change in a "control field." The example in Figure 4-12 illustrates a detailed printed report in which minor totals are taken for Gross Pay and Net Pay when there is a change in Department Number. Final totals are also printed after all the records have been processed.

```
10/26/75                 P A Y R O L L   R E P O R T                PAGE 01

         DEPARTMENT        EMPLOYEE NAME      GROSS PAY      NET PAY


            02          ADAMS, THOMAS P.        276.42       192.43
            02          BURNSTEIN, AARON F.     193.52       165.87
            02          CARLIN, DOROTHY         159.63       128.91
            02          ELLIOT, CALVIN A.       154.12       107.28
            02          FRENCH, WILLIE Z.       174.73       142.04

                        TOTAL DEPARTMENT 02    $958.42*     $736.53*


            03          ARAR, JACOB B.          100.00        80.00
            03          CHAMPION, ARTHUR G.     298.49       224.63
            03          GILLIS, DORIS J.        251.71       202.17
            03          HOBBS, BOB A.           187.38       125.49
            03          INDICA, RAFAEL U.       226.25       173.41

                        TOTAL DEPARTMENT 03   $1,063.83*    $805.70*


            07          UMAMM, JULUIS S.        286.96       248.99
            07          PERATA, HUBERT T.       192.76       151.17

                        TOTAL DEPARTMENT 07    $479.72*     $400.16*


            09          QUINN, ALBERT F.        371.44       288.50
            09          REYNOLDS, MARTE E.      172.38       140.01
            09          SIMPSON, OMAR A.        288.41       206.22
            09          SOLOMON, NANCY F.       144.06        98.73

                        TOTAL DEPARTMENT 09    $976.79*     $733.46*


            10          TELLER, STANLEY G.      226.00       177.48
            10          THOMPSON, GILBERT F.    163.05       136.19
            10          VINCENT, CARLA T.       177.29       126.31
            10          YELLEN, BOB C.          288.10       202.72

                        TOTAL DEPARTMENT 10    $854.44*     642.70*


                        FINAL TOTAL   $4,333.20**  $3,318.55**
```

Figure 4-12 Example of Minor Totals and Final Totals

Note in the report the constant TOTAL DEPARTMENT is printed followed by the appropriate department number on the line identifying the minor total. It is important that the analyst have enough knowledge of computer programming and the capability of the computer to realize that it is possible to print identifying information of this type when a total is being taken.

Multi-Level Control Breaks and Totals

It is possible to have more than one control field within the data which is processed and therefore several levels of control breaks on a report. This is illustrated in Figure 4-13, where plant number has been added to the data and totals are taken for both plant number and department number. Higher level totals are commonly called "intermediate" totals and "major" totals.

```
   10/26/75              P A Y R O L L   R E P O R T                    PAGE 01

   PLANT NO   DEPARTMENT       EMPLOYEE NAME      GROSS PAY    NET PAY

     103        02          ADAMS, THOMAS P.         276.42     192.43
                02          BURNSTEIN, AARON F.      193.52     165.87
                02          CARLIN, DOROTHY          159.63     128.91

                           TOTAL DEPARTMENT 02     $629.57*    $487.21*


                03          ARAR, JACOB B.           100.00      80.00
                03          CHAMPJON, ARTHUR G.      298.49     224.63

                           TOTAL DEPARTMENT 03     $398.49*    $304.63*


                     TOTAL PLANT NO 103          $1,028.06**   $791.84**


     107        02          ELLIOT, CALVIN A.        154.12     107.28
                02          FRENCH, WILLIE Z.        174.73     142.04

                           TOTAL DEPARTMENT 02     $328.85*    $249.32*


                03          GILLIS, DORIS J.         251.71     202.17
                03          HOBBS, BOB A.            187.38     125.49
                03          INDICA, RAFAEL U.        226.25     173.41

                           TOTAL DEPARTMENT 03     $665.34*    $501.07*


                     TOTAL PLANT NO 107           $994.19**    $750.39**


     121        07          OMAMM, JULUIS S.         286.96     248.99
                07          PERATA, HUBEPT T.        192.76     151.17

                           TOTAL DEPARTMENT 07     $479.72*    $400.16*


                09          QUINN, ALBERT F.         371.44     288.50
                09          REYNOLDS, MARTE E.       172.88     140.01
                09          SIMPSON, OMAR A.         288.41     206.22
                09          SOLOMON, NANCY F.        144.06      98.73

                           TOTAL DEPARTMENT 09     $976.79*    $733.46*


                10          TELLER, STANLEY G.       226.00     177.48
                10          THOMPSON, GILBERT E.     163.05     136.19
                10          VINCENT, CARLA T.        177.29     126.31
                10          YELLEN, BOB C.           288.10     202.72

                           TOTAL DEPARTMENT 10     $854.44*    $642.70*


                     TOTAL PLANT NO 121          $2,310.95**  $1,776.32**


                           FINAL TOTAL           $4,333.20*** $3,318.55***
```

Figure 4-13 Example of Multi-Level Control Break Report

In the example in Figure 4-13, it can be seen that there are two control fields—the plant number and the department number. Thus, totals are taken whenever the department number changes and whenever the plant number changes. In addition to the minor total which is taken when the department number changes, an "intermediate" total is taken when the plant number changes.

There is no limit on the number of control breaks which can be taken and totals accumulated for changes in control values. Thus, there may be reports in which four or five levels of totals are printed. It is up to the analyst and the user to determine what totals would be useful on a report and which ones should be accumulated on any control breaks which may be possible on the input data.

From the report in Figure 4-13 it can be seen that the plant number is printed for only the first employee in the given plant. It is not printed for each subsequent employee. The department number is printed for each employee. The printing of the control field for only the first record belonging to the control group is called "group indication" and is commonly used on reports which contain control breaks so that it is easier to see where the control breaks occur and thus easier to read the report.

Whenever a report is to be printed using control breaks, it should be noted that the data being processed must be in a sequence so that all of the data belonging to one group is together, that is, all of the payroll records for employees in department 02 must be processed together. It would be impossible to group the records for department 02 if they were randomly interspersed within the data processed. Therefore, when the analyst designs reports which are to be group-indicated or which are to contain minor or intermediate totals, the data must be sorted in such a manner as to place all records pertaining to one group together.

In the report in which minor totals were taken by department number, the records were arranged in sequence by department number (see Figure 4-12). Note that all records relating to department 02, department 03, etc. are grouped together. Thus, the records must be sorted on department number prior to producing the report on the computer.

In the report in which "minor" totals are taken on department number and "intermediate" totals are taken on plant number, the records must be sorted by department number **within** plant number to produce the report illustrated in Figure 4-13. Thus, it can be seen that the analyst must be aware of the requirements concerning the sequence of the data to be processed when designing the output to be produced from the system. Chapter 5 and Chapter 6 will contain additional information concerning the design of input data so that it can be sorted for proper processing.

Another important characteristic of all of the reports which have been illustrated thus far is the use of headings, that is, information which is contained at the top of each page and is used to identify the data which is found on the report. With very few exceptions, all reports which will be produced from a system will contain heading information of some kind for identification purposes. The wording and placing of the headings is normally a part of the analyst's function when he designs the output formats.

Summary Reports

It should be noted also from the example in Figure 4-13 that there is certain information which lends itself to summarizing. For example, it may be useful to have a "summary report" which merely indicates the gross and net pay for each department. An example of a summary report is illustrated below.

```
 10/26/75      P A Y R O L L   S U M M A R Y   R E P O R T          PAGE 01

              PLANT NO         GROSS PAY              NET PAY


                 103           1,028.06               791.84

                 107             994.19               750.39

                 121           2,310.95             1,776.32
```

Figure 4-14 Example of Summary Report

Note from Figure 4-14 that the gross pay and the net pay for each plant are printed. This data is "summarized" from the "detail" data which is contained on the report in Figure 4-13. The gross pay printed for plant 103 is the sum of the gross pay for the departments in the plant. Thus, summary reports are used to accumulate data and to present it in a summarized format.

A summary report is normally used when the reader of the report is not concerned with all of the detail data and is only interested in totals. For example, the vice-president of the company may not be interested in what each of the employees is being paid, but he might be interested in the payroll total for each plant within the company. Therefore, he would receive the summary report for the payroll as illustrated in Figure 4-14 rather than a detailed report as illustrated in Figure 4-13. Normally, the higher in the business organization a report is directed, the more condensed and summarized the report should be.

DESIGNING REPORTS

When it is determined that printed reports are to be a form of output, the analyst will normally use a standardized "printer spacing" chart to assist in laying out the format of the report. The basic purpose of the printer spacing chart is to provide for a standardized method of documenting and designing reports indicating the content and spacing of the data to appear on a page. The printer spacing chart serves as the document which will be used by the programmer to program the computer to produce the report as indicated.

The Printer Spacing Chart is a document which simulates the page which will be printed on the printer. Figure 4-15 illustrates a blank printer spacing chart. Each vertical column represents each print position available on the computer in exact proportion to the actual printing which will occur when a report is produced. From the printer spacing chart illustrated, it can be seen that there are 132 print positions represented on the chart. This is the number of print positions commonly available on high speed printers. It should be noted, however, that not all printers have 132 print positions. Printers are also available which range from 80 to 144 print positions. It is, therefore, important for the analyst to know the characteristics of the available computer hardware because he must design the reports to be produced within the capabilities of the system.

The horizontal lines numbered 1-50 represent the lines on the report. In the printer spacing chart illustrated there are six lines to the inch. The "standard form" which is illustrated in Figure 4-15 has a printing area of 11'' x 14''. The actual size of the computer paper used in the computer may vary considerably and a variety of paper sizes, styles, and colors are available.

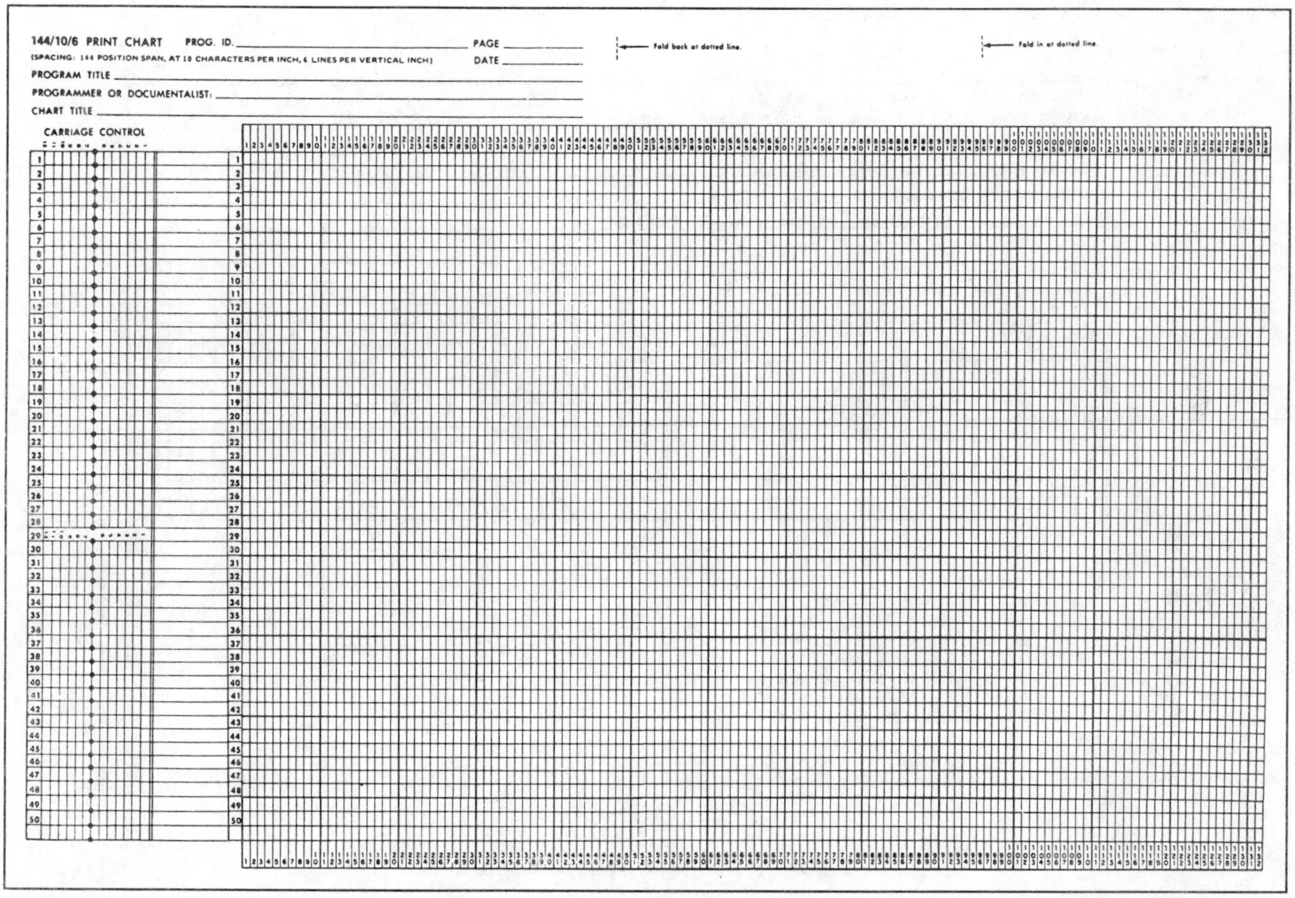

Figure 4-15 Printer Spacing Chart

To design a report, the analyst merely selects the print positions on the printer spacing chart where he desires the specific fields to print and places X's in the selected positions. Figure 4-16 illustrates a printer spacing chart and a related report. X's are used on the printer spacing chart to indicate the printing of "variable" information. Variable information is information that will be changing as the report is produced. For example, the name field on the report illustrated in Figure 4-16 will vary with each printed line; therefore, X's are used to indicate that this is where the names are to print.

Note that the X's begin on line 6 indicating that there is to be approximately a 1-inch top margin on the paper. The two rows of X's indicate that the report is to be single spaced. It can be seen that the department number is to be placed in print positions 6 and 7 and that the name is to print in positions 13 through 32. The entries (DEPT) and (NAME) are not to be printed and are merely used to identify the fields that are to be printed. It is recommended data that is not to be printed be placed in parentheses. It should again be noted that the basic purpose of the printer spacing chart is to provide a method for designing a report, and it is important to note that the method used must be sufficiently detailed to allow the programmer to "program" from the information provided.

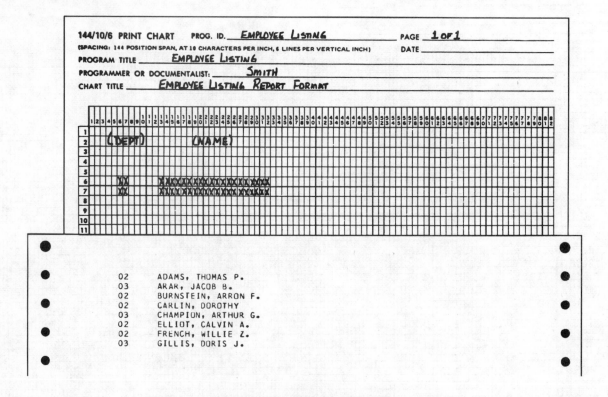

Figure 4-16 Single Spaced Report

Figure 4-17 illustrates a printer spacing chart which specifies that a double spaced report is to be prepared. Double spacing is indicated by leaving a blank line between the rows of X's.

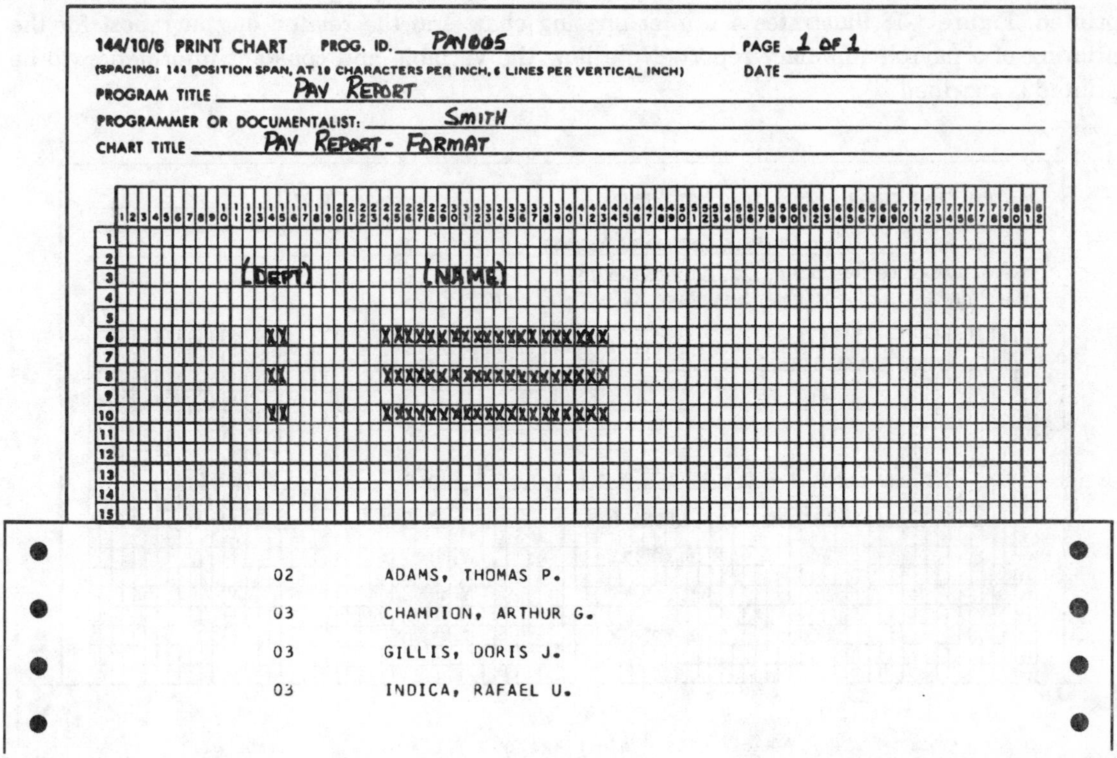

Figure 4-17 Double Spaced Report

Note in the example in Figure 4-17 that the department number is to print in print positions 14 and 15 and the name is to print in positions 24 through 43.

Another type of information that will be printed on a report is "constant" information. Constant information is that type of data that does not change with each line. Report titles and column headings are examples of constant information printed on a report. Constant information is indicated on the printer spacing chart by specifying the constant as it is to be printed. Figure 4-18 illustrates a printer spacing chart and the related output report for the printing of a payroll summary report. Note how the variable and constant information to be printed is specified.

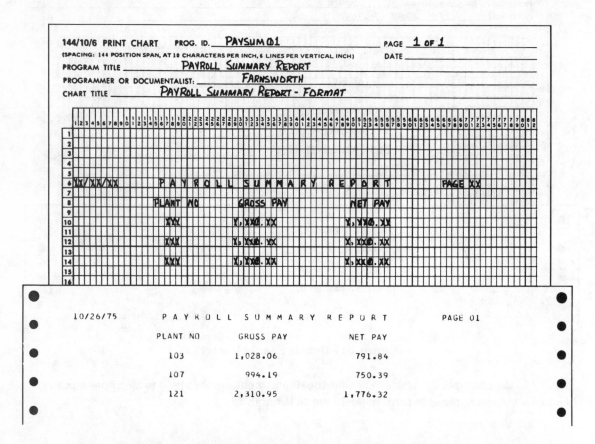

Figure 4-18 Printer Spacing Chart with One Line Variable and Constant Data

The printer spacing chart in Figure 4-18 also illustrates the method used to specify "editing" of a printed report. Editing refers to the process of zero-supressing a field, that is, the suppression of the printing of non-significant leading zeros, and insertion of special characters such as the decimal point, comma, and dollar sign in a field.

In most business applications it is desirable to edit numeric fields, especially fields representing dollars and cents, to make the report more readable. It should be noted in Figure 4-18 that the Gross Pay and the Net Pay fields are to be edited with a decimal point and a comma, and that the fields are to be zero-suppressed.

Editing is indicated on the printer spacing chart by inserting the special characters to be printed, such as the comma and the decimal point, in the exact positions where they are to be printed on the report. Zero-suppression is indicated by placing a zero (∅) in the position where zero-suppression is to stop. In the example in Figure 4-18 the entries on the printer spacing chart indicate that zero-suppression is to occur up to and including the print position before the decimal point in the Gross Pay and Net Pay fields. Thus, if the Net Pay field contained the value 105, it would print as 1.05; and if the field contained the value 05, it would print as .05.

The following examples illustrate how editing is specified on the printer spacing chart and how the edited field will appear on the output report.

Data To Be Edited	Printer Spacing Chart	Edited Results
205.21	XX∅.XX	205.21
000.43	XX∅.XX	.43
071.04	XX∅.XX	71.04
1237.04	$X,XX∅.XX	$1,237.04
0038.75	$X,XX∅.XX	$ 38.75
0008.49	$X,X∅X.XX	$ 08.49
0007.50	X,XX.XX	$7.50

Figure 4-19 Example of Editing

In the examples above, the effect on the printed report when numeric data is edited prior to printing can be seen. The last example illustrates editing with a "floating" dollar sign, that is, a dollar sign which is printed to the left of and adjacent to the first significant digit in the numeric field. It is suggested that a floating dollar sign be specified on the printer spacing chart by placing a dollar sign ($) to the left of the decimal point, as illustrated in Figure 4-19.

Reports with Special Characters

One consideration in report design which is quite significant relative to the efficient preparation of the report is the use or "mis-use" of special symbols or characters in the report. Excessive use of special characters, such as the dollar sign, can significantly slow down the speed at which the report is printed on the computer. If, for example, there are ten amount fields to be printed on a report and each of the amount fields are printed with a dollar sign, the utilization of the dollar sign in each of the fields could reduce printing speeds as much as 50 percent as compared to printing the fields without the dollar sign. As a general rule, therefore, when there are numerous adjacent fields to be printed, the analyst should use as few special symbols in each of the fields as possible. Thus, it can again be seen that the systems analyst should have a good technical background relative to the use and operation of computer hardware.

Reports with Totals

When designing more complex reports, such as those requiring totals, the analyst must specify sufficient detail on the printer spacing chart so that the programmer is totally aware of the desired spacing. For example, when taking totals the programmer must know how many blank lines are to be placed before the total and how many lines are to be spaced after the total is printed and before the next detail line is printed. The printer spacing chart in Figure 4-20 illustrates the report design for taking minor totals when there is a change in department number. Note that there is one blank line before the total is printed and two blank lines after the total is printed.

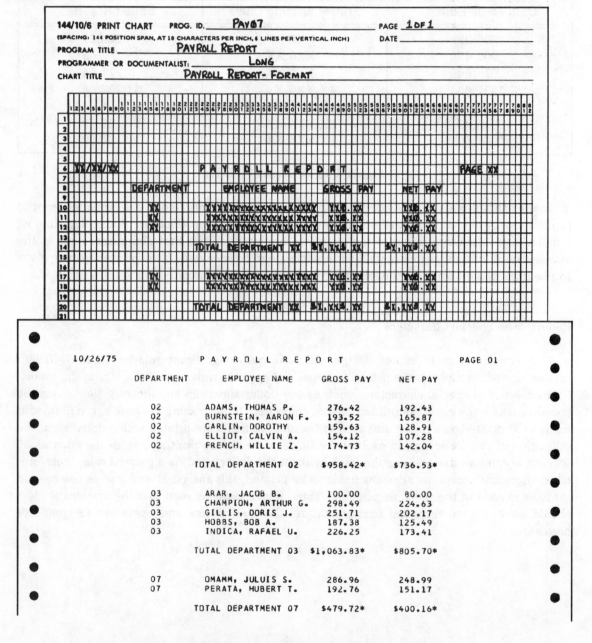

Figure 4-20 Printer Spacing Chart for Report with Totals

Another important area which will concern the analyst when designing output reports is determining the size of the fields to be used for totals. Without careful planning it is possible that totals may overlap adjacent fields. Thus, it is the responsibility of the systems analyst when designing reports to determine the maximum size fields that will appear on the report and the maximum size totals that may develop by reviewing historical records and making inquiries during interviews.

In the example in Figure 4-20, it can be seen that the Gross Pay field for each detail record has been allotted 5 digits. Thus, the analyst has determined that no gross pay will ever exceed the value 999.99. In addition, it should be noted that six characters have been allotted for the Department Total for Gross Pay. This allows for a maximum total of $9,999.99. The printer spacing chart has been designed so that this is the maximum amount which can be printed, and this is the maximum amount that the programmer will allow for when he writes the program to produce the specified output. Thus, when designing the report the analyst must be certain that the print positions specified are of sufficient number to print the maximum size answer which can develop.

It is also good policy, when possible, to allow for futher expansion of the fields if this is required at a later time. For example, it can be seen in Figure 4-20 that if it is required to enlarge the size of the departmental total field for Gross Pay, the field would overlap the employee name field, thus necessitating a complete redesign of the report. Therefore, it perhaps would have been desirable to allow additional blank spaces between the Name Field and the Gross Pay Field in the layout illustrated in Figure 4-20 to allow for future expansion.

Continuous Forms

It will be noted in the previous examples of the printer spacing chart that a single page was illustrated to show the contents of the report. When a report is prepared on the computer, however, the report is not printed on individually separate pages but rather is printed on continuous forms. An example of continuous forms is contained in Figure 4-21.

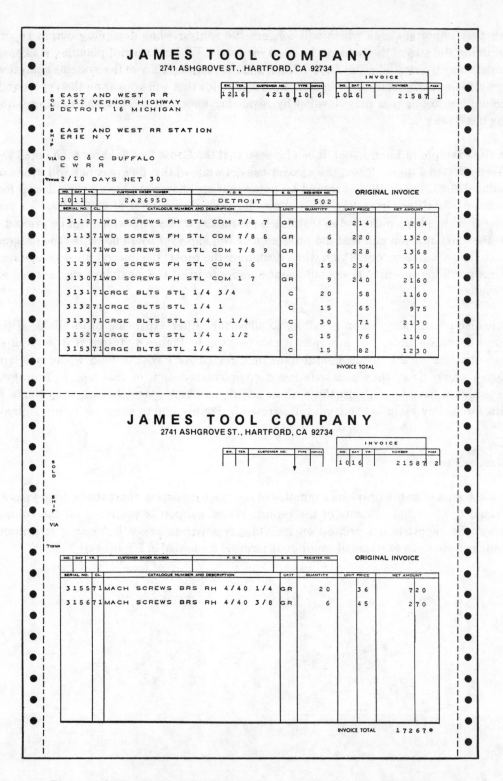

Figure 4-21 Example of Continuous Forms with Pre-printed Headings

Computer reports are always prepared using continuous forms. When the forms are run on the printer, the pages are connected, or continuous. It should be noted, however, that each page is separated by a page perforation. Thus, after the reports have been printed, each page can be separated. It will be noted also that there are sprocket holes along each side of the form. These sprocket holes are used to mount the forms on the printer and will normally be removed from the final report by separating them at the perforation, as illustrated.

The form in Figure 4-21 is an example of a pre-printed form, that is, there is printing already on the form and the data to be printed by the printer will fit into the blanks and spaces on the form. A pre-printed form is used when it is determined that certain information on the form will be more easily identified by printing on the form rather than by headings which will be printed by the computer. In many cases, pre-printed forms will be used for external documents, that is, documents which will be distributed to outside users or customers.

Pre-printed forms are normally expensive because the form must be specially ordered from the paper company and the forms must be printed by the paper company. Therefore, in many applications, especially those which are for internal use, a "standard" form will be used. This is a continuous form on which there is no pre-printed information. An example of a "standard" form is illustrated in Figure 4-22.

Figure 4-22 Example of Standard Continuous Forms

Note from the example in Figure 4-22 that the standard form is merely lined and has no pre-printed information. Thus, any heading or identification information which is to be on the report must be printed by the computer. Note also that there are no perforation lines down the side of the form. In some cases this is desirable so that the reports generated on the standard form can be stored in a folder using the sprocket holes on the form. The standard form can, however, be ordered with perforations along the sides in the same manner as the form illustrated in Figure 4-21.

The use of standard forms without preprinted headings is often desirable. In addition to reducing the cost of the form itself, an even greater savings may be obtained by reducing computer operations time. If, for example, thirty jobs are to be run during the day and all require the use of different special forms, thirty minutes to an hour of computer time could easily be ''lost'' each day when the operator is changing the forms, aligning the paper, etc. Although the printing of report and column headings requires some extra ''print'' time by the computer, this loss is quickly offset by the frequent forms changing and aligning. The analyst, therefore, should give careful consideration to which applications require the use of pre-printed forms and which applications may use computer-generated headings.

Space Control on Continuous Forms

Another consideration when designing output reports, especially when dealing with pre-printed forms, is the use of the carriage control tape which is used in conjunction with the computer printer. A carriage control tape can be used to determine where on a page a line is to be printed because the computer program can command the printer to skip to a particular ''punch'' in the tape. A carriage control tape which has been designed to be used with an invoice is illustrated in Figure 4-23.

Across the top of the carriage control tape are the numbers 1-12, which are called ''channels'' in the carriage control tape. These channels correspond to sense brushes on the printer and when holes are punched in the channels the skipping of the paper or the positioning of the paper for printing is controlled by the sensing of the holes in each channel of the carriage control tape.

Two of the more commonly used channels are channel 1 and channel 12. Channel 1 is used as the ''first printing line stop.'' A punch in this channel indicates where printing is to begin on a page. The punch in channel 12 normally indicates the overflow point on a page, that is, where the last line of print is to appear on a page. Punches in channel 2, 3, 4, etc. can be used to cause printing to occur at specific locations on a page (see Figure 4-23).

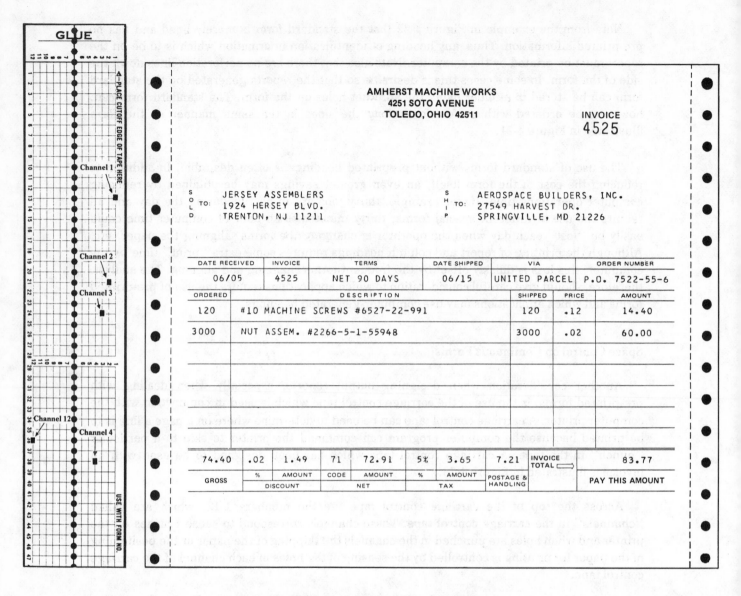

Figure 4-23 Example of Carriage Control Tape

Note from the example in Figure 4-23 that a hole is punched in the carriage control tape corresponding to the position where each of the beginning lines on the invoice are to be printed, and that each of the holes is in a different column on the carriage control tape. Thus, the printer would be instructed (by means of the computer program) to skip to channel 1 to print the "Sold To" information, to skip to channel 2 to print the "Data Received" line, to skip to channel 3 to print the detail lines, and to skip to channel 4 to print the totals. The punch in channel 12 would indicate the last line for the detail lines and if this punch was sensed, the detail lines would be continued on a second invoice since all of the detail lines could not fit on a single page.

After the carriage control tape has been designed and punched for a particular application, it is glued together in a circle and inserted into a special unit on the printer. This is illustrated in Figure 4-24.

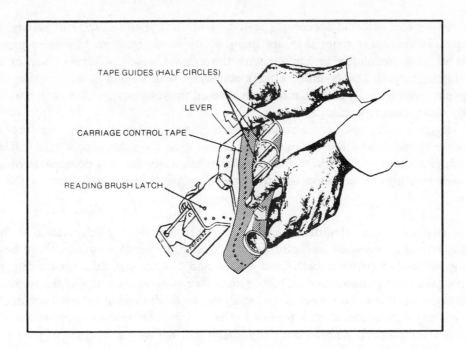

Figure 4-24 Inserting Carriage Control Tape

It should be noted that although most high-speed printers used with computers function in basically the same way, there may be variations as far as the number of columns which can be printed, the format of the carriage control tape, etc. Therefore, as noted previously, it is important that the analyst be familiar with the hardware on which the system is to be implemented during the design phase of the systems project.

USES OF OUTPUT INFORMATION

As noted previously, in many cases the format and type of report which is to be produced is dependent upon the use to which the report is to be put. In general, there can be two broad classifications of reports which are generated from a data processing system: internal documents and external documents.

Internal documents are those which are used within the company and which normally are used by individuals in the daily performance of their jobs. The only major requirements which are placed on internal documents is that they are designed to be capable of being used for the job for which they were intended. In most cases, there are no types of restrictions which are placed on these documents except for their usefulness.

External documents, on the other hand, are documents which are used outside the company, that is, they are designed for use by persons other than those who perform daily functions within the company. An example of external documents would be invoices which are sent to customers of a company or paychecks which are distributed to the employees. Where internal documents are designed primarily for usefulness, in many applications external documents must be designed not only for usefulness but also to satisfy some other types of requirements, such as formats dictated by company policy, legal requirements, or other restrictions.

The reason that external documents require more attention in terms of satisfying certain restrictions is that many times they are going to be used, read, or otherwise processed by persons who are completely unfamiliar with the system which produces them or with data processing in general. For example, a bank statement which contains information concerning the deposits, withdrawals, and balance of an account must be designed in such a way that it is perfectly clear what has transpired concerning the account. A bank statement which is not clearly designed will have two undesirable effects: it will confuse the bank customer as he will not be able to understand the statement and it will give the bank a poor public image. This public image is one very important consideration when designing any documents which will be distributed outside the company and the analyst must be very cognizant of the company policies.

Since computers are being used more and more in distributing information to the public, especially billing information and information concerning charge accounts, there has recently been a great deal of criticism both from within and outside the data processing profession concerning the inaccuracies found on billing and other statements and also the format found in these documents. It is to be expected that analysts, when designing external documents in the future, will pay a great deal more attention to the design of these documents so that the public can utilize them properly and the company image will not suffer because of poorly designed forms and documents.

An especially difficult document to be designed by analysts is the so-called "turnaround" document, that is, the one which is prepared on the computer and then is sent out to customers or other outside persons with the expectation that the form will be returned to the data processing department for further processing. Since the person must not only try to read and comprehend the information on the form, but must also place information on the form, the design must be thoroughly thought out prior to implementing the system which uses the document. Chapter 5 has a detailed discussion of the design of source documents and much of that discussion pertains to turnaround documents also.

Another type of external output which must be specifically designed to conform to given standards are those documents which are produced from the computer to fulfill some type of legal requirements of the company. Such documents are many times rigidly controlled so that the analyst has no options in the design—they must be designed to conform to regulations imposed by the government or other regulatory agencies. An example of this is the W-2 form which must be prepared at the end of the year for tax purposes showing the amount of money earned by an employee and the amount of money deducted for taxes. The format of this is established by the government and the analyst must use the given format.

In other types of documents, however, the analyst may have some leeway in the design as long as the document conforms to some standards. An example of this may be the payroll check. Banks may have certain limitations as to the size of a check and where on the check certain values such as the account number are to be placed. The analyst, however, would be able to design any specific wording pertaining to the company or other information as long as the check followed the prescribed limitations.

DISPOSITION OF OUTPUT

As has been noted, one of the major tasks of the systems analyst is to determine what information is to be contained in the output of the computer system and the format in which the data is to be presented to the user. Another important area of concern for the analyst, and one which is sometimes relegated to a minor role although it should not be, is the disposition of the output once it has been produced. It is critical that the analyst be able to determine who is to receive the output and also in what form they are to receive the output.

The disposition of the output will many times be determined by when the reports or other output are produced. Normally, reports are prepared on some type of fixed timetable. For example, there may be daily reports, weekly reports, monthly reports, quarterly reports, and semi-annual or annual reports. The time when these reports are to be produced is normally determined by the use to which the reports are put. For example, if the stock in an inventory is updated daily, the updated stock list should probably be produced daily so that the inventory people are working with current data. On the other hand, the payroll is run once a week because employees are paid once a week. The run to produce the W-2 forms for a company is made once a year because this is the only time the reports are required.

Another type of report which may be produced by a data processing system are "on-request" reports, that is, reports which are produced only when requested by someone in a position to request them. In many instances, "on-request" reports are exception reports which will report a fact which is inconsistent with what should be happening and an executive or manager wished to see what the exception report indicates.

The analyst must determine which types of reports are required for the system which is being designed and he must also determine where the reports will be sent to and who will see them. When determining these things, the analyst must take into consideration some features of computer printed reports which have not yet been discussed. Primary among these is the fact that continuous forms need not be only one copy. Continuous forms can be purchased with up to 8 or 9 copies. It should be noted, however, that with the exception of certain types of paper, any more than four copies is somewhat difficult to produce clearly and if possible, the analyst should try to use four or less parts to the paper.

After it has been determined who is going to receive the report, and therefore, how many copies may be required, and also how often the report is to be produced, the analyst must determine the best method of presentation of the report to the user. Of course, reports can be given to the user in the same status as they are when they are taken off the printer, that is, as continuous forms. In some cases, especially when single part paper is being used, this is perfectly satisfactory.

In other cases, however, usually when there is multiple part paper or when the report is to be stored in some type of binder, the carbon of the multiple part paper will be removed (decollated) and the pages will be separated from one another on the perforations of the continuous forms (bursted). These two functions can be performed automatically by machines which are designed to decollate and burst reports. The example in Figure 4-25 illustrates a burster and a decollator.

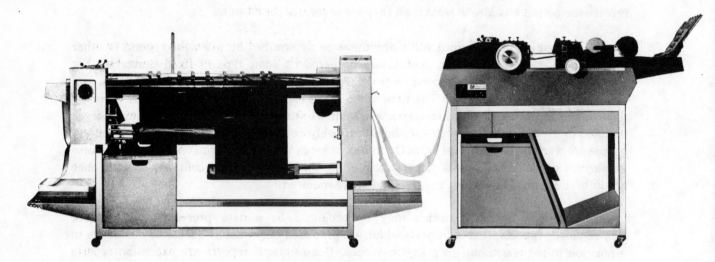

Figure 4-25 Example of Decollator and Burster

The decollator will separate the carbon paper from the paper containing the report so that each part of the report can be dispersed to different users. The burster is used to separate each page of the report so that it can be placed in some type of binder to be dispersed to the user. In actual practice, the report would first be decollated and then bursted and bound for the user.

When designing or recommending various types of reports the analyst must carefully consider specialized hardware which may be needed in order to implement the production of the reports. For example, if six copies of a report are needed, it is essential that a decollator capable of separating six-part forms be obtained when the system is implemented. Or, in a system where hundreds or even thousands of payroll checks are produced, it is necessary that some type of automatic check signing machine be obtained. There have been cases cited where the system (and the analyst) have ''gotten into trouble'' because when the first six-part report consisting of 5,000 pages was produced, there was no decollator so it was necessary to separate the report by hand!—or when the first computerized payroll was implemented it was necessary for the treasurer to sign 10,000 checks by hand!

Thus, the analyst must not only consider the proper design of the output but also the many problems related to the effective disposition of the output.

SUMMARY

The design of the output of a system is critical to the successful implementation of a data processing system because it is the output which will provide the information justifying the system. Therefore, the analyst must spend considerable time and effort in determining what information must be produced from the system, the format in which the output will be presented to the user, and the methods to be used to place the output in the users' hands.

CASE STUDY - JAMES TOOL COMPANY

In light of the various formats and uses of reports, the analyst must decide which ideas to incorporate into the system for which he is designing the output. In the case study, the output for the payroll system of the James Tool Company must be designed.

When designing the output, the analyst, as noted, must take into consideration a number of factors including: what data is to be output from the system and in what format this data is to be produced. In the payroll system for James Tool Company, it was noted in Chapter 3 that the five output reports which are produced from the manual system are the Payroll Checks and Employee Stubs, the Payroll Register, the Employee Compensation Records, the Union Deduction Register, and the Overtime Report. When designing a new system, especially one which is to be processed on a computer, it is necessary that the analyst analyze the output from the manual system to ascertain if the output is necessary in terms of the processing which will be accomplished in the new system and also if there is any new output which can be produced from the computer system which was not possible or practical in the manual system.

It should be obvious that any payroll system is going to have to produce payroll checks for the employees. Thus, the payroll checks and the accompanying employee check stubs must be produced from the new system. The payroll register which is produced in the manual system must also be produced because it is the only report which indicates the total payroll of the company and is used by several departments in evaluating and otherwise processing the payroll of the company. The Union Deduction Register must still be produced because it is a requirement of the union that they have a listing of the deductions made for union dues. Therefore, these three reports will be produced by the new computer system containing essentially the same information as in the manual system.

The Employee Compensation Record was also produced in the manual system. It will be recalled, however, that this record was primarily a historical record of what had transpired concerning each employee's pay during the year. The report did not contain any data which was used on a week-to-week basis in reporting or preparing the payroll. Due to the storage capabilities of a computer, that is, the ability to record data on magnetic tape or direct-access devices (see Chapter 6), the information which is recorded on the Employee Compensation Record can just as easily be recorded on tape or disk and held and updated there. Therefore, it was decided for the James Tool Company that it was not necessary to create the Employee Compensation Record each week since this information could be stored and used through another medium.

As was noted in Chapter 3, the Overtime Report was used infrequently by Arnold Henderson, the Vice-President of Financial Affairs. Therefore, the analysts decided that it would be more appropriate for this report to be an ''on request'' report rather than a report which was automatically created each week. It should be noted, however, that it will still be necessary to design the format of the report and, in addition, it will be necessary to be sure that the information necessary to create the report is available.

Case Study - Payroll Register

After determining which reports are to be generated, it is necessary to determine what information is to be contained on each report, the format of this information on the report, and the source of the data to be on the report. In analyzing the payroll register which is produced in the manual system, it was decided that all the information on the register was required. In addition, it was determined that a unique identifier was required for each employee within the company. This identifier will be an employee number.

The employee number consists of seven digits—the first two digits are the number of the department where the employee will work. The next four digits are a sequence number which uniquely identifies the employee and also indicates seniority within a department, that is, the first employee hired will be number 0001, the second will be 0002, etc. The last digit in the employee number is a ''check digit'' which ensures that the employee number has been recorded correctly. The methods of processing check digits are discussed in Chapter 8.

Once the fields for an output report have been determined, the formats of the fields must be designed. The formats of the fields for the payroll register are illustrated on the report analysis form in Figure 4-26.

SYSTEM DOCUMENTATION			
NAME OF SYSTEM Payroll		**DATE** November 20	**PAGE** 1 **OF** 1
ANALYSTS D. Mard H. Coswell		**PURPOSE OF DOCUMENTATION** Payroll Register - Report Analysis	

FIELD	DESCRIPTION	NO. OF CHAR.	COMMENTS
Detail			
Employee No.	Alphanumeric	8	7 numeric digits with hyphen between second and third digits
Name	Alphabetic	30	
Social Security No.	Alphanumeric	11	Print: XXX-XX-XXXX
Regular Earnings	Numeric	5	Edit with decimal point
Overtime Earnings	Numeric	5	Edit with decimal point
Total Earnings	Numeric	6	Edit with comma and decimal point
Federal Withholding	Numeric	5	Edit with decimal point
FICA	Numeric	4	Edit with decimal point
State Withholding	Numeric	5	Edit with decimal point
Credit Union Deduction	Numeric	5	Edit with decimal point
Union Dues	Numeric	4	Edit with decimal point
Net Amount	Numeric	6	Edit with dollar sign, comma, and decimal point
Department Totals			
Regular Earnings	Numeric	7	Edit with comma and decimal point
Overtime Earnings	Numeric	7	Edit with comma and decimal point
Total Earnings	Numeric	7	Edit with comma and decimal point
Payroll Totals			
Regular Earnings	Numeric	8	Edit with dollar sign, comma, and decimal point
Overtime Earnings	Numeric	8	Edit with dollar sign, comma, and decimal point
Total Earnings	Numeric	8	Edit with dollar sign, comma, and decimal point

General Comments

1. Headings and the page number should be printed as per the printer spacing chart.
2. The period-ending date should be the Sunday date of the previous week.
3. Each department should begin on a new page.
4. A maximum of 50 lines, including headings, should be printed on one page.

Figure 4-26 Report Analysis Form

After the data has been determined, the analyst must design the format of the payroll register which will be produced on the computer. It was decided that the placing of the information should be in the same sequence as on the manually prepared report. However, it was also decided that it would be useful to have minor totals for the payroll paid to the employees of each department and also to have final totals of all of the payroll paid weekly. Therefore, the format illustrated in Figure 4-27 was designed for the payroll register report.

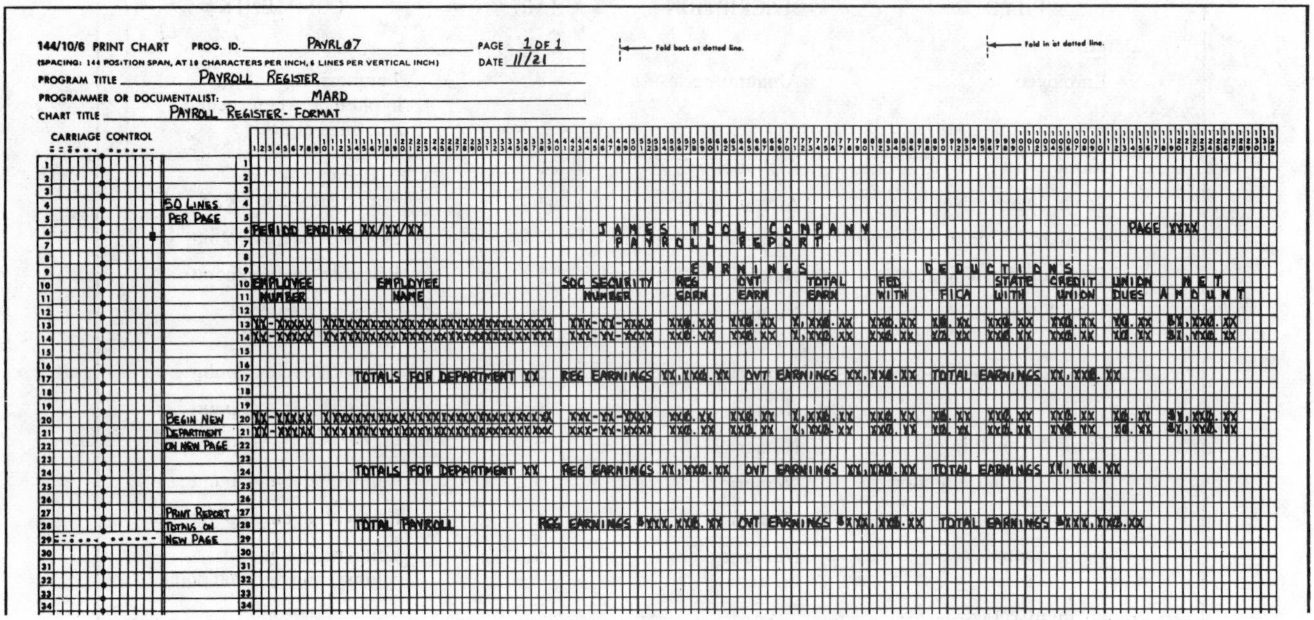

Figure 4-27 Printer Spacing Chart - Payroll Register

Case Study - Union Deduction Register

A second report which is required for the James Tool Company Payroll System is the Union Deduction Register. In analyzing the Union Deduction Register which is produced in the manual system and in talking with the unions, the analysts Mard and Coswell determined that the same information which was on the manually prepared report should be on the report produced by the new system because the unions were happy with the report and there was no reason for change. Therefore, they first made up the Report Analysis form and then designed the report on the printer spacing chart. These forms are illustrated in Figure 4-28 and Figure 4-29.

SYSTEM DOCUMENTATION			

NAME OF SYSTEM
Payroll

DATE
November 24

PAGE 1 **OF** 1

ANALYSTS D. Mard
H. Coswell

PURPOSE OF DOCUMENTATION
Union Deduction Register Report Analysis

FIELD	DESCRIPTION	NO. OF CHAR.	COMMENTS
Detail			
Employee No.	Alphanumeric	8	7 numeric digits with hyphen between second and third digits
Employee Name	Alphabetic	30	
Social Security No.	Alphanumeric	11	Print: XXX-XX-XXXX
Total Earnings	Numeric	6	Edit with comma and decimal point
Dues-This Week	Numeric	4	Edit with comma and decimal point
Dues-This Period	Numeric	4	Edit with comma and decimal point
Union Totals			
Dues-This Week	Numeric	6	Edit with dollar sign, comma and decimal point
Dues-This Period	Numeric	6	Edit with dollar sign, comma and decimal point
Final Totals			
Dues-This Week	Numeric	7	Edit with dollar sign, comma and decimal point
Dues-This Period	Numeric	7	Edit with dollar sign, comma and decimal point

General Comments

1. Headings and the page number should be printed as per the printer spacing chart.

2. The period is the beginning and ending date of the pay period to the unions. These dates are fixed yearly.

3. The week number is the number of the week within the given period.

4. The week-ending date should be the Sunday date of the previous week.

5. The union is to be determined from the union code in the employee's master record.

6. The local number is fixed.

7. A new page must be begun when a new union is started.

8. The final totals must be printed on a separate page.

Figure 4-28 Report Analysis Form - Union Deduction Register

144/10/6 PRINT CHART   PROG. ID. _PAYRL10_   PAGE _1 OF 1_
(SPACING: 144 POSITION SPAN, AT 10 CHARACTERS PER INCH, 6 LINES PER VERTICAL INCH)   DATE _11/25_
PROGRAM TITLE _____ _UNION DEDUCTION REGISTER_
PROGRAMMER OR DOCUMENTALIST: _____ _MARD_
CHART TITLE _____ _UNION DEDUCTION REGISTER- FORMAT_

```
                                40 LINES
                                PER PAGE

PERIOD XX/XX/XX - XX/XX/XX          JAMES TOOL COMPANY                           PAGE XXX
WEEK NUMBER X                     UNION DEDUCTION REGISTER
WEEK ENDING XX/XX/XX      UNION XXXXXXXXXXXXXXXXXXXX LOCAL XXXX

EMPLOYEE              EMPLOYEE                   SOC SECURITY    TOTAL     DUES-THIS   DUES-THIS
NUMBER               NAME                       NUMBER          EARNINGS  WEEK        PERIOD
XX-XXXXX  XXXXXXXXXXXXXXXXXXXXXXXXXXXXXXX       XXX-XX-XXXX  X,XXX.XX   XX.XX        XX.XX
XX-XXXXX  XXXXXXXXXXXXXXXXXXXXXXXXXXXXXXX       XXX-XX-XXXX  X,XXX.XX   XX.XX        XX.XX
                                                      UNION TOTALS   $X,XXX.XX   $X,XXX.XX

         EACH UNION
         MUST START
         ON A NEW PAGE  XX-XXXXX  XXXXXXXXXXXXXXXXXXXXXXXXXXXXXXX  XXX-XX-XXXX  X,XXX.XX  XX.XX  XX.XX
                        XX-XXXXX  XXXXXXXXXXXXXXXXXXXXXXXXXXXXXXX  XXX-XX-XXXX  X,XXX.XX  XX.XX  XX.XX
                                                      UNION TOTALS   $X,XXX.XX   $X,XXX.XX

         FINAL TOTALS
         MUST BE
         PRINTED ON
         A NEW PAGE                                   FINAL TOTALS  $XXX,XXX.XX  $XX,XXX.XX
```

Figure 4-29 Printer Spacing Chart - Union Deduction Register

In the Union Deduction Register, unlike the payroll register, the totals are printed in the same columns as the detail data which comprise the totals. For example, the "dues this week" field is placed in columns 68-72 and the totals for the "dues this week" for each union and for all of the unions are in the same columns. When this occurs, it is important for the analyst to be aware that the total fields will contain more digits than the detail fields and therefore, the detail fields may have to be spaced further apart than would normally be necessary if the totals were not being printed in the same columns. The following example illustrates what happens if this is not taken into consideration.

144/10/6 PRINT CHART   PROG. ID. _PAYRL10_   PAGE _1 OF 1_
(SPACING: 144 POSITION SPAN, AT 10 CHARACTERS PER INCH, 6 LINES PER VERTICAL INCH)   DATE _11/25_
PROGRAM TITLE _____ _UNION DEDUCTION REGISTER_
PROGRAMMER OR DOCUMENTALIST: _____ _MARD_
CHART TITLE _____ _UNION DEDUCTION REGISTER- FORMAT_

```
                                40 LINES
                                PER PAGE

PERIOD XX/XX/XX - XX/XX/XX          JAMES TOOL COMPANY                           PAGE XXX
WEEK NUMBER X                     UNION DEDUCTION REGISTER
WEEK ENDING XX/XX/XX      UNION XXXXXXXXXXXXXXXXXXXX LOCAL XXXX

EMPLOYEE              EMPLOYEE                   SOC SECURITY    TOTAL     DUES-   DUES-
NUMBER               NAME                       NUMBER          EARNINGS  WEEK    PERIOD
XX-XXXXX  XXXXXXXXXXXXXXXXXXXXXXXXXXXXXXX       XXX-XX-XXXX  X,XXX.XX   XX.XX   XX.XX
XX-XXXXX  XXXXXXXXXXXXXXXXXXXXXXXXXXXXXXX       XXX-XX-XXXX  X,XXX.XX   XX.XX   XX.XX
                                                      UNION TOTALS   $X,XXX.XX$X,XXX.XX

         EACH UNION
         MUST START
         ON A NEW PAGE  XX-XXXXX  XXXXXXXXXXXXXXXXXXXXXXXXXXXXXXX  XXX-XX-XXXX  X,XXX.XX  XX.XX  XX.XX
                        XX-XXXXX  XXXXXXXXXXXXXXXXXXXXXXXXXXXXXXX  XXX-XX-XXXX  X,XXX.XX  XX.XX  XX.XX
                                                      UNION TOTALS   $X,XXX.XX$X,XXX.XX

         FINAL TOTALS
         MUST BE
         PRINTED ON A
         NEW PAGE                                     FINAL TOTALS  $XX,XXX.XX$XX,XXX.XX
```

NOTE: Fields Overlap

Figure 4-30 Example of Improper Column Allocation

Note from Figure 4-30 that even though the detail columns for ''Dues-Week'' and ''Dues-Period'' have four spaces between them, which is adequate for clarity on the report, when the grand totals are printed, the totals run into one another and there would be no indication where one total stopped and another started. In addition, there is no room for the insertion of a dollar sign as is done on the actual report as illustrated in Figure 4-29. Thus, the analyst must be aware of which fields are to be used as totals and to plan the columns for the detail data accordingly.

In addition to the planning of the horizontal columns on a report, the analyst must be concerned with the vertical spacing on a report. This is true when it is determined how many lines will be printed on a page but is also true when determining the spacing which is to take place between each detail line and between the detail lines and the total lines. In the payroll register illustrated in Figure 4-27, there are to be two blank lines between the last detail record for a department and the department total. The first detail record for a new department is to begin on a new page, regardless of the number of detail records which have been printed for the previous department. This is indicated by the note in the margin.

In the Union Deduction Register (Figure 4-29), each time the union changes, a total will be taken and then a new page will be begun for the new union. These skipping methods are taken so that the report can be separated and distributed to the proper union. As with the horizontal spacing layout, there are no hard and fast rules which are applicable to vertical spacing, but as a general rule it is usually good to space at least one blank line between detail and total lines. In many cases, two or three lines should be spaced but again the primary consideration is the legibility of the report. It will be noted that there is also spacing between some of the lines within the heading and between the heading lines and the detail lines. This spacing is again determined by clarity and legibility requirements.

Case Study - Payroll Checks

As was noted previously, reports which are to be used for internal documents, such as the payroll register, and even reports which are to be used as external documents which can be clearly identified by the computer generated headings, such as the union deduction register, do not require pre-printed forms. On the other hand, documents which are to be utilized for legal purposes or for certain external documents which are to be distributed to the public may require pre-printed forms. Such is the case with the payroll checks and the employee stubs which must be printed for the James Tool Company Payroll System. As a general rule, all payroll checks or other checks made in a computer system are pre-printed.

When pre-printed forms are called for, such as when payroll checks are being designed, the analyst must determine both what information is to be on the form and the format in which the data is to be presented, in the same manner as with standard or non pre-printed forms. In addition, he must prepare the form so that it can be printed by a paper or printing company to conform with the design specifications. The payroll checks which are used in the James Tool Company payroll system are illustrated in Figure 4-31

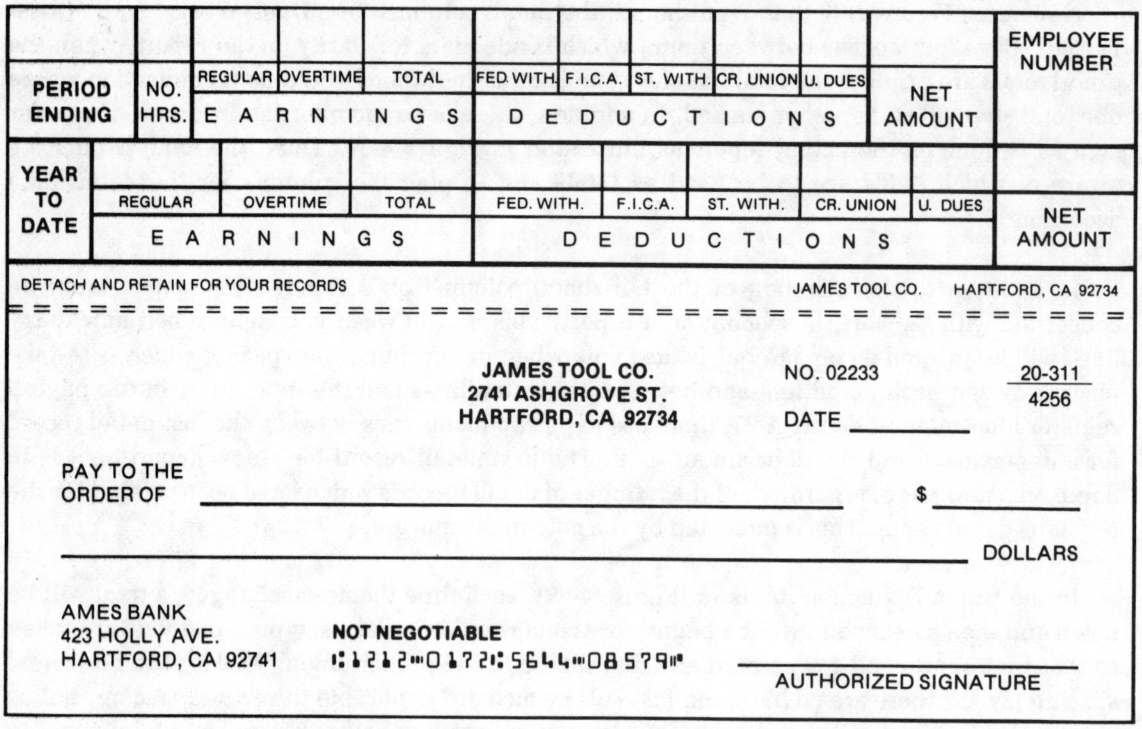

PERIOD ENDING	NO. HRS.	REGULAR	OVERTIME	TOTAL	FED. WITH.	F.I.C.A.	ST. WITH.	CR. UNION	U. DUES	NET AMOUNT	EMPLOYEE NUMBER
		E A R N I N G S			D E D U C T I O N S						
YEAR TO DATE		REGULAR	OVERTIME	TOTAL	FED. WITH.	F.I.C.A.	ST. WITH.	CR. UNION	U. DUES	NET AMOUNT	
		E A R N I N G S			D E D U C T I O N S						

DETACH AND RETAIN FOR YOUR RECORDS   JAMES TOOL CO.   HARTFORD, CA 92734

= =

JAMES TOOL CO.   NO. 02233   20-311
2741 ASHGROVE ST.   ——————
HARTFORD, CA 92734   DATE _____   4256

PAY TO THE
ORDER OF   _____ $ _____

_____ DOLLARS

AMES BANK
423 HOLLY AVE.   **NOT NEGOTIABLE**
HARTFORD, CA 92741   ⑈1212⑈0172⑈5644⑈08579⑈   _____
                                                   AUTHORIZED SIGNATURE

Figure 4-31 Payroll Checks

Note from the example in Figure 4-31 that the check and the employee stub are pre-printed with all of the identification needed and all of the areas on the check stub lined off. Thus, during the preparation of the checks in the computer system, only the data called for will have to be printed—no headings or other identification material will be printed on the computer printer.

When a pre-printed form is prepared by the analyst, he must be as much aware of the spacing on the form and the reservation of the required space for all of the values in the fields as when he is designing a form which is not pre-printed. Therefore, again, the normal way in which the form is laid out is on a printer spacing chart. The printer spacing chart for the payroll checks is illustrated in Figure 4-32.

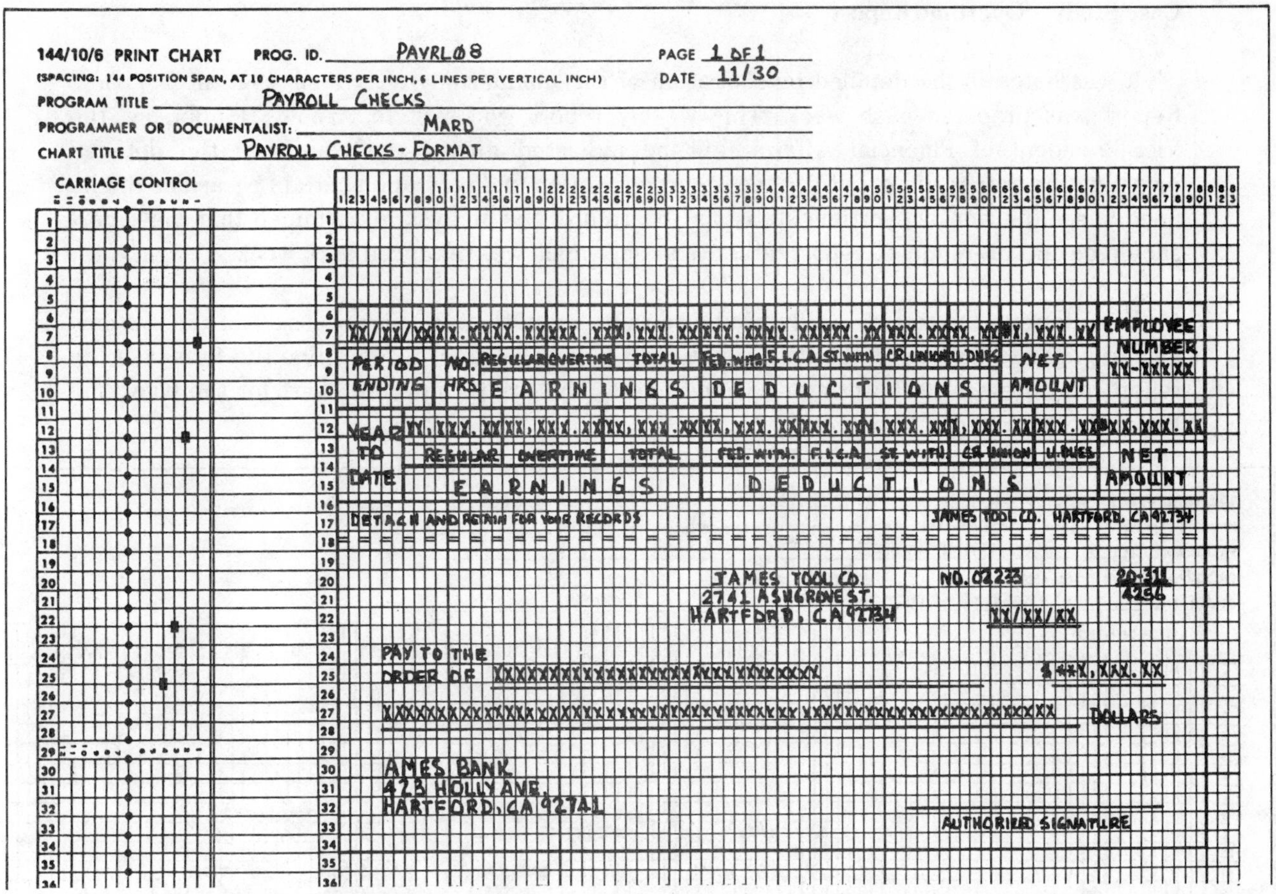

Figure 4-32 Example of Printer Spacing Chart and Pre-printed Form

From the example in Figure 4-32 it can be seen that the layout of the pre-printed checks is done in the same manner as with the blank forms used for the payroll register and the union deduction register. Every vertical line accounts for one print position found on the printer. There must be, as discussed previously, allowance for totals being larger than detail figures. Thus, the earnings and deduction fields for the year-to-date information are larger than those for the weekly earnings and deductions fields.

As with the other reports which are to be generated from the system, the analyst must be concerned with determining what data will appear on the report, in this case the checks. As can be seen, most of the information on the checks is the same information which is contained on the payroll register. Therefore, there must be consistency between the payroll checks and the payroll register in terms of the number of digits reserved for each field on each report. Thus, the same amount of space is allocated for the information on the payroll checks as on the payroll register.

Case Study - Overtime Report

It was noted in the detailed investigation of the manual payroll system that an Overtime Report was prepared each week. This weekly report was sent to Arnold Henderson, the Vice-President of Financial Affairs, but he indicated during the course of the detailed investigation that he did not utilize the report very often and only upon certain occasions did he need to see the weekly overtime statistics. Therefore, the analysts determined that the report should be an "on-request" report rather than one which was produced each week.

Even though the report is only produced upon request, it is still necessary in the systems design phase to determine the data which is to be on the report, determine the format of the report, and draw it on the printer spacing chart. The printer spacing chart for the Overtime Report is illustrated in Figure 4-33.

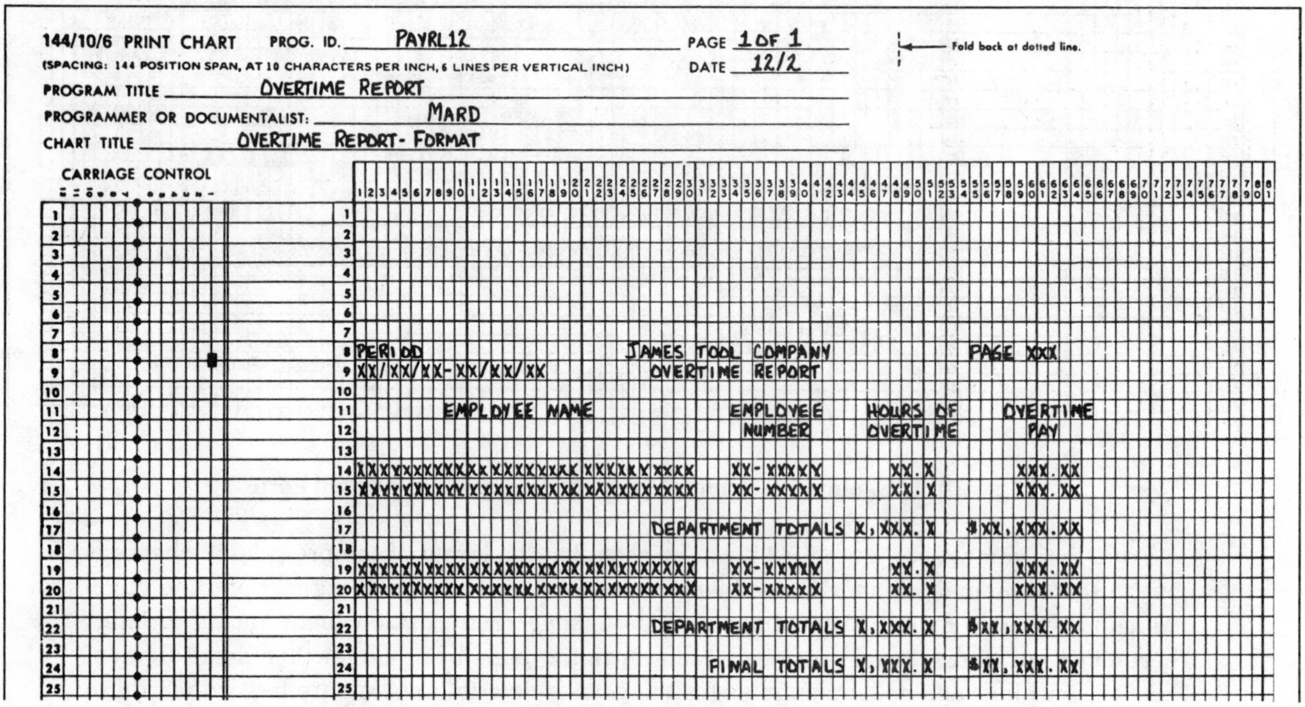

Figure 4-33 Printer Spacing Chart - Overtime Report

It can be seen that the Overtime Report which is to be produced by the new computer system contains the same information as on the manually prepared report (see Figure 3-16). This is because it was stated by Henderson that the report was complete and useful as prepared in the manual system. The only additions are the employee number, the department totals, and the final totals.

Case Study - Time Cards

The payroll system for the James Tool Company also has use for Card Output. It will be recalled that in the manual payroll system, the time cards were prepared manually by the timeclerks each week. The employees then used these time cards to clock in and clock out each day. In the new system, however, it is not necessary to prepare the time cards manually since all the information for each employee will be contained in the files of the system. Thus, the time cards which will be used by each employee to clock in and clock out will be prepared on the computer system. The format of the time card to be punched is illustrated in Figure 4-34.

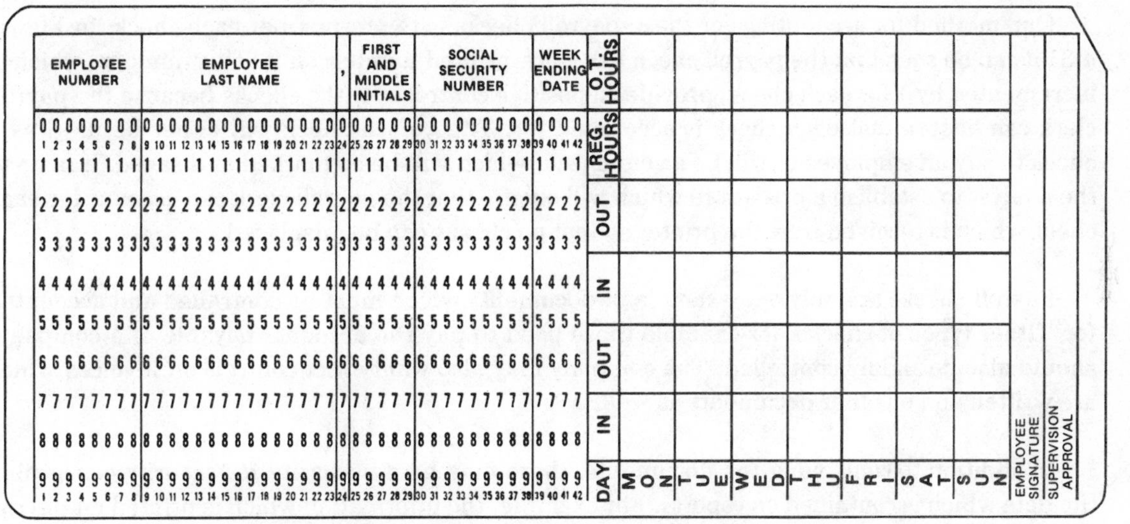

Figure 4-34 Format of Time Card - Output from Payroll System

In the example in Figure 4-34 it can be seen that the employee number, the last name of the employee, the first and middle initials of the employee, and the social security number are contained on the card. It should be noted that this information is contained on the card merely for identification purposes, that is, so the cards can be distributed to the proper departments and each employee can identify his time card. Therefore, only the last name, which the analysts determined could be contained in 15 characters, is placed on the card, together with the initials of the employee.

This card will NOT be used as an input record when it is returned to the payroll department. Instead, the completed time card will be used as a source document from which the data entry department will prepare an input card to be read into the payroll system. It will also be noted that the payrate and the total hours worked are not contained on this card as they were on the time card used in the manual system (see Figure 3-10). This is because the payrate will be contained in the employee master records stored on disk or tape and the totals for the hours worked can be calculated by the payroll programs from the times punched in the input card. Thus, since all calculations pertaining to pay can be made using the computer, there is no need for this information to be calculated manually and recorded on the time card.

Thus, it can be seen that even though punched cards do not provide the mass of information which is possible with printed reports, they can provide an important output medium, especially when the card is to be used as input or as a source document to the system itself.

Case Study - Output Control

As was noted previously, it is important for the analyst to consider restrictions and controls which are placed on the system output in terms of forms design and report layout. Another area in which controls and restrictions may be found is in controlling the forms themselves. For example, in the James Tool Company Payroll System, payroll checks, which are pre-printed forms, are to be used. Control of these payroll checks is quite critical because if a payroll check is lost, there is a possibility that it could be forged or otherwise mishandled. Therefore, the payroll department must account for every paycheck which is printed for it.

One method for accounting for these payroll checks is to pre-number each check. In Figure 4-31 it can be seen that the payroll check has a pre-printed number on it. This number, which is incremented by 1 for each check, provides a positive control over the checks because the payroll clerk can ensure that each check is accounted for. Thus, each check will either be used as a check to pay an employee or will be a void check, which will be so marked and saved. It is up to the analyst to establish a procedure which will ensure that the payroll clerk can account for each check which is received from the printer so that no checks can be misplaced.

Payroll checks are only one example of documents which must be controlled and accounted for. Other types of checks, for example those used to pay the accounts payable of a company, should also be rigidly controlled. The company may also want strict controls on invoices which are written and on other documents as well.

In addition to controlling the documents, there may be a necessity for security concerning the data which is contained on reports. For example, the information which is printed on payroll reports such as the payroll register is not normally available for everyone to see because an employee's pay should not be known to other employees. Therefore, there are normally security procedures, designed by the analyst, which are established so that only designated computer operators and designated payroll department employees are allowed to see the information on these reports.

The reports which indicate the company's sales and profits normally contain sensitive information and this information should not be available to many people. In some cases, these reports will be "coded" so that only a person who is familiar with the coding procedure can read the information on the report.

SUMMARY

The first step in the systems design phase is to design the system output, including the data to be contained in the output, the determination of the formats of the output, the medium on which the output is to be produced, and any security measures which must be taken. This work must be completed prior to further work on the system. Once the output has been designed, the analyst can then procede to determine what input must be available to produce the desired output.

STUDENT ACTIVITIES—CHAPTER 4

REVIEW QUESTIONS

1. List the sequence of steps involved in the Systems Design Phase of the Systems Project.

   1)

   2)

   3)

   4)

2. List the steps which should be undertaken when designing the systems output.

   1)

   2)

   3)

   4)

   5)

3. When analyzing the output requirements of a system, what questions should be posed by the analyst?

   1)

   2)

   3)

   4)

   5)

4. What is meant by the term ''Exception Report''?

5. Who normally receives summary reports in a business organization: top management or line personnel?

6. What elements determine if pre-printed forms should be used for computer-prepared reports?

7. What is the difference between an ''internal document'' and an ''external document''?

8. What is a "turnaround document"? What difficulties in design does it present?

9. What are "on-request" reports? When are they useful?

10. What is the purpose of a decollator? A burster?

DISCUSSION QUESTIONS

1. A systems analyst was quoted as saying "Input must be designed before output because if output is designed first, in consultation with the user, the output wanted by the user may have information which is not available or is too expensive to capture. Then when you tell the user he can't have the information requested he gets mad and data processing loses face."

   What do you think of this position?

2. Some systems analysts argue that "you give a user what he wants. If he wants long reports with reams of data, that is what you give him. Otherwise, the user will be unhappy with data processing because he feels that the analyst is trying to tell him his own business."

   Others say that "it is the function of the analyst to dictate to the user what information can be obtained from a system. If you listen to a user, you will never get anywhere because the user really doesn't know what he wants and, besides, he doesn't understand data processing."

   What do you think of these arguments? Why?

3. It has been said by some analysts that " it is absolutely necessary to keep on top of all the technological developments in data processing and use only the latest equipment for producing output from a system. Otherwise, a system will be designed which will be obsolete in a year."

   Other analysts feel that "it is always best to use the tried and true methods in data processing. By being a 'pioneer' with new equipment, you are only inviting systems which do not work properly and will cost many times what they should."

   What is your opinion?

CASE STUDY PROJECT—CHAPTER 4
The Ridgeway Company

INTRODUCTION

In the review of the billing system, Gunther Swartz pointed out that not only were incorrect monthly statements being sent to many customers, but that he also had no detailed information relative to sales in the restaurant, bar, and pro shop.

Mr. Swartz indicated that he would like to have the following reports on a monthly basis, in addition to the monthly statements which are to be sent to the members:

1. A report listing daily sales in the restaurant, bar, and pro shop.

2. A report listing total sales to each of the members for the month.

3. A report listing all members who made no purchases during the month.

In reviewing similar billing systems, the analyst obtained the following sample forms to serve as guidelines for the design of the new statement:

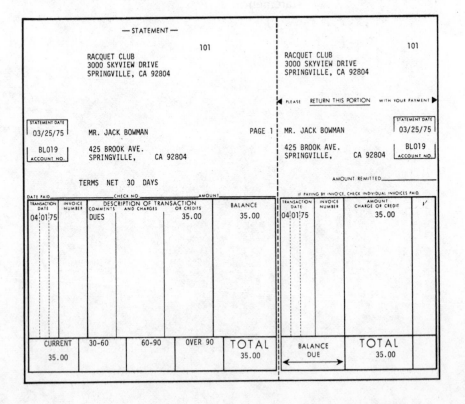

STATEMENT

The Tennis Club
200 RIDGEWOOD ROAD, HILLCREST, CALIFORNIA 92656

L014
TO: MR. & MRS DAVID ROWE
8034 MAPLE ST.
HILLCREST, CA. 92656

DATE   07/25/75

BILLS ARE DUE AND PAYABLE ON RECEIPT
PLEASE IDENTIFY YOUR REMITTANCE BY CLUB NUMBER   DETACH TOP PORTION AND RETURN WITH REMITTANCE

DATE	DESCRIPTION OF CHARGE		CHARGES	CREDITS	BALANCE
					36.66
06/25/75	BALANCE FORWARD				
07/25/75	DUES		45.00		
06/20/75	BAR	02181	.55		
06/26/75	BAR	02351	.55		
06/28/75	BAR	02436	1.15		
07/02/75	REST.	2752	4.30		
07/02/75	PRO SHOP	4522	14.35		

	TOTAL
	102.56

The Tennis Club   PLEASE PAY THIS AMOUNT

— STATEMENT —

RACQUET CLUB 101	RACQUET CLUB 101
3000 SKYVIEW DRIVE	3000 SKYVIEW DRIVE
SPRINGVILLE, CA 92804	SPRINGVILLE, CA 92804

◄ PLEASE   RETURN THIS PORTION   WITH YOUR PAYMENT ►

STATEMENT DATE			STATEMENT DATE
03/25/75	MR. JACK BOWMAN	MR. JACK BOWMAN	03/25/75
BL019	425 BROOK AVE. PAGE 1	425 BROOK AVE.	BL019
ACCOUNT NO.	SPRINGVILLE, CA 92804	SPRINGVILLE, CA 92804	ACCOUNT NO.

TERMS NET 30 DAYS   AMOUNT REMITTED_____

DATE PAID_____CHECK NO._____AMOUNT____   IF PAYING BY INVOICE, CHECK INDIVIDUAL INVOICES PAID.

TRANSACTION DATE	INVOICE NUMBER	DESCRIPTION OF TRANSACTION COMMENTS AND CHARGES	CHARGES OR CREDITS	BALANCE		TRANSACTION DATE	INVOICE NUMBER	AMOUNT CHARGE OR CREDIT	✓
04/01/75		DUES	35.00	35.00		04/01/75		35.00	

CURRENT	30-60	60-90	OVER 90	TOTAL		BALANCE DUE	TOTAL
35.00				35.00		◄——►	35.00

STUDENT ASSIGNMENT 1
DESIGN OF CUSTOMER STATEMENTS

1. Define the information which must be contained on the customer statements and prepare a report analysis form for the customer statements.

2. Design the format of the customer statements using a printer spacing chart.

3. Designate the disposition of the customer statements after they have been printed, including the number of forms to be produced and where these forms are to be sent. Document this procedure in a narrative form.

STUDENT ASSIGNMENT 2
DESIGN OF MANAGEMENT REPORTS

1. Define the information which must be contained on each of the three management reports requested by Gunther Swartz. Prepare a report analysis form for each of the reports.

2. Design the format for each of the three management reports using a printer spacing chart.

3. Designate the disposition of the management reports. Document this procedure in a narrative form.

   NOTE: The report listing daily sales in the restaurant, bar, and pro shop is to be in sequence by date and the report listing total sales to each member of the club is to be in member number sequence, as per the request from Gunther Swartz.

4. Define and design any other reports which you feel would be of assistance to management and which can be cost-justified.

DESIGN OF SYSTEMS INPUT

CHAPTER 5

DESIGN OF
SYSTEMS INPUT

5

INTRODUCTION

As noted in the previous chapter, once it is determined what is to be produced from the system, that is, what the output of the system is to be, it must be determined what data is necessary to generate the required output. As an additional consideration the analyst must also be concerned with the method of data entry that is to be used to input data into the system.

Historically, the punched card with the related keypunch machines and punched card verifiers have been the most widely used form of input and input preparation equipment used for computer systems. During the past decade, however, a number of "data entry" devices have become available which should be reviewed by the analyst in determining the type of input media to be used for the system being designed.

When the punched card was the primary form of data entry, the basic problem in designing input was merely one of preparing a well designed source document and punched card layout to facilitate rapid keypunching. Today with numerous devices for preparing input data for the computer, ranging from portable card punches to sophisticated remote terminals and optical character recognition devices, the analyst is faced not only with the problem of proper input design but also with the problem of equipment selection.

Perhaps even more important the analyst must consider the method of data entry as a part of a total data processing system that provides for the most efficient method of input preparation, sorting and processing of data, and the ultimate preparation of the desired output so that the data generated at the source can be processed with a minimum of human intervention and cost.

TYPES OF DATA ENTRY DEVICES

The numerous data entry devices now available can be grouped into the following basic categories:

1. Card Punches and Verifiers
2. Key-To-Tape Units
3. Key-To-Disk Units
4. Optical Character Readers
5. Terminals

In addition to these basic types of data entry devices there are numerous specialized units, such as point of sale recorders for the retail industry, specialized terminals for the banking industry, etc.

Card Punches and Verifiers

Card punches, commonly called ''keypunches,'' and verifiers are the oldest, and for many years, the most widely used method of preparing data for input to the computer. It is estimated that there are over 500,000 keypunches in use throughout the data processing industry. For many years the basic operation of the keypunch was to record, in the form of punched holes on a card, data which could be efficiently processed by machines. Figure 5-1 illustrates an IBM 29 keypunch.

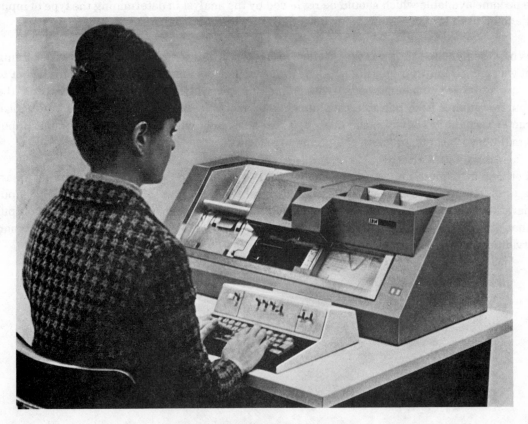

Figure 5-1 IBM 029 Keypunch

When using the keypunch it is necessary to have some method to ensure the accuracy of punching. This is normally accomplished through the use of the verifier. To insure the accuracy of punching, the newly punched cards are inserted into the verifier and the data just punched is rekeyed. If the holes in the newly punched card do not correspond to the data as it is keyed in on the verifier, the machine stops, the keyboard locks and an error is indicated to the operator. If the discrepancy exists after three tries in keying in the information on the verifier, a notch is recorded over the card column in error. If no errors are detected in the verifying processing a correction notch is placed on the right edge of the card. Figure 5-2 illustrates a card that has been correctly punched and a card that has been incorrectly punched. Note the notches in each of the cards.

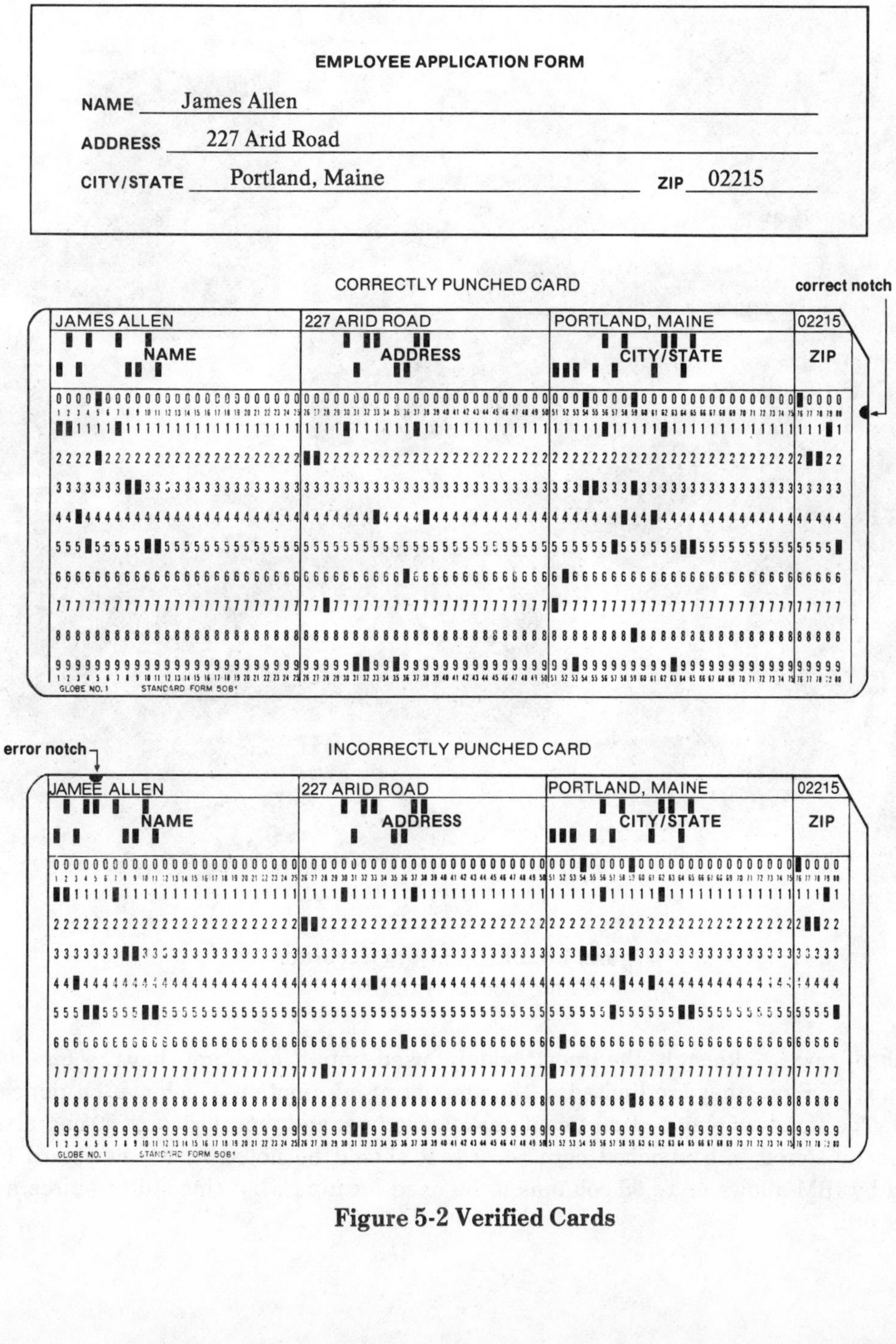

Figure 5-2 Verified Cards

In the late 1960's, the verifying keypunch was introduced. With this type of device, punching and verifying was performed on the same unit. The initial data is entered into an electronic memory. Punching is not performed until after all 80 columns of information have been placed in memory and verified. Verification is performed by rekeying the information and comparing the information to the data stored in memory. If a discrepancy exists, corrections may be made to the data stored in memory prior to punching. When the data has been verified as being correct, the entire card is punched at one time. This type of punching and verifying permits much faster error correction and reportedly results in a 20 to 30 percent increase in productivity over the conventional keypunch. Figure 5-3 illustrates the IBM 129 "data recorder."

Figure 5-3 IBM 129 Data Recorder
(Photo Courtesy of IBM)

Punched cards, although the most widely used input medium, have some distinct disadvantages. First, they are limited in the amount of information which can be punched in one card. The standard punched card contains 80 columns, and therefore only 80 characters or digits can be stored in a punched card record. It should be noted that a new card format designed by IBM allows up to 96 columns to be used for input, but this still requires a fixed length record.

Cards are also more difficult to store and to handle than other types of input which may be used with a system. Since cards are relatively bulky, they require more storage space for the same amount of data than disk or tape. In addition, they are more susceptible to errors in handling because they can be dropped or otherwise mishandled. Therefore, punched cards are normally used when the input volume is to be relatively small, thus reducing the probability of errors when handling and also reducing the amount of storage space required.

It must be noted, however, that the analyst must always consider the hardware available when making decisions concerning the medium to be used for input. For example, if the company has keypunch machines for creating punched cards and the application does not totally justify additional expenditures to acquire the machines required to prepare data on another medium, punched cards will normally be used simply because it is most economical to use them.

KEY-TO-TAPE SYSTEM

From a systems point of view the keypunch is a very slow means of data entry and requires extensive personnel resources. The need to increase the efficiency of the keypunch operation and lower the data entry costs without significantly changing the data preparation procedures led to the development of a number of keypunch replacement systems during the past 10 years.

One of the first types of keypunch replacement units was the key-to-tape stand alone system. With the key-to-tape system data is entered via a keypunch-type keyboard and stored directly onto magnetic tape using magnetic tape reels or cartridges as the form of storage. Key-to-tape systems provide for the verification of data using the same machine. Figure 5-4 illustrates a Mohawk Data Sciences 6401 Data-Recorder.

Figure 5-4 Key-To-Tape Unit
(Photo Courtesy of Mohawk Data Sciences Corp.)

Magnetic Tape as Input

Magnetic tape is composed of a plastic material normally one-half inch wide and is coated on one side with metallic oxide on which data may be recorded in the form of magnetic spots. The data recorded on magnetic tape may include numbers, letters of the alphabet, or special characters. Data is recorded in a series of parallel channels or tracks along the length of the tape. The example in Figure 5-5 illustrates how characters are stored on magnetic tape.

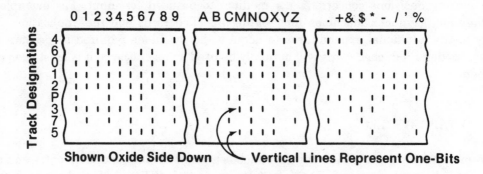

Figure 5-5 Magnetic Tape

When using magnetic tape as input, there is no requirement that the records be of any given length, such as 80 column cards; therefore, if input records longer than 80 columns are required, there is no restriction on record length with tape, whereas there would be with cards. For example, if an 85 character record is required for a particular application and punched cards are used as the input medium, two cards would be required for each record, one card containing 80 characters and a second card containing the remaining 5 characters. Thus, if there were 100,000 records 85 characters in length to be created, 200,000 cards would be required. If tape were used the 85 characters could be recorded in adjacent positions for each tape record, considerably simplifying systems design, programming, and operations.

Additional advantages of using magnetic tape as input include the fact that tape records can be read much faster than card records because of the computer hardware which is associated with the reading of records and much more information can be stored on magnetic tape than on cards. In most computer systems 1600 characters can be stored on one inch of tape, that is, the same amount of data which can be stored on twenty 80-column cards can be stored on one inch of magnetic tape. Thus, since much more data can be stored on tape and the tape can be read faster, magnetic tape is very useful as an input medium, especially if a large volume of input data is to be prepared and processed.

In summary, then, the advantages of the key-to-tape units include:

1. A single unit can perform both data entry and verification.

2. The operator is capable of correcting errors as soon as detected.

3. Data entry speed is reportedly 25-30 percent faster than punching cards using the keypunch, resulting in the need for fewer machines and people.

4. Magnetic tape as a form of input is reusable.

5. Magnetic tape as a form of input provides a greater density for storing data and a faster means of input to the computer system than punched cards.

6. When recorded on magnetic tape, records cannot be lost or put out of sequence, once recorded.

7. The devices are virtually silent, tending to decrease "operator fatigue."

There are, however, some disadvantages of key-to-tape input, including:

1. Higher equipment costs as compared to the keypunch.

2. The loss of the ability to access, manipulate, and read individual records as compared to punched cards.

3. Some retraining is required if personnel has previously used punched card equipment.

4. Switching from punched cards to magnetic tape as a form of input may require some reprogramming of existing programs if an electronic data processing system is currently in operation within the company.

KEY-TO-DISK SYSTEMS

As data recorders gained acceptance, key-to-disk systems called "shared processor systems" began to emerge. In these systems, a number of keyboards (usually varying from 2 to 64) are connected to a common processor. All of the keyboards, under control of the processor, enter data onto one disk, with the information received from each of the keyboards allocated to different portions of the disk. Verification is made possible by comparing data which is rekeyed with data which has been stored on the disk.

In addition to being verified, the data to be stored on the disk may be edited, formatted, accumulated, and otherwise processed under control of the common processor. The disk on which the data is stored can be removed to be used as input to a large computer system or, in some key-to-disk systems, data can be transferred directly from the disk to the large computer system via data communications equipment.

The system also allows for close monitoring of the operators since the control processor can accumulate detailed statistics relative to the number of jobs completed, the number of keystrokes per hour performed by each operator at each station, and other such information. Figure 5-6 contains an illustration of a key-to-disk system.

Figure 5-6 Key-To-Disk System
(Photo Courtesy of Mohawk Data Sciences Corp.)

In addition to the features which are available with key-to-disk systems, they offer economy of operation as compared to punched card data entry systems. Economy is due to increased productivity which is realized from these systems. For example, in one study it was reported that eight key-to-disk stations replaced fourteen keypunch stations.

The primary disadvantage, in addition to the fact that there is a higher initial cost for the key-to-disk systems when compared to card data entry systems, is the fact that with most key-to-disk systems currently available any expansion of the system must be done in increments ranging up to four stations at one time, while keypunches can be implemented one at a time.

Floppy Disk Systems

A relatively new type of data entry device to emerge on the market makes use of a flexible disk, commonly called a "floppy disk." The floppy disk is approximately the size of a 45 rpm record and in a typical system is capable of holding up to 3,000 card images. The floppy disk is receiving some consideration as a replacement for punched cards as input. A single disk costs approximately eight dollars, the same as 6,000 to 8,000 cards, but has the additional advantages of being reusable, providing a much more compact form of storage, providing much easier handling, and is a faster method of input.

The reading devices used for floppy disks are not very complicated and are relatively inexpensive when compared to disks commonly used with computer systems. Figure 5-7 illustrates the IBM 3740 data entry system which utilizes the floppy disk.

Figure 5-7 "Floppy Disk" Data Entry System
(Photo Courtesy of IBM)

OPTICAL CHARACTER READERS

Optical character readers represent an attempt to change the way in which source data is captured and generated by providing a means for a source document to be directly read by a machine. One of the earliest attempts at optical reading or scanning was in test scoring types of applications in which pencil marks on an answer sheet were electronically "read" and interpreted. Today this basic concept of Optical Mark Reading (OMR) has been expanded to include Optical Character Reading (OCR) so that it is now possible to read not only pencil marks recorded on preprinted forms but a variety of type fonts, and even hand-printed characters as well. Figure 5-7 illustrates an IBM 3886 Optical Character Reader.

Figure 5-8 Optical Character Reader
(Photo Courtesy of IBM)

The example in Figure 5-9 illustrates a form designed to handle personnel file updating that utilizes the principles of optical mark reading and optical character reading of hand-printed characters.

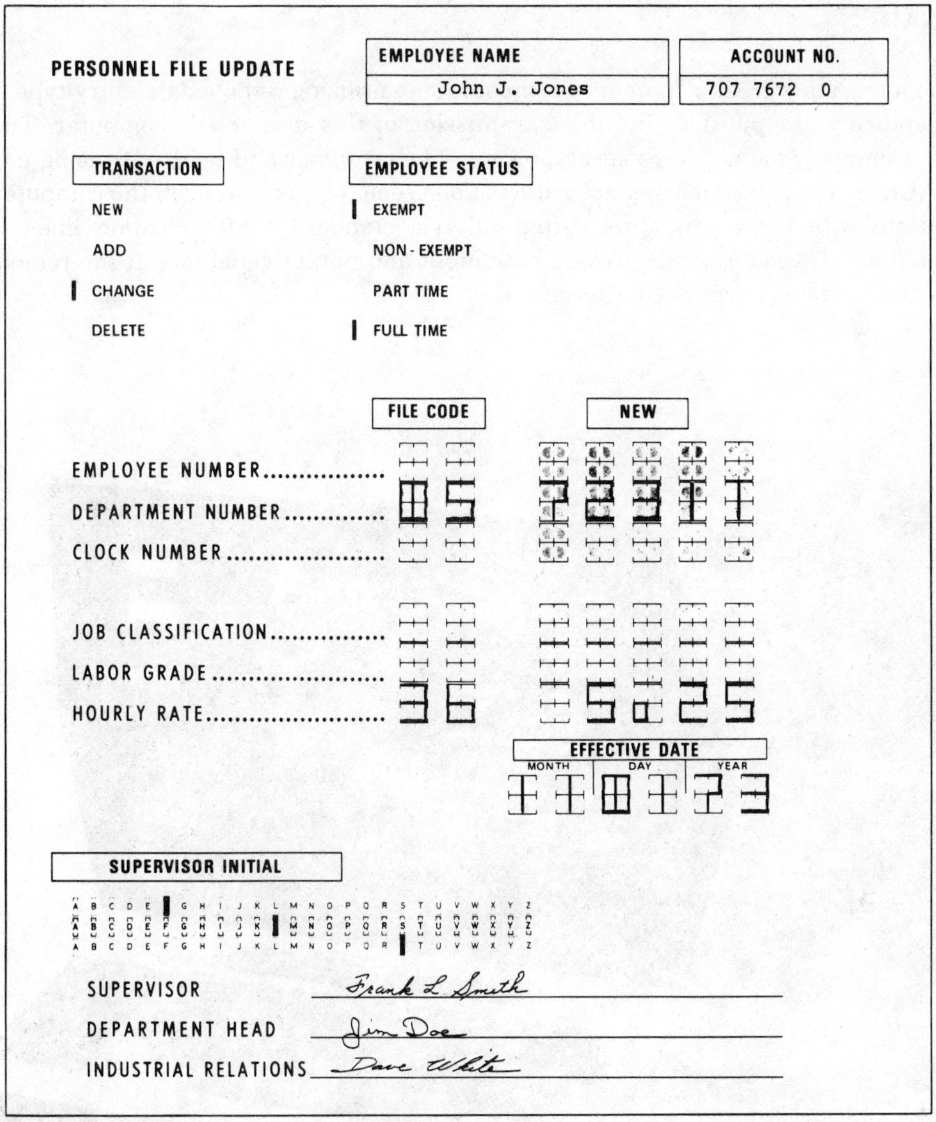

PERSONNEL FILE UPDATE

EMPLOYEE NAME: John J. Jones

ACCOUNT NO.: 707 7672 0

TRANSACTION: NEW / ADD / | CHANGE / DELETE

EMPLOYEE STATUS: | EXEMPT / NON - EXEMPT / PART TIME / | FULL TIME

FILE CODE — NEW

EMPLOYEE NUMBER................

DEPARTMENT NUMBER.............

CLOCK NUMBER.....................

JOB CLASSIFICATION...............

LABOR GRADE.......................

HOURLY RATE........................

EFFECTIVE DATE: MONTH DAY YEAR

SUPERVISOR INITIAL

A B C D E F G H I J K L M N O P Q R S T U V W X Y Z

SUPERVISOR Frank L. Smith

DEPARTMENT HEAD Jim Doe

INDUSTRIAL RELATIONS Dave White

Figure 5-9 Example of Document for Optical Character Reader

The output from optical character readers can take a variety of forms, that is, it is possible to read the forms and take the information directly into the computer; or, the form can be read and the data can be transferred to punched cards, magnetic tapes, or magnetic disks. Optical Character Readers can typically accept a variety of form sizes and process from 300 to 3000 sheets per hour.

TERMINALS

With increasingly greater frequency companies are utilizing remote data entry-type stations for the capturing of input data and the transmission of this data to the computer. Two basic types are commonly found—keyboard typewriter-like terminals and terminals using a cathode ray tube (CRT). These terminals, placed at locations remotely located from the computer, often communicate with the central processing unit via standard communication lines such as telephone lines. These terminals provide both input and output capability at the remote sites. Figure 5-10 illustrates a typical CRT terminal.

Figure 5-10 Example of CRT Terminal

Intelligent Terminals

The first terminals used with computer systems were referred to as transaction or conversational terminals. These terminals had no processing or editing capabilities within themselves and were merely used to key information into the computer or to receive and display information sent from the computer.

One of the newest concepts relative to the utilization of terminals is the development of INTELLIGENT TERMINALS. The intelligent terminal is designed to be capable of performing some processing within its own internal circuitry such as the editing and validating of data. The advantage of the intelligent terminal is that it relieves the main computer of some of the processing required to prepare the input data for processing. In determining whether a data processing system should utilize intelligent terminals, the analyst must weigh the increased cost of providing "intelligence" at the terminal against the reduced cost and usage of the central processing unit of the computer system because of the processing capabilities of the terminal.

The example in Figure 5-9 illustrates an intelligent terminal. Note that the unit illustrated contains two tape cassette input/output units that further increase the flexibility of the device.

Figure 5-11 Intelligent Terminal

As noted, an intelligent terminal is able to accept, store, and execute a variety of instructions which can be used to control operations which must be performed on data entered into the terminal. For example, in a payroll application, the intelligent terminal could be used to perform the following data checking operations as data from the time card is entered into the terminal.

TIME CARD							
EMPLOYEE NO. 10005			PERIOD ENDING 1/31				
NAME ANTHONY A. ALLEN							
DEPT NO. 05			SOCIAL SECURITY NO. 525-66-1824				
DATE	IN	OUT	IN	OUT	TOTAL REGULAR HOURS	TOTAL O/T HOURS	OVERTIME APPROVAL
1/27	7:58	5:01			8.0		
1/28	7:49	5:00			8.0		
1/29	7:54	5:02			8.0		
1/30	7:57	5:45			8.0		
1/31	7:59	5:00			8.0		
		TOTALS		40.0			
		PAYRATE		$ 4.00 HR.			
		TOTALS		160.00	+	= $160.00	
UNION CODE 01						TOTAL EARNINGS	

CHECKING
TO BE
PERFORMED

1. Department Number must be between 01 and 26.

2. The Union Code must be 1, 2, 3, 4, or 5.

3. Regular hours for each employee cannot be less than 1 or more than 40.

Figure 5-12 Example of Data Checking by Intelligent Terminal

In the example above, the intelligent terminal could be programmed to perform the checks indicated as the data is entered by the operator at a site remote from the central computer. Any data that did not meet the specified conditions would cause an error message to be displayed on the screen. This editing and validating of the input data results in less need for computer time at the central computer.

Additional Data Entry Hardware

It should again be pointed out that there are currently available numerous other types of data entry devices for specialized application areas such as point of sale devices for the retail industry, data collection systems for manufacturing industries, etc. It thus becomes apparent that the analyst must be well versed in the latest developments in computer technology if systems are to be designed that take into consideration the newest developments in computer hardware.

DESIGN OF SYSTEM INPUT

Although the analyst will have uncovered many of the characteristics of the current system and how it operates in the previous phases of the systems project, it is necessary to review the information obtained to ensure a thorough understanding of the method of input used in the current system prior to beginning the design of the input for the new system. Facts to be reviewed related to the current system include:

1. Who enters data in the current system?

2. What data is entered?

3. Where does the data enter the system?

4. When does the data enter the system and how often is the input data generated?

5. How does the data enter the system?

It is on the basis of a thorough understanding of the sources and methods of input in the current system that the analyst can design more efficient and improved systems for the handling of the input in the proposed system.

Other factors of concern to the analyst include the volumes of data entering the system, the source documents used, the length and size of data fields on each of the source documents, and special coding systems which are currently in use to represent data.

After reviewing the input of the current system, the basic steps in the design of the input for the new system include the following:

1. Determine what input data is needed to produce the required output.

2. Determine the source of the data needed to produce the output.

3. Determine the method of data entry to be used.

4. Design the format of the input data and related source documents.

DETERMINING THE INPUT DATA
NEEDED TO PRODUCE THE REQUIRED OUTPUT

In defining the input requirements, the analyst is merely defining the data that is needed as input to the system to generate the required output. In order to determine the required output, the analyst should consult the Report Analysis forms and printer spacing charts which were designed in the design of the system output. It will be recalled that the Report Analysis form contained an analysis of the data to be on the report, including the names of the fields, a description of the fields, that is, if the fields are alphabetic, alphanumeric, or numeric, and the number of characters in each of the fields. For example, if a Name and Address Listing were to be prepared, the Report Analysis form and printer spacing chart illustrated in Figure 5-13 could be prepared.

```
                    S Y S T E M   D O C U M E N T A T I O N

NAME OF SYSTEM                          DATE
             Personnel                            6/5        PAGE  1  OF  1

ANALYSTS                                PURPOSE OF DOCUMENTATION   REPORT
             Jones                      ANALYSIS FORM NAME AND ADDRESS LISTING
```

| | | NO. OF | |
FIELD	DESCRIPTION	CHAR.	COMMENTS
Employee Number	Numeric	5	
Employee Name	Alphabetic	20	
Employee Address	Alphanumeric	25	
City/State	Alphabetic	25	
ZIP Code	Numeric	5	

GENERAL COMMENTS

1. Report should include report and column headings with date and
   page number.

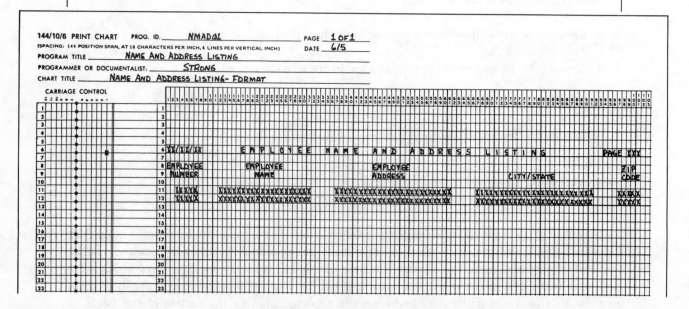

Figure 5-13 Example of Report Analysis Form and Printer Spacing Chart

It will be noted from the Report Analysis form that there is no indication of the source of the data for the output since in the output design phase the analysts merely determined what the output was to be and its format. Therefore, after ensuring that all output for the system is known, the analysts must determine the source of the data which is to be input to the system.

DETERMINING THE SOURCE OF DATA NEEDED TO PRODUCE THE OUTPUT

After the analyst has determined the input data required to produce the output, he is faced with the more difficult task of determining the source of the data. In any data processing system, data can originate from four basic sources. These sources are:

1. Data originating from a single source document.

2. Data that is calculated or originated by the computer program.

3. Data extracted from several sources outside the program.

4. Data originating from tables that serve as input to a computer program.

Data Originating from a Single Source Document

The easiest input design problem occurs when all the data for an output report originates from a single source document. For example, in an application it may be necessary to change the addresses of employees within the company. Thus, an Employee Name and Address Change Form could be used. The data from this source document would be input to the computer program and then the changes to the addresses would be listed on the printed report. This process is illustrated in Figure 5-14.

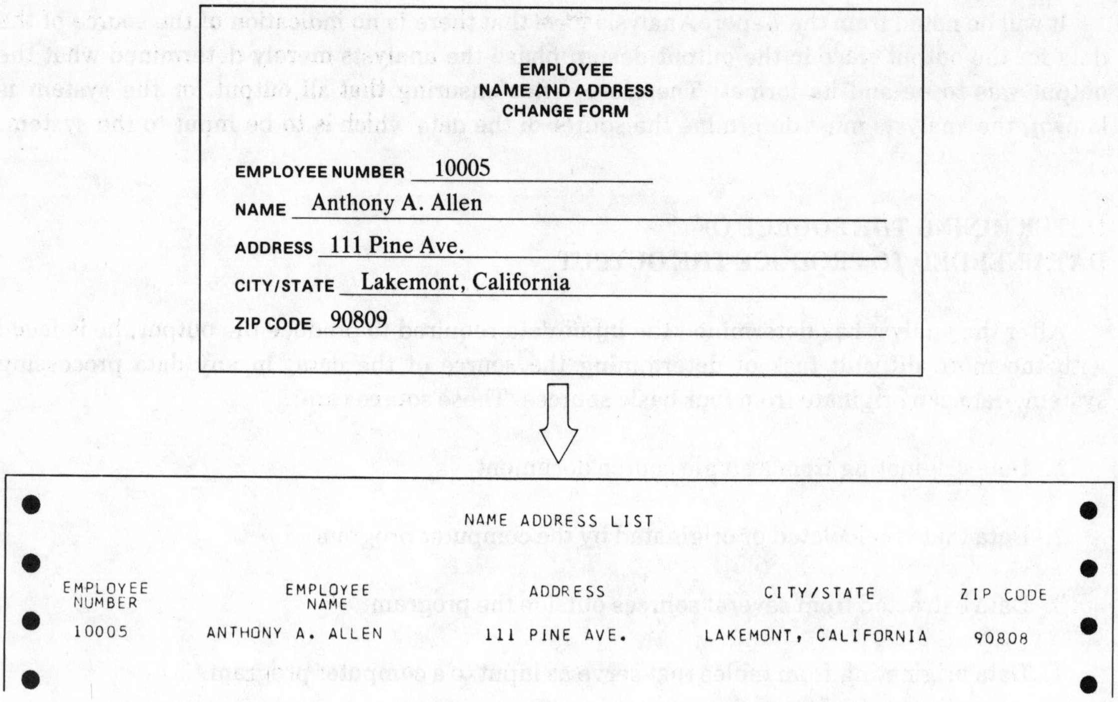

```
                    ┌──────────────────────────────────────┐
                    │            EMPLOYEE                   │
                    │       NAME AND ADDRESS                │
                    │         CHANGE FORM                   │
                    │                                       │
                    │  EMPLOYEE NUMBER  10005               │
                    │                                       │
                    │  NAME   Anthony A. Allen              │
                    │                                       │
                    │  ADDRESS  111 Pine Ave.               │
                    │                                       │
                    │  CITY/STATE   Lakemont, California    │
                    │                                       │
                    │  ZIP CODE  90809                      │
                    └──────────────────────────────────────┘
```

```
                              NAME ADDRESS LIST

  EMPLOYEE          EMPLOYEE            ADDRESS          CITY/STATE         ZIP CODE
   NUMBER            NAME
   10005         ANTHONY A. ALLEN      111 PINE AVE.   LAKEMONT, CALIFORNIA  90808
```

Figure 5-14 Example of Single Source Document

Note from the example in Figure 5-14 that all of the information for the computer-generated report originates from the single source document. When the analyst has determined the source of the input data, in this case the Employee Name and Address Change Form, it is then possible to complete an Input Analysis Form which includes the source of the input data. This is illustrated in Figure 5-15.

<div align="center">**S Y S T E M D O C U M E N T A T I O N**</div>				
NAME OF SYSTEM Name and Address Changes		**DATE** 5/12		**PAGE** <u>1</u> **OF** <u>1</u>
ANALYSTS Longview		**PURPOSE OF DOCUMENTATION** Input Analysis Form		

FIELD	DESCRIPTION	NO. OF CHAR.	SOURCE	COMMENTS
Employee Number	Numeric	5	Employee Name And Address Change Form	Must be all numeric
Employee Name	Alphabetic	20	Employee Name And Address Change Form	
Employee Address	Alphanumeric	25	Employee Name And Address Change Form	
Employee City/State	Alphabetic	25	Employee Name And Address Change Form	
ZIP Code	Numeric	5	Employee Name And Address Change Form	Must be all numeric

Figure 5-15 Input Documentation Form

In the example in Figure 5-15 it can be seen that there is a similarity between the information on the Report Analysis Form and the Input Analysis Form. It differs in two important aspects: the source of the input, that is the source document, is specified and in the comments area the requirements of the field are noted. In the example, the source of the input is specified as the Employee Name and Address Change Form for all of the fields. The Employee Number and ZIP Code are to contain numeric information and they must be checked to ensure that they do contain numeric data. Techniques for field and data checking will be discussed in a later chapter.

Data that is Calculated or Originated by the Computer Program

In many types of output reports and other forms of output, the data that appears will originate not only from the source documents but also from constants within the program or calculations made by the program as the input data is processed. Although this creates no serious systems design problems, the analyst should note this fact on the input analysis form, as in subsequent phases of the systems project the analyst must define the programming specifications for the computer programmer. These programming specifications must indicate the calculations which are to occur and other special types of information such as the size of the calculated field, etc.

The example in Figure 5-16 illustrates a report of gross pay prepared from the information contained on employee time cards. Note in the example that the hours worked is multiplied by the pay rate to produce the gross pay.

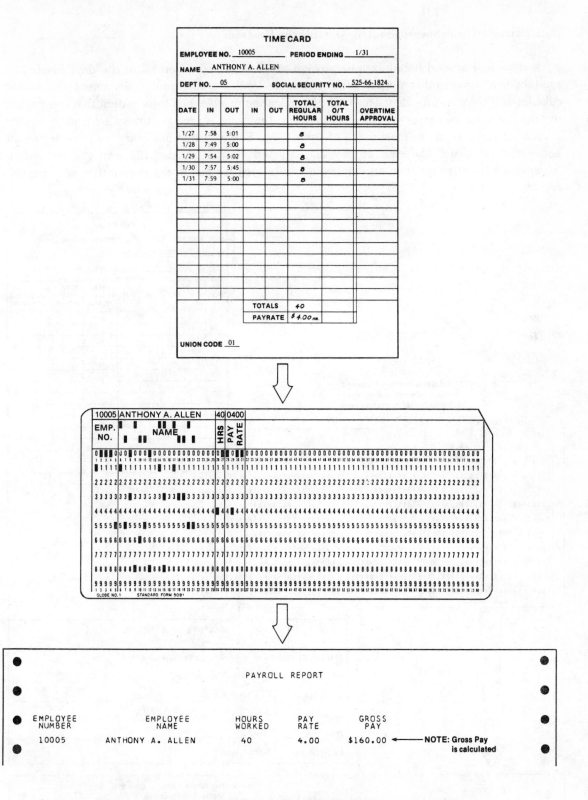

Figure 5-16 Example of Calculation Required for Output Data

Data Extracted from Several Sources Outside the Program

A third and more difficult systems design problem occurs when all of the data needed to produce a report or other output is not available from a single source document and is not calculated within the program. This type of application exists when information that is a part of the output must be extracted from either more than one source document or from a source document and a separate master file of some type. The example in Figure 5-17 illustrates an application in which the rate of pay is extracted from a master file and the remaining information is extracted from an employee's time card for use in the preparation of a payroll report.

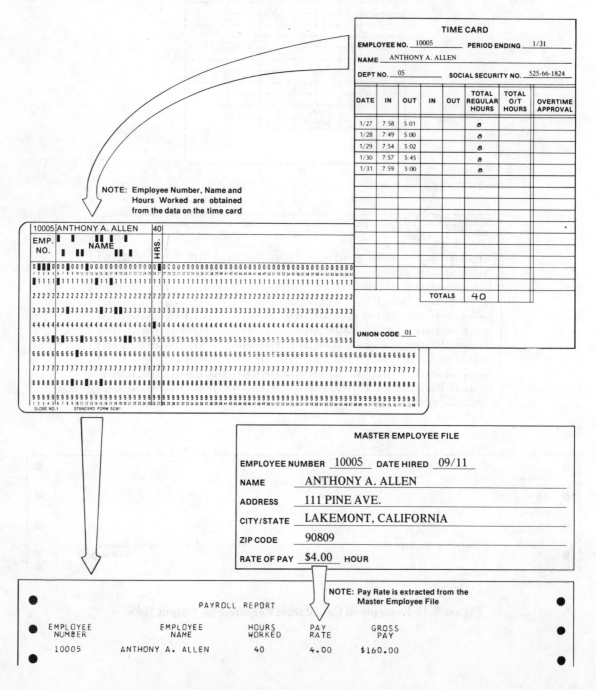

Figure 5-17 Example of Multiple Sources of Data

Note in the example illustrated in Figure 5-17 that the pay rate is **not** contained on the time cards but must be extracted from a Master Employee File to provide for the calculating of the gross pay. Thus, in this example, the analyst is faced with the task of not only designing the input record consisting of the information from the time card, but he is also faced with the task of designing a master employee record and integrating this into a system which provides for the pay rate to be extracted from the master record to produce the needed output report.

It should be noted that a number of methods exist for integrating this master data into a data processing system. Master file design and many of the methods which may be considered will be discussed in Chapter 6.

Data Originating from Tables

In some applications the analyst may want to consider the use of TABLES as a form of input. Using tables as a part of the input process can sometimes simplify the systems design process by eliminating the need for multiple input files.

In the previous example, the pay rate was not included on the time card but instead was extracted from a master file. Often confidential information such as pay rates are not placed on documents which can be viewed by other employees. A solution to problems of this type could be the use of a pay rate table. For example, assume that a company has 100 different pay rates for all employees. These pay rates could be designed into a table and the table could be used as input to the program. The example in Figure 5-18 illustrates a pay rate table in which each of the 100 different pay rates are assigned a code. This code is placed on the time card and transferred to the input data card for subsequent processing.

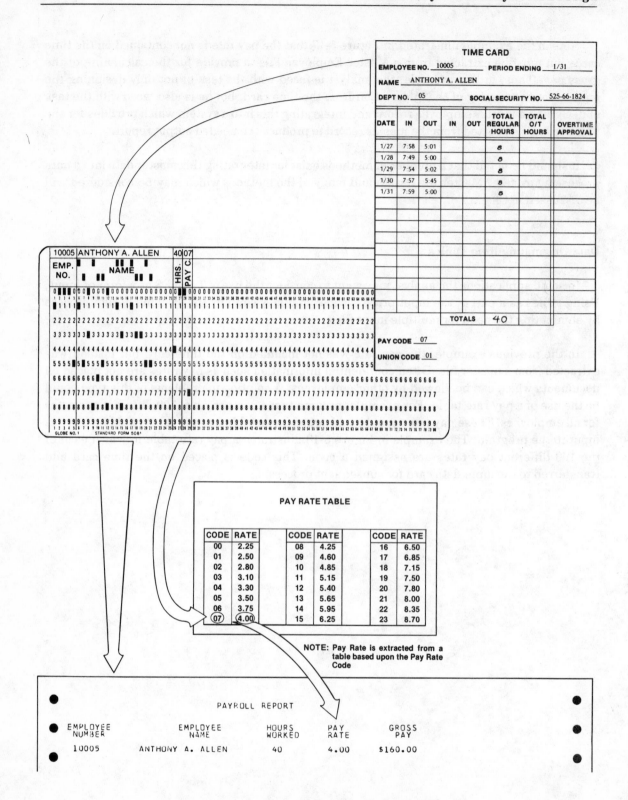

Figure 5-18 Example of Extracting Data from a Table

Note from the example above that a pay code is recorded on the time card and is transferred to the punched card which will be input to the computer program. This code is then used to "search" the table and when the proper code is found, the corresponding pay rate is extracted from the table, printed on the report, and used in the calculation to determine the gross pay.

It should be noted that a table can be stored as a part of the computer program itself or the table could be stored on an external device such as magnetic disk and referenced as required. The use of tables as a form of input has several significant advantages that should be apparent to the systems analyst. For example, in the application just illustrated, if there were 10,000 employees and frequent changes in pay rates for all employees, a change in the 100 entries in the table would provide an accurate method of providing for the pay rate changes for all 10,000 employees. If the actual pay rate was recorded in a master record, a change in pay would neccessitate a change in all 10,000 master records.

Tables also provide a means of condensing data on input data records and provide for increased data entry speed when large amounts of alphabetic data must be entered to satisfy output requirements. For example, assume a company has 10 salesmen who account for all of the company sales. Rather than transcribing the salesman name on an input record used for a sales report it is possible to place a Salesman Code in the input record and the code and related salesman name in a table and then extract the name when required to produce the output report. This processing is illustrated in Figure 5-19.

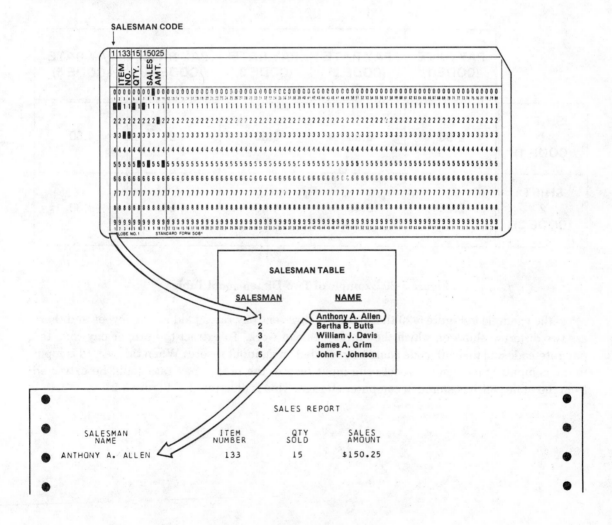

Figure 5-19 Example of Table Usage

Note in the example in Figure 5-19 that a salesman code is recorded in the punched card which is input to the program and the code and related salesman name is recorded in the table which is stored in computer storage. When the output report is produced, the computer program to produce the output will be read the salesman code from the input record and extract the proper name from the salesman table to produce the proper output report. Again, it should be noted that if there are large volumes of input records to be prepared, the entry of a salesman code in the input record rather than the salesman name can result in significant savings in data entry time.

Two-Dimensional Tables

Another type of table which is commonly used in data processing applications is the two-dimensional table. A two-dimensional table that could be used in a payroll application is illustrated in Figure 5-20.

	PAY RATE (CODE 1)	PAY RATE (CODE 2)	PAY RATE (CODE 3)	PAY RATE (CODE 4)	PAY RATE (CODE 5)
SHIFT 1 (CODE 1)	2.50	3.00	3.50	4.25	4.80
SHIFT 2 (CODE 2)	2.75	3.40	4.00	4.65	5.30

Figure 5-20 Example of Two-Dimensional Table

In the example in Figure 5-20 there are five different pay rates paid to employees and there are two different shifts on which the employees can work. To extract the proper pay code the pay rate code and the shift code could be recorded on an input record. When this record is input to the computer program to process the input record, the proper pay rate could be extracted from the table which is stored in computer storage. This is illustrated in Figure 5-21.

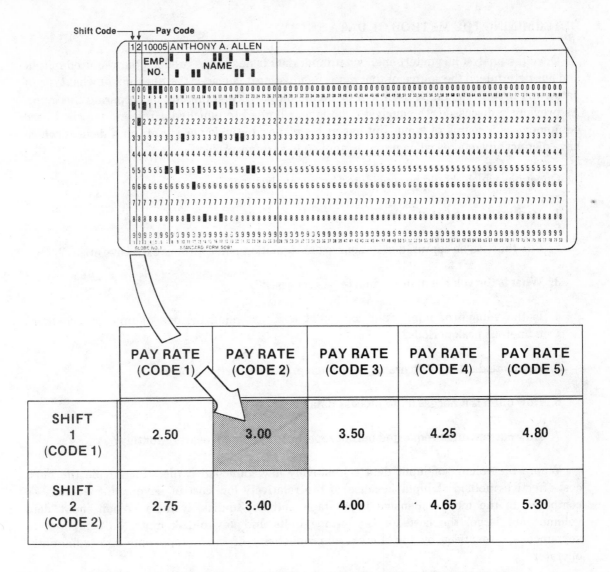

Figure 5-21 Example of Data Retrieval in Two-Dimensional Table

Note in the example above that Shift Code 1 and Pay Code 2 are recorded in the input record. When the record is processed, the pay rate corresponding to Shift Code 1 and Pay Code 2, which is $3.00, would be extracted from the table.

Tables of all types are commonly used in business data processing applications; therefore, it is essential that the analyst have a thorough understanding of the uses and applications of tables when designing the input for a computer system.

DETERMINING THE METHOD OF DATA ENTRY

Once the analyst has determined what input data is needed to produce the required output and has determined the source of that data, he must next consider the question of what type of data entry system should be used to provide the most efficient method of preparing the input. As previously discussed, the analyst is no longer limited to punched cards and punched card equipment as a source of input, but has at his disposal a wide variety of input devices which have distinct advantages when properly applied to systems projects.

Questions which should be considered by the analyst include:

1. How is the data to enter the system?

2. Can the data be captured at its source or origination point without excessive errors?

3. What is the volume of input data to be processed?

4. Is the volume of input data consistent over extended periods of time or are there numerous peak periods?

5. Is remote data entry desirable or is batch processing satisfactory?

6. How often is it necessary to process data?

7. How current must requested output reports be? Daily? Weekly? Monthly?

Where the volume of input data is relatively low, punched cards still provide the most cost-effective method of input because of the relatively low cost of keypunch machines as compared to the more expensive key-to-tape and key-to-disk systems. When input data volumes are large, the costs of key-to-tape units and key-to-disk units, as compared to keypunches, are offset by the increased productivity of key-to-tape and disk data entry operators.

Perhaps the most efficient method of data entry is to capture the data at its source through optical character reading devices or by having the originator of the data enter the data into the remote terminals. Again, the primary disadvantage of optical character reading devices is the high initial cost of the input devices. Similarly the use of remote terminals requires a relatively large computer system to support remote data entry.

In any case, with the design of a new system, the analyst must give serious consideration to the method of data entry because once the initial input records are designed and the method of data entry selected, subsequent changes in the future can become very costly.

DESIGNING THE FORMAT OF THE INPUT DATA

The actual design of the input records is one of the most important steps in the systems design process. At this step, the analyst must consider not only the design of each individual input record that is to be a part of the system, but the analyst must also analyze existing source documents from which the input is to be derived and redesign the source documents if necessary.

Design of Source Documents

The source document is the initial form on which data is transcribed. The source document may be a "master record" comprising permanent information related to a system or the source document may be a transaction record reflecting the daily business activities.

Although in any systems project many of the source documents may already be in existence, it is the job of the analyst to review existing documents to determine their effectiveness in supplying the system with the information needed to produce the desired output. Often a form designed for a manual system will be totally inefficient when applied to an electronic data processing system with a related method of data entry.

The transcribing of information from a source document to an input record through the use of a data entry device is subject to a great variation in production rate since this is a manual operation; therefore, anything which simplifies this procedure will tend to ensure a faster and more accurate operation. The basic rules related to the design of source documents and related input records are summarized below:

1. Source documents should be designed to allow for the most rapid recording of data using a data entry device.

2. Source documents should be designed to allow for reading and related recording of the data by reading the source document from left to right and top to bottom.

3. Fields that are to be duplicated, that is, fields which will contain the same information in more than one record, should be grouped together and placed in the high-order (leftmost) positions in the record.

4. Fields which contain variable information for each record and therefore must be manually keyed should not be interspersed with fields that are to be duplicated or skipped.

5. Fields should be grouped together on the input media by type of data, that is, all numeric fields should be grouped together and all alphabetic fields should be grouped together.

6. Unused portions of a record that are to be skipped, such as unused card columns on a card, should be placed in the low-order (rightmost) positions of the input record.

The analyst, by following these basic rules of source document and input record design, can significantly increase data entry speed. To illustrate this, the example in Figure 5-22 contains a source document and related input record which are poorly designed.

SOURCE DOCUMENT

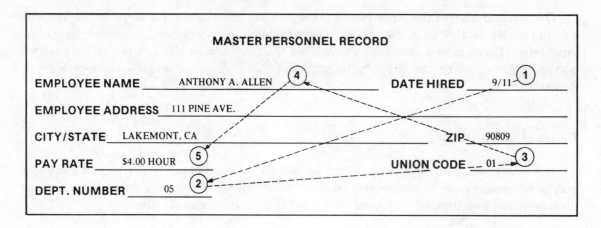

INPUT RECORD

Figure 5-22 Example of Poor Design of Source Document and Input Record

Note in the example in Figure 5-22 that the design of the source document and related input record does not lend itself to rapid data entry operations. The data entry operator must skip around the source document in order to find the data to be entered in the input record. In addition, numeric and alphabetic fields have not been grouped together on the input record and there are fields which are skipped within the data. Thus, this source document and input record do not follow the basic rules for good design.

The example in Figure 5-23 illustrates a redesign of both the source document and the input record to facilitate data entry.

SOURCE DOCUMENT

MASTER PERSONNEL RECORD

DATE HIRED _09/11_ ①----- DEPARTMENT NUMBER --- _05_ ---②

UNION CODE _01_ ③ PAY RATE _$4.00 HOUR_ ④

EMPLOYEE NAME _ANTHONY A. ALLEN_ ⑤

EMPLOYEE ADDRESS _111 PINE AVE._

CITY_LAKEMONT_ STATE _CA_ ZIP _90809_

INPUT RECORD

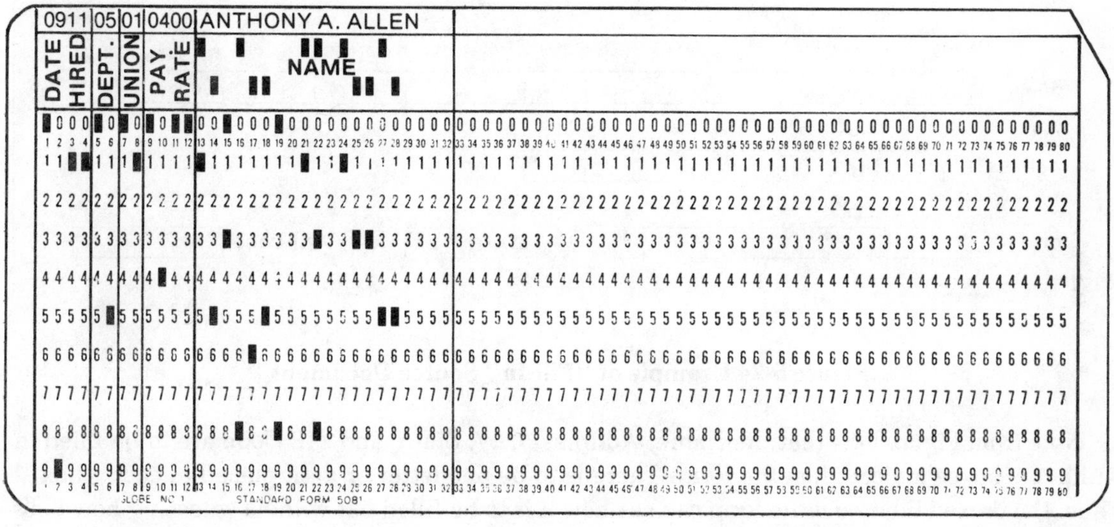

Figure 5-23 Example of Good Design for Source Document and Input Record

Note in the example above that the source document has been designed to allow reading of all data from left to right, top to bottom, and that the input record has been designed to provide punching in this manner. Also note in the input record that all numeric data has been grouped together rather than having the pay rate follow the employee name.

In addition to coordinating the design of the source document with the input record to provide for rapid data entry, there are a number of specific methods for designing source documents that are effective for improving the accuracy of the data that enters the system.

One consideration which must be of concern to the analyst is how the data is to be recorded on the source document, that is, who is to complete the source document. In many applications the source document will be completed by trained clerical personnel. In other cases, however, the design of the system will provide that a source document be completed by the originator of the information, such as a customer completing a bank deposit slip or a student completing a registration form. In those cases where the person completing the form is outside the company organization, it is essential that the source document be designed so that all individuals completing the form do so in an identical manner. There are several techniques that can be used to ensure consistency in the completing of source documents. These techniques are described in the following paragraphs.

The basic requirement of a source document in terms of ensuring accurate information is that the form must be kept as simple as possible so that it is easily filled out. When a document is to be used to provide variable information, many times the "fill-in" type of form is best. An example of a form of this type is illustrated in Figure 5-24.

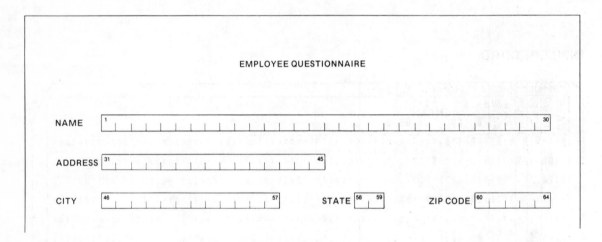

Figure 5-24 Example of "Fill-In" Source Document

Note from Figure 5-24 that the Name, Address, City, State, and ZIP Code are to be filled in within the boxes indicated on the form. There is to be one letter or digit placed in each "box." Thus, the person filling out the form can see how it is to be filled out without guessing how long the name is to be or where it should be placed. This form illustrates one of the important "maxims" of form design, that is, the form should be designed to force the person using the form to accurately record data on the form. By arranging the boxes as illustrated, the user of the form will find it difficult to make a mistake.

It should also be noted that there are numbers associated with each box on the form. These numbers correspond to the character positions in the input records which will be prepared from the source documents. Thus, the Name entry would begin in position 1, the Address entry would begin in position 31, and so on. This feature of the source document allows easy transcription by the data entry operator who is to prepare the computer input.

When the person filling out the source document is to make a choice out of a given number of entries, there are several techniques which can be used to enable that person to easily make the choice and allow it to be easily transcribed by the data entry operator. One of the most useful is the "check-box" approach, which is illustrated in Figure 5-25.

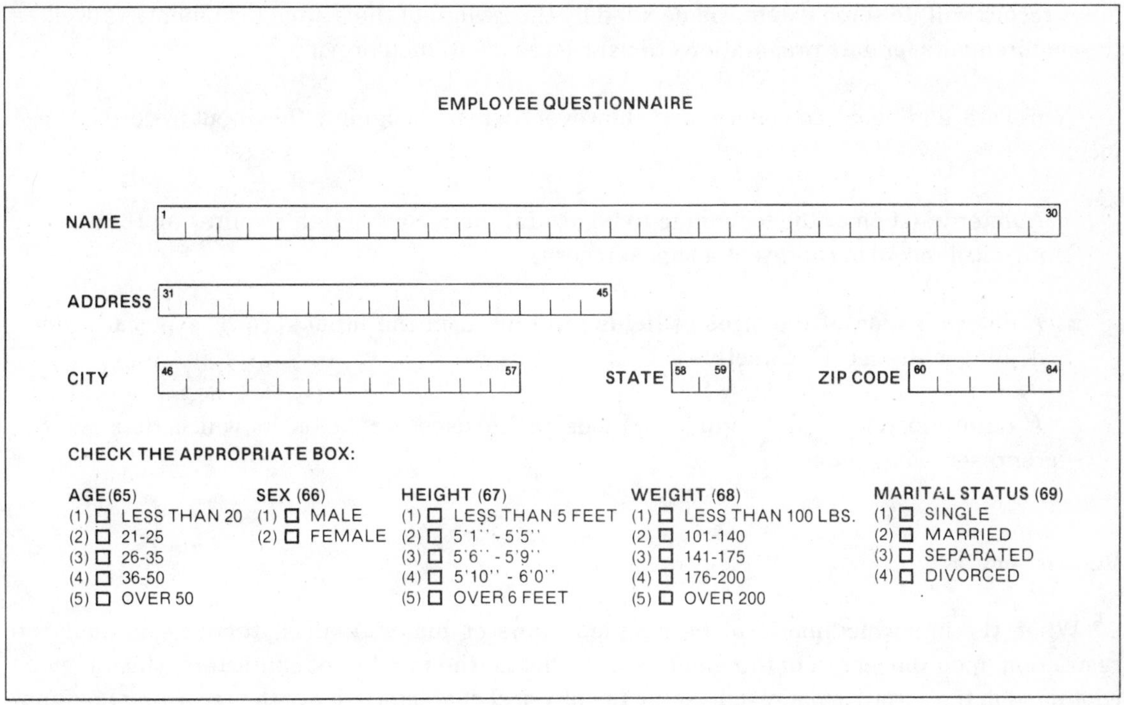
EMPLOYEE QUESTIONNAIRE

NAME (1–30)

ADDRESS (31–45)

CITY (46–57) STATE (58–59) ZIP CODE (60–64)

CHECK THE APPROPRIATE BOX:

AGE (65)
(1) ☐ LESS THAN 20
(2) ☐ 21-25
(3) ☐ 26-35
(4) ☐ 36-50
(5) ☐ OVER 50

SEX (66)
(1) ☐ MALE
(2) ☐ FEMALE

HEIGHT (67)
(1) ☐ LESS THAN 5 FEET
(2) ☐ 5'1" - 5'5"
(3) ☐ 5'6" - 5'9"
(4) ☐ 5'10" - 6'0"
(5) ☐ OVER 6 FEET

WEIGHT (68)
(1) ☐ LESS THAN 100 LBS.
(2) ☐ 101-140
(3) ☐ 141-175
(4) ☐ 176-200
(5) ☐ OVER 200

MARITAL STATUS (69)
(1) ☐ SINGLE
(2) ☐ MARRIED
(3) ☐ SEPARATED
(4) ☐ DIVORCED

Figure 5-25 Example of Check Boxes on Source Document

Note from the example in Figure 5-25 that the person completing the form can check the applicable box on the form in answer to the questions posed. The data entry operator can then record, in the columns indicated, the number associated with the checked box. The computer program will then interpret the number by relating the number with the associated meaning. This use of codes in input records will be discussed later in this chapter.

It should be noted that the source documents from which computer input is created may have more than one purpose. For example, if the questionnaire in Figure 5-24 and Figure 5-25 were for an insurance policy, it may well be that the questionnaire would also act as a contract between the customer and the insurance company, binding the company to writing the policy. Therefore, there may be legal requirements or restrictions placed on source documents by the company which must be considered by the analyst when designing the forms. As with the output documents discussed in Chapter 4, there are always the restrictions which may be required by management which dictate, to a certain extent, the limits for design which the analyst has.

When designing a new system the analyst should consider all aspects of the systems design process. Far too often there is a tendency for the analyst to design the input records from the existing source documents with little consideration given to the need for maximizing the speed and accuracy of data entry. Yet, in most installations, data entry can account for a significant amount of company resources and is one of the more costly of the data processing functions.

DESIGN OF INPUT RECORDS

The design of the input record is normally performed concurrently with the design of the source document so, as noted previously, the preparation of the input record from the source document can take place as efficiently as possible. Thus, the positioning of the fields in the input record will, to some extent, be decided by the design of the source document, as well as the requirements for data preparation efficiencies as mentioned previously.

There are additional considerations, however, when designing the input record. These include:

1. A determination of the technique to be used if the record length requires more than one physical record to represent a logical record.

2. A determination of the sizes of fields to be used in the input record, especially when dealing with special characters.

3. A determination of the coding methods to be used for fields in which data will be represented by codes.

Record Length

When the input medium is to be magnetic tape or magnetic disk, there is normally no restriction upon the length of the input record, that is, the number of characters which may be contained in the record is not restricted by the physical characteristics of the recording medium. When, however, the record is to be stored on a punched card or, in some instances, is to be entered from a CRT or other type of remote device, there may be restrictions on the length of the input record. If this situation occurs, it may be necessary to have more than one physical record, that is, more than one card, in order to contain all of the information which is to be a part of the input record.

Because of the restrictions placed on record lengths by the physical attributes of certain recording media, such as punched cards, it is best if the logical record length can be kept within the length of the physical record. If this is not possible, then two or more physical records will be required for each logical record. If this is necessary, a control number, such as an employee number, must be recorded on each physical record so that the records can be identified as belonging to the same logical record. When this is necessary, the control field must be in the same location within each record so that the records can be sorted and otherwise processed properly.

Two specialized forms are normally utilized by the analyst in documenting input records. One is a card layout record for use when designing a single punched card record and the other is a multiple card layout form utilized when a series of related records are to be designed. These forms are illustrated in Figure 5-26.

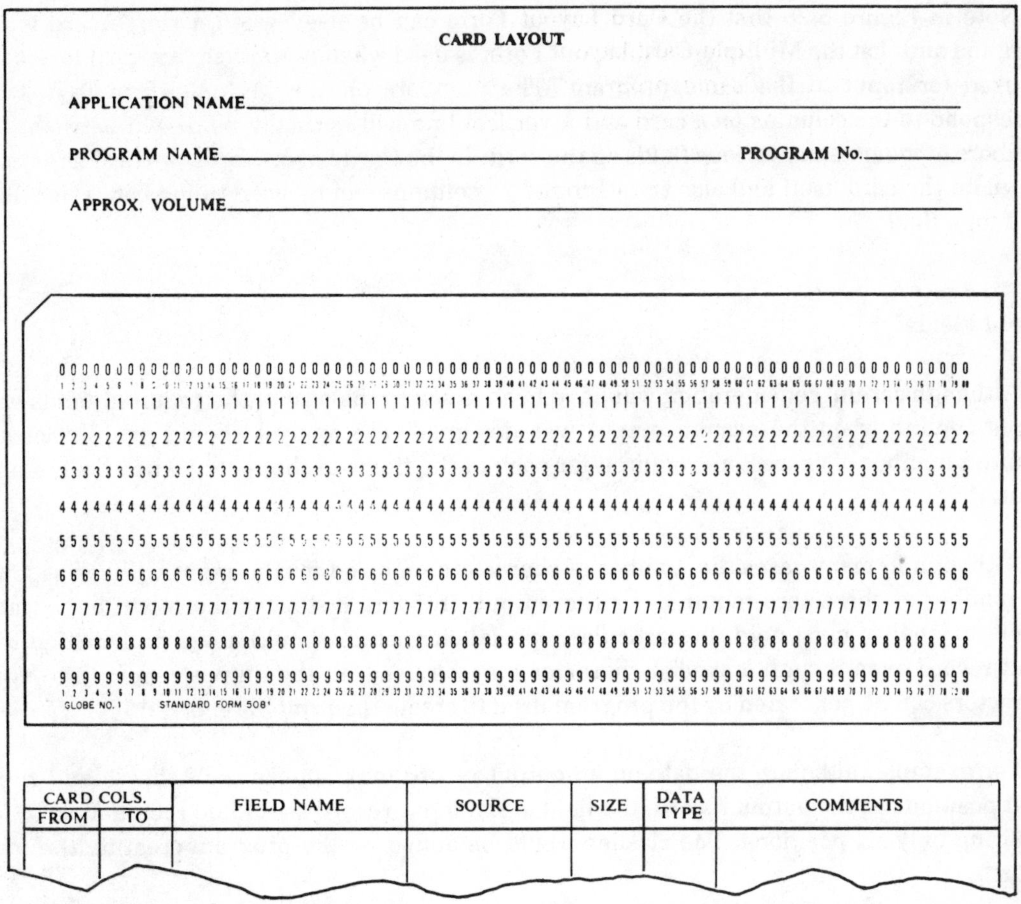

CARD LAYOUT

APPLICATION NAME_____

PROGRAM NAME_____ PROGRAM No._____

APPROX. VOLUME_____

GLOBE NO. 1   STANDARD FORM 508

CARD COLS.		FIELD NAME	SOURCE	SIZE	DATA TYPE	COMMENTS
FROM	TO					

IBM

INTERNATIONAL BUSINESS MACHINES CORPORATION

MULTIPLE-CARD LAYOUT FORM

Form X24-6599-0
Printed in U.S.A.

Company _____

Application _____ by _____ Date _____ Job No. _____ Sheet No. _____

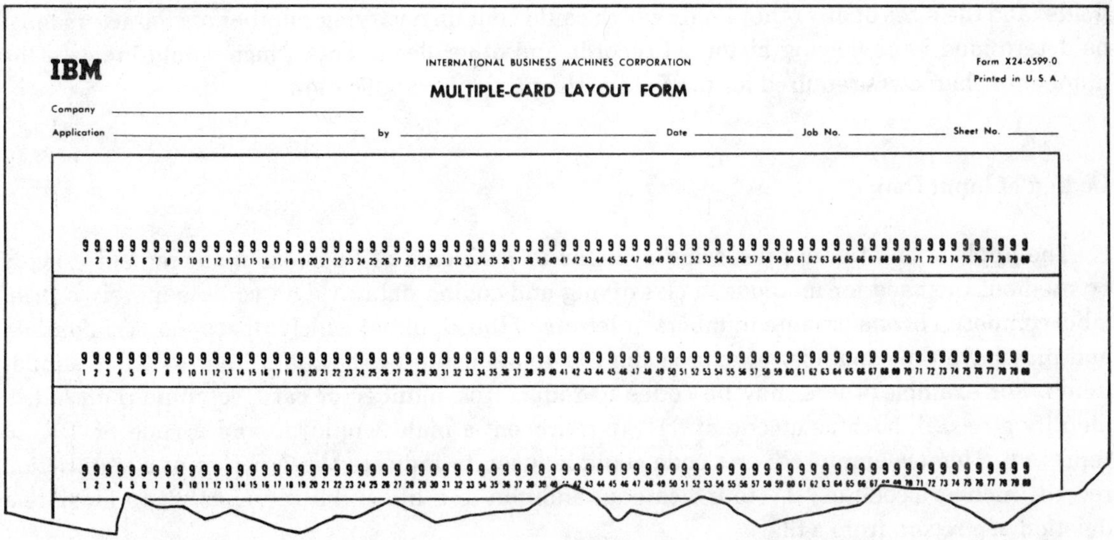

Figure 5-26 Example of Card Layout Forms

Note in Figure 5-26 that the Card Layout Form can be used when a single card is to be designed and that the Multiple-Card Layout Form is used when more than one card format is to be used for input to the same program. The numbers on the Multiple-Card Layout form correspond to the columns on a card and a vertical line will normally be drawn between these numbers to indicate the various fields on the card. In the Card Layout form, vertical lines can be drawn on the card itself and also the "From-To" columns can be used to indicate the columns used for a field.

Size of Fields

Although careful consideration will have been given to the size of the required fields on the output reports when the system output was designed, the analyst should again review the required number of characters required for each of the fields of the record when designing the input.

Although in many cases the number of characters required for the input will correspond to the number of characters required for the output, in some instances there will be variations which the analyst must note. In particular, special characters should never be included in the input record even though a special character may be used on the output report. The special characters can be generated by the program used to create the printed report.

For example, although the date on an output record may appear as 09/11/75 and require eight positions on the output record, the field in the input record should be recorded as 091175, requiring only six positions. The slashes would be added by the program creating the output report.

It should also be noted again that the size of amount fields, the size of name and address fields, and the sizes of any other fields which could contain a varying number of characters must be determined by reviewing historical records and other documents which would indicate the number of characters required for the field in the particular application.

Coding of Input Data

The application of data processing systems to problems of business management quickly pointed out the need for methods of classifying and coding data. A data code is merely a brief label composed of one or more numbers or letters of the alphabet which identify an item of data and may have the capability of expressing the relationship of that data item to other similar items. For example, there may be codes to reduce the number of card columns required to identify a record, such as a code of "1" to represent a male employee and a code of "2" to represent a female employee; or a code could be used to indicate the function of a transaction record, such as a code of "1" to indicate an addition to a file and a code of "2" to indicate a deletion of a record from a file.

Codes have long been a part of our lives even before data processing applications. We have codes for highways—Route 66, codes for airplane flights—flight 452, codes for classes at school—business 1A, and a social security number which is a code. In fact, much of one's life is arranged around codes of some type. The systems analyst should have an understanding of the basic types and methods of coding data, for codes are frequently used as a method for identifying or specifying certain types of data elements in a computerized system.

Before discussing specific coding techniques there are some basic characteristics of a good coding system that should be understood. These techniques are summarized below.

1. THE CODE SHOULD BE UNIQUE - The coding method selected should ensure that only one value of the code with a particular meaning can be assigned to one particular type of record or value within a record.

2. THE CODE SHOULD BE EXPANDABLE - In developing a coding structure provision must be made to provide for additional entries in the structure so that the new items will fit within the structure within the proper sequence.

3. THE CODE SHOULD BE SIMPLE TO APPLY - In any coding structure the assignment and use of codes may be performed by persons within the organization who are not familiar with data processing principles. Therefore, the code must be simple to apply and use.

4. THE CODE SHOULD BE CONCISE - The code developed should be as concise as possible to adequately describe each item being represented.

5. THE CODE SHOULD LEND ITSELF TO SORTING - In many applications the code developed should lend itself to sorting to allow the arrangement of the items reflected in the code in an ascending or decending sequence.

With the concept of large, shared data bases it is essential that a systemic approach to the coding of data be undertaken by the analyst. The following paragraphs describe several of the common methods used in coding data.

TYPES OF CODES

One of the more common methods for coding data is the "sequence" code wherein numeric values are placed in a sequence to represent different conditions or to indicate different meanings. For example, if it is desired to represent sex and marital status on an application form, the numeric codes "1" and "2" could be used for sex and the numeric codes "1," "2," "3," and "4" could be used for marital status.

CODE	SEX	CODE	STATUS
1	- Male	1	- Married
2	- Female	2	- Single
		3	- Divorced
		4	- Widowed

Note in the examples above that a numeric code which is in numeric sequence is used to represent sex and marital status. In some cases, the numeric values used to represent data can be arbitrarily chosen. In most cases, however, it is advantageous to consider which values are to be used to represent data. In general, the most frequently found conditions of the data should be represented by the lowest numeric value. In the example above the majority of application forms are filled out by married men. Therefore, the code value "1" is assigned to males and the code value "1" is applied to married persons. The next most likely group are single persons so they are given a code of "2," and so on.

The reasons for this type of arrangement are two-fold: First, in the computer program which processes this data, it is most economical in terms of processing time to perform a minimum number of compare operations to determine the proper data; thus, if the code which occurs most often is compared first, there will be a minimum number of comparisons in the program. Secondly, it is much more logical and makes the program easier to understand if the sequence of comparison in the program is, for example, "1,2,3,4" rather than, for example, "3,1,4,2." Therefore, in order to get the benefit of both a logical program and the most efficient processing within the program, the most frequently found condition of the data should be represented by the lowest numeric value, the next most frequent the next lowest value, and so on.

The advantage of codes of the type illustrated above is that they do condense the number of characters on an input record required to represent data, and through programming the descriptive names may be printed on the output reports by referencing the numeric code.

The numeric sequence code does not, however, provide for classification of groups of similar data types and is best used where there are a small number of coding classifications and where it is known that there will not be the need to insert or add classifications within the list.

CLASSIFICATION CODES

Another common coding technique makes use of codes which are designed to represent major and minor classifications of data by succeeding digits within a number. For example, to generate an employee number using a classification code there could be a code for store number, a code for the department number, a code for the type of worker, and a unique employee number.

STORE	DEPARTMENT	TYPE OF EMPLOYEE
1 - Los Angeles	10 - Auto Parts	100 - Salesman
2 - Long Beach	20 - Warehouse	200 - Warehouseman
		300 - Shipping Clerk

Thus, an employee with an assigned employee number of 003 that worked in the Los Angeles store, in the Auto Parts Department as a salesman could be assigned the following employee identification number:

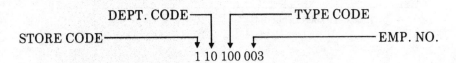

In the coding structure illustrated above there could be up to 9 stores, 99 departments, and 999 classifications of employees. Thus, as well as covering the current requirements, this coding structure has allowed for future expansion.

In many applications, the development of a code or identification for a given element of the system is more complicated than assigning an employee number or a status code. For example, it may be necessary to classify over a million parts in an inventory with many more factors than store and department. Regardless of the complexity, however, the general requirements are that the code be able to uniquely identify the part or the element of the system to be identified, without any duplication; and that the length of the number be reasonable so that an excessively long number is not used to identify the element. The requirement of uniqueness is important because when processing the data, it will be vital that two different parts are not identified by the same part number.

As pointed out there are few universal standards that may be applied to the coding of data, and although there are some suggested techniques, the analyst is responsible for devising a coding structure that lends itself to the most efficient processing for the system under study.

SUMMARY

The design of the input data to be processed within a computer system will, in most instances, depend upon the data which will be required to produce the required output. The source document is a very critical document because its ease of use will many times improve accuracy within the system and will lead to a more easily used system. After the input and the output of the system have been designed, it remains for the analyst to specify the processing sequence which will produce the required output from the designated input and to indicate the permanent files in which data will be stored.

CASE STUDY - JAMES TOOL COMPANY

In the sample payroll system for the James Tool Company it was found that the master data, that is, the data which describes each employee and his pay rate, number of exemptions, etc. was contained on the Employee Master Record Form. Changes in employee status, such as a change in pay rate, was recorded on a separate Deduction and Pay Rate Change Form, and historical data such as year-to-date earnings was contained on an Employee Compensation Record. These forms are again illustrated below.

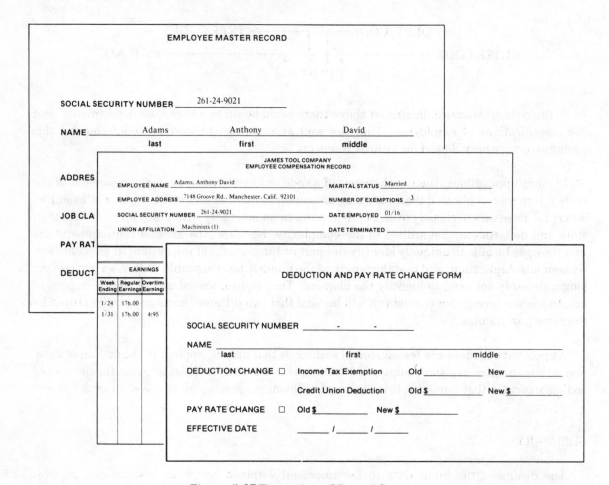

Figure 5-27 Forms from Manual System

In reviewing the forms from the manual payroll system and after preparing input analysis forms, the analysts noted that most of the information required for the new computer system was contained on these forms. They noted also, however, that the data was not arranged in a manner which was easily transferable to punched cards for input to the new system. Thus, it was decided that it was necessary to redesign the forms.

In addition, the analysts noted that the Deduction and Pay Rate Change form contained entries to change only the deductions and pay rate for an employee. There was no form which was used in case the name or address had to be changed or if other information on the Employee Master Record had to be changed. They had found out in their investigation that these changes were made manually by merely crossing out one value and inserting another in the manual system.

Because of these missing values which were not required in the manual system but which would be required in the computer system, and because of the need that the source document be easily used when preparing computer input, the analysts concluded that two forms were required—one for new employees and one for changes to be made to the employee master records. The forms which were designed are illustrated in Figure 5-28 and Figure 5-29

NEW EMPLOYEE FORM

DATE___/___/___

FIRST CARD

FUNCTION (Col 1)

[1] New Employee

EMPLOYEE NUMBER (Col 2-8)

NAME (Col 9-38)

ADDRESS (Col 39-57)

CITY/STATE (Col 58-73)   ZIP CODE (Col 74-78)   SEQUENCE (Col 80)

[1]

SECOND CARD

FUNCTION (Col 1)

[1] New Employee

EMPLOYEE NUMBER (Col 2-8)

SOCIAL SECURITY NO. (Col 9-17)   JOB CLASSIFICATION (Col 18-19)   STARTING DATE (Col 20-25)

PAY RATE (Col 26-29)   UNION (Col 30-31)   EXEMPTIONS (Col 32-33)

MARITAL STATUS (Col 34)   CREDIT UNION (Col 35-39)   SEQUENCE (Col 80)
(1) ☐ Married   (3) ☐ Separated
(2) ☐ Single   (4) ☐ Divorced   [2]

Figure 5-28 New Employee Form

EMPLOYEE CHANGE FORM

DATE____/____/____

CHECK THE APPROPRIATE FUNCTION (Col 1)

(2) ☐ Name and / or Address Change   (6) ☐ Exemptions Change
(3) ☐ Classification and / or Union Change (7) ☐ Marital Status Change
(4) ☐ Employee Termination   (8) ☐ Credit Union Deduction Change
(5) ☐ Pay Rate Change

EMPLOYEE NUMBER (Col 2-8) All Functions

NAME (Col 9-38) Function 2

ADDRESS (Col 39-57) Function 2   CITY / STATE (Col 58-73) Function 2   ZIP CODE (Col 74-78) Function 2

JOB CLASSIFICATION (Col 18-19) Function 3   TERMINATION DATE (Col 20-25) Function 4

PAY RATE (Col 26-29) Function 5   UNION (Col 30-31) Function 3   EXEMPTIONS (Col 32-33) Function 6

MARITAL STATUS (Col 34) Function 7   CREDIT UNION (Col 35-39) Function 8

(1) ☐ Married   (3) ☐ Separated
(2) ☐ Single   (4) ☐ Divorced

Figure 5-29 Employee Change Form

Note from Figure 5-28 that the New Employee Form allows space for two separate cards—one with the Name, Address, City, State and ZIP Code, and a second card for the other information required for the new employee. The form makes use of two techniques in designing a source document, that is, the "fill-in" type of entry and the "check-box" type of entry. Where variable information must be entered, such as the Name and Address, fill-in entries are to be made, but where a choice must be made from a given number of possibilities, such as Marital Status, the check-box approach is used.

Another consideration in source document design is whether the form is to be hand-written or typewritten. If the document is to be filled in by hand, it is normally necessary to make the "fill-in" boxes larger than if the form is to be filled in using a typewriter. When a typewriter is to be used, it will normally be the best policy to space the "boxes" either 10 or 12 characters per inch so that the typist can type normally and fill in the boxes. Since the New Employee Form is to be filled out by the personnel department using a typewriter, the form was designed with all boxes a tenth of an inch in width so that the typewriter can be used.

As noted, the form is completed by the personnel department whenever a new employee is hired or a change is to be made to the master record of an employee. Figure 5-30 contains an example of the New Employee Form with the data filled in.

NEW EMPLOYEE FORM

DATE 01 / 16 /

FIRST CARD

FUNCTION (Col 1)

|1| New Employee

EMPLOYEE NUMBER (Col 2-8)

|01 - 25423|

NAME (Col 9-38)

| Adams, Anthony David . . . . |

ADDRESS (Col 39-57)

| 7148 Groove Road . . |

CITY/STATE (Col 58-73)	ZIP CODE (Col 74-78)	SEQUENCE (Col 80)
Manchester, CA .	92101	1

SECOND CARD

FUNCTION (Col 1)

|1| New Employee

EMPLOYEE NUMBER (Col 2-8)

|01 - 25423|

SOCIAL SECURITY NO. (Col 9-17)	JOB CLASSIFICATION (Col 18-19)	STARTING DATE (Col 20-25)
261 - 24 - 9021	3F	01 - 21 - .

PAY RATE (Col 26-29)	UNION (Col 30-31)	EXEMPTIONS (Col 32-33)
0440	01	03

MARITAL STATUS (Col 34)	CREDIT UNION (Col 35-39)	SEQUENCE (Col 80)
(1) ☒ Married (3) ☐ Separated (2) ☐ Single (4) ☐ Divorced	00000	2

Figure 5-30 Example of Completed New Employee Form

It should be noted that the form was designed so that the person completing the form would have no question concerning where the data was to be entered and the person preparing the input record from this source document would know which columns in the input record are supposed to be used. The column specifications which are included with the name of the field are for the convenience of the data entry operator. It should be noted also that these column specifications could not be made unless the format of the input record had been designed. Thus, it should be seen that the design of the source document and the design of the input record normally are closely coordinated efforts.

Case Study - Time Card-Source Document

There is an additional source document in the new system which was not used in the manual system. It will be recalled that a time card is to be punched from the computer system. This time card is used as a source document to prepare the input record which indicates the hours worked for each employee. The time card which is punched for use by the employee and which serves as a source document is illustrated below.

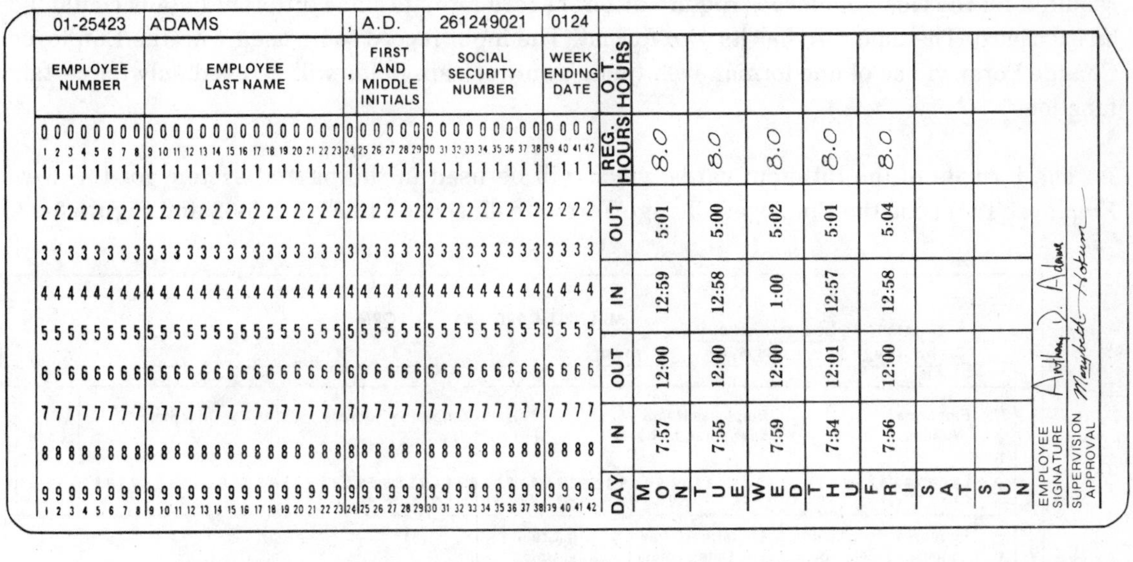

EMPLOYEE NUMBER	EMPLOYEE LAST NAME	FIRST AND MIDDLE INITIALS	SOCIAL SECURITY NUMBER	WEEK ENDING DATE	DAY	REG. HOURS	OUT	IN	OUT	IN
01-25423	ADAMS	, A.D.	261249021	0124		O.T. HOURS				
					MON	8.0	5:01	12:59	12:00	7:57
					TUE	8.0	5:00	12:58	12:00	7:55
					WED	8.0	5:02	1:00	12:00	7:59
					THU	8.0	5:01	12:57	12:01	7:54
					FRI	8.0	5:04	12:58	12:00	7:56
					SAT					
					SUN					

EMPLOYEE SIGNATURE: Anthony D. Adams
SUPERVISION APPROVAL: (signature)

Figure 5-31 Example of Time Card as Source Document

Case Study - Input Record Format Design

Once the source documents have been designed, the analyst must design the input records which are to be used as input to the computer system. In the sample problem of the payroll system for the James Tool Company, the determination was made that card input should be used because there were keypunches in the data processing department and the volume of input transactions for the payroll system was not deemed large enough to justify purchasing or leasing key-to-tape devices or a more sophisticated data entry method.

By designing the source documents, the analyst has, for the most part, identified that data which is to be contained in the input records. In some cases, there may be additional identification codes and other data that is not directly used as source data which may be included within the input records but, in general, by examining the source documents to the system, the analyst will be able to design the input data.

In the payroll system it will be recalled that there are three major source documents—the New Employee Form, the Employee Change Form, and the Time Card. The two input records required for the New Employees require different card formats since different data is contained on each card. The time card has its own format. The input record to be used with the Employee Change Form will be of one format even though only certain fields will be used only for certain functions.

The formats of the different cards which will be used in the payroll system for the New Employee Form and the Employee Change Form are illustrated in Figure 5-32 and Figure 5-33.

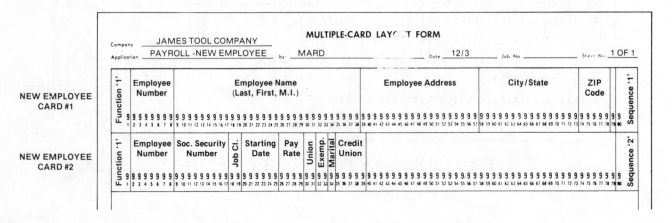

Figure 5-32 Card Format for New Employee Input Record

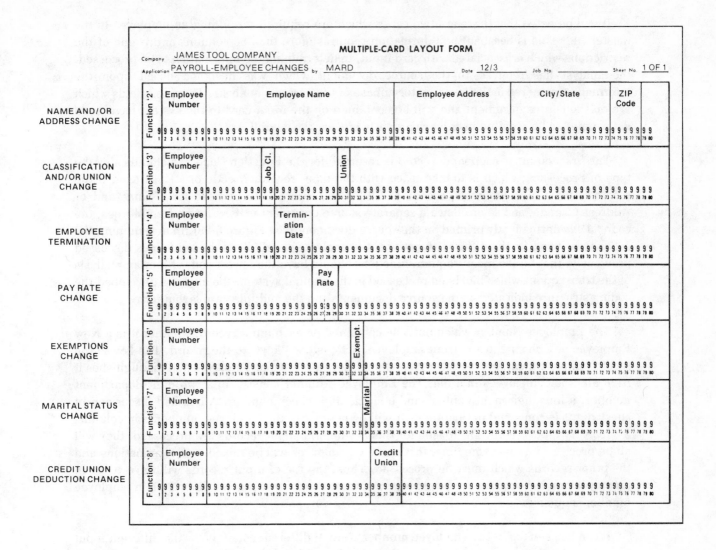

Figure 5-33 Card Format for Employee Change Record

In the examples in Figure 5-32 and Figure 5-33 it can be seen that the Multiple-Card Layout Form is used to depict the format of the card. Each of the numbers on the form correspond to a column on the card. A line is drawn on the form between columns to indicate the location of the different fields on the card. Thus, the analyst can graphically illustrate how the cards will be prepared for processing on the computer.

It will be noted that there are two cards which are required for adding an employee to the master file. This is because the information requires more than 80 columns and is one of the restrictions which is associated with card input. In determining the information to be processed, the analyst has referred to the source document which was designed based upon the information gathered in the investigatory phase of the systems analysis. Thus, the fields which are on the source document and will be contained on the input card to the system have been determined to be required in order to properly process the payroll.

The first column of each input record is reserved for the Function Code, which indicates the type of processing which is to take place with the input record. A code of "1" is used if the record is for a new employee, a code of "2" if the record is to be used for a Name and/or Address Change, and so on. Since a separate source document is used for new employees, the code "1" is permanently printed on the source document (see Figure 5-28). In determining the codes to be used for the different functions, the analyst should follow the guidelines discussed previously, that is, the most frequent functions should have the lowest code. From an analysis of the transactions which had been processed in the manual system, the analysts developed the coding structure indicated on the source documents and the multiple-card layout forms.

The Employee Number which must be contained on each input record, whether it is a New Employee or a change to a current employee, acts as the "key" to the record. The key to the record uniquely identifies the record as belonging to a particular employee and distinguishes it from all other employees in a file. The Employee Number consists of a two digit department number, a four digit man number, and a single digit check number. A detailed discussion of check digits is contained in Chapter 8. It will be noted that the employee number is in columns 2-8 in every input record. The reason for this is when the input records are processed, they will all be processed at the same time, that is, new employees will be mixed with terminations and the other records which must be processed. Thus, the file of input records will have to be in employee number sequence to be processed since the employee master file will be in employee number sequence.

It can be seen that if the employee number were in different columns for the different input records, it would be impossible to sort the records and have them all be in employee number sequence. For example, if the employee number for the new employee card was in columns 2-8 and the employee number in the termination input record was in columns 7-13, the cards could not be sorted because there would be no common columns on which the sort could take place. Therefore, when designing the input data, it is important for the analyst to consider those fields which may be involved in the sequencing of the file so that in different formats, these control fields are always located in the same relative positions of each record.

Another code which is required is the "sequence code" in column 80 of the new employee records. As noted previously, two cards are required for new employees, with the first card containing the name, address, city/state, and ZIP code and the second card containing the remainder of the information for a new employee. The first card has a sequence number of one and the second card has a sequence number of two (see Figure 5-32). The addition of an employee to the master file cannot take place without both of these records, that is, both the card with the name and address and the card with the social security, etc. must be present in order to place the employee on the master file. Without either of these cards, the transaction is invalid. Therefore, the sequence number can be used to ensure that both cards have been read prior to adding the record.

In addition, the cards used to add a new employee to the master file must be sorted in such a manner that the card with the sequence number "1" will be first and the card with the same employee number and a sequence number of "2" will be second. Therefore, the input data must be sorted on sequence number within employee number, which will place it in the desired sequence. It should be noted that the sequence number is placed in column 80 and for the input records other than new employees, there are no values to be entered in column 80. This is done so that when the input records are sorted on the values in column 80, only the new employee records will be rearranged, that is, there will be blanks in column 80 for the other records and their sequence will not be altered. This is illustrated in Figure 5-34.

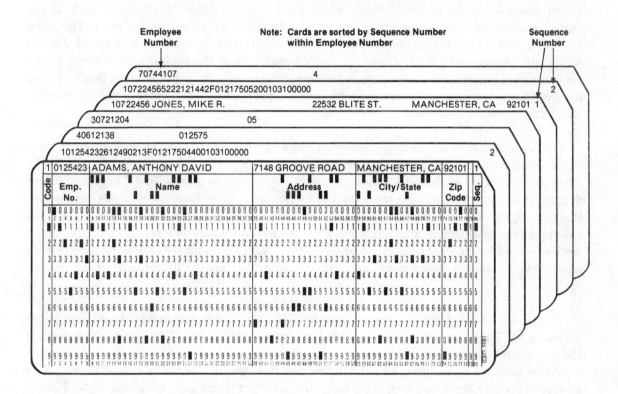

Figure 5-34 Example of Sorting on Sequence Number within Employee Number

Note from Figure 5-34 that the input cards are sorted by Sequence Number within Employee Number. Whenever it is necessary to sort some input data on a certain value, such as the Sequence Number for the new employees, and not sort other data on this number, the columns selected for this value to be sorted must either be columns which are not used in other input records or columns which will contain constant data for all the other records. This will avoid any problems when sorting the data.

When determining what information is to be contained within the input records, it is also important, as noted previously, to determine the size of the field required to contain the information. When the analyst is dealing with fields which will contain a known number of characters or digits, then only that number of card columns needed to contain the digits need be allocated. For example, it is known that the employee number contains seven digits and the ZIP code always contains five digits; thus, this is the number of columns allocated to these fields on the card.

On the new employee card and the change card, there have been thirty columns allocated for the employee name, nineteen for the street address, and sixteen for the city and state. When determining the length of fields which can contain variable length data, such as the name, address, and city-state, there are no exact standards which can be used. The main concern of the analyst is that there be enough length allocated to contain all of the characters which can appear in the field. Thus, the analyst may look at the typical names found in the company or at typical street addresses to determine the maximum length of these fields. Again, the overriding requirement is that there be enough space allocated to contain any information which can appear in these fields.

On the second card for the new employees and on certain of the change cards there are also fields which may contain variable length information. The pay rate field is one of these fields. After examining all of the rates at which people were paid in James Tool Company, it was decided that no one would make more than 99.99 per hour. Therefore, the analysts determined that four digits were sufficient for the pay rate within the input record. In addition, no one had ever taken out over 999.99 in a single week for their credit union deduction. Thus, five digits were allocated for the weekly credit union deduction. Note that both these determinations were based upon historical data which was accumulated in the investigation phase of the system project.

It may also be noted that the starting/termination date field (columns 20-25) contains spaces for six digits. Thus, the date will be recorded in a month (MM), day (DD), year (YY) format. If the date must be printed on an output report, some punctuation such as slashes (MM/DD/YY) would be inserted by the program. Whenever possible, the size of fields should be kept to a minimum on both input and master files and the program which prints the data should insert the required punctuation.

Another feature of the cards which make up the input records which will be noted is that, where possible, the same columns in the cards are used for the same fields, regardless of the function of the input record. Thus, for example, in the payrate change card (code=5) the payrate field is placed in columns 26-29 which is the same as new employee cards even though the payrate could have been placed in some other columns. There are several reasons for this. First, it enables the programmer to define a single field within the program which will process the input records that will be used for the payrate and no matter which input record is being processed, this field will contain the payrate. Therefore, the programmer is able to establish common routines within his program to process the data.

Secondly, it allows the data entry operator to make a program for the data entry device which will work for any of the input records which are to be prepared. For example, some field in the card may have to be "zero-filled," that is, the leading non-significant digits within the field may have to contain zeros instead of being blank. The operator can make a program for the data entry device to automatically cause this zero filling to take place. If the same columns were not used for a given field in each input record, the operator would not be able to make a program which could be used for all of the input formats and this would require the operator to take more time in preparing the input data.

Case Study - Time Cards

As noted previously, the time cards which are punched from the system will be used as a source document (see Figure 5-31). These time cards will not, however, be used as input to the system; rather, other input cards will be prepared using the time cards as source documents. This process, together with the input card which has been designed, is shown in Figure 5-35 and Figure 5-36.

Card Layout Form

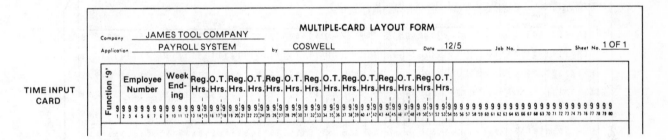

Figure 5-35 Format of Time Input Card

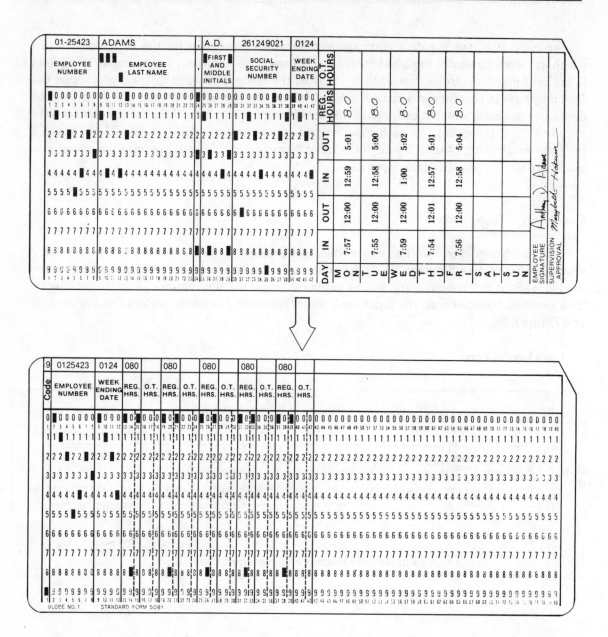

Figure 5-36 Time Card Preparation

Note from the examples in Figure 5-35 and Figure 5-36 that the time card which is punched as output from the system will act as the source document in the preparation of the input card containing the hours worked by an employee. Only the total of regular and overtime hours are punched in the input card. The computer program will then total the hours and determine the weekly pay for the employee.

Case Study - Employee Compensation Records

When a system such as the payroll system is converted from a manual system to a computer system it is normally necessary that certain information which is found in the manual system be converted from data in the format used for the manual system to data in a format which is usable in the computer system. Such is the case in the James Tool Company with the information found on the Employee Compensation Records in the manual system.

It will be recalled that in the manual system the history of the employee's pay was kept on the Employee Compensation Record form. It is necessary that this information be contained in the computer system for two primary reasons: first, the year-to-date earnings and deductions must be printed on the weekly paycheck stubs; and, secondly, at the end of the year, W-2 forms must be prepared from the yearly earnings for tax purposes.

Whenever it is necessary to convert data to a computer system, the analyst must be concerned with the format of the records which will be used to contain the data which is input to the computer system. The format of the card input to be used to place the employee compensation information into the system is illustrated in Figure 5-37.

JAMES TOOL COMPANY
EMPLOYEE COMPENSATION RECORD

EMPLOYEE NAME __Adams, Anthony David__   MARITAL STATUS __Married__

EMPLOYEE ADDRESS __7148 Groove Rd., Manchester, Calif. 92101__   NUMBER OF EXEMPTIONS __3__

SOCIAL SECURITY NUMBER __261-24-9021__   DATE EMPLOYED __01/16__

UNION AFFILIATION __Machinists (1)__   DATE TERMINATED _____

WEEKLY PAYROLL											YEAR-TO-DATE								
	EARNINGS			DEDUCTIONS					NET PAY		EARNINGS			DEDUCTIONS					NET PAY
Week Ending	Regular Earnings	Overtime Earnings	Total Earnings	Federal Tax	FICA	State Tax	Credit Union	Union Dues	Net Amount	Check Number	Regular Earnings	Overtime Earnings	Total Earnings	Federal Tax	FICA	State Tax	Credit Union	Union Dues	Net Amount
1/24	176.00		176.00	21.50	10.30	3.52		3.00	137.68	02015	176.00		176.00	21.50	10.30	3.52		3.00	137.68
1/31	176.00	4.95	180.95	23.30	10.59	3.80		3.00	140.26	02232	352.00	4.95	356.95	44.80	20.89	7.32		6.00	277.94

MULTIPLE-CARD LAYOUT FORM

Company __JAMES TOOL COMPANY__

Application __PAYROLL SYSTEM__   by __MARD__   Date __12/10__   Job No. _____   Sheet No __1 OF 1__

COMPENSATION INPUT CARD

Code	Employee Number	Reg. Earn. YTD	O.T. Earn. YTD	Total Earn. YTD	Fed. With. YTD	FICA YTD	State With. YTD	Credit Union YTD	Union Dues YTD	Net Amount YTD		Sequence '3'

Figure 5-37 Employee Compensation Record

Note from the example in Figure 5-37 that the Employee Number and all of the Year-To-Date information from the Employee Compensation Record is to be contained in the card input record. The Employee Number is to be used to enable the information to be placed in the proper master record within the system and would be determined when an Employee Number is assigned to each employee prior to system implementation.

Note that a code is contained in column 1 of the input record. This code is to be used to indicate whether the employee represented by the record is currently employed by the company. Since the information found in these records will be used to prepare W-2 forms and a W-2 form must be prepared for every employee which was employed within a company for the past year, all year-to-date information for both current employees and past employees must be kept. A code of ''1'' in column 1 will indicate that the employee is currently employed and a code of ''2'' will indicate that the employee has been terminated within the year. A sequence number is contained in column 80 so that these cards can be sorted with the Addition cards when loading the Employee Master File.

SUMMARY

The overall design of the source documents and input records to a system is critical since many errors which are encountered in the operation of a data processing system can be directly attributed to invalid input data entering the system. It is only through careful design of the input records based upon the source documents which are to be used with the system that accurate and efficient data entry methods may be used to ensure a system which operates smoothly and error-free.

After the input and output have been designed, it is necessary that the analyst turn his attention to the files which must be designed in order to store and process the data through the system. This is the subject of the next chapter.

STUDENT ACTIVITIES—CHAPTER 5

1. Explain how verification is accomplished when using keypunch machines as a method of data entry.

2. Explain the operation of the "verifying keypunch."

3. What are the advantages of key-to-tape data entry systems as compared to keypunching?

4. What are the disadvantages of key-to-tape data entry systems?

5. What are advantages of data entry systems using the "floppy disk"?

6. What is meant by the term "Intelligent Terminal"?

7. List the basic steps in the design of systems input.

   1)
   2)
   3)
   4)

8. List the basic sources from which input data may originate.

   1)

   2)

   3)

   4)

9. List the factors which should be considered when determining the method of data entry.

   1)

   2)

   3)

   4)

   5)

   6)

   7)

10. Summarize the basic rules related to the design of source documents.

11. What factors should be considered when designing input records?

12. List five characteristics of a good coding system.

    1)

    2)

    3)

    4)

    5)

13. Explain the use of "classification codes."

DISCUSSION QUESTIONS

1. It has been stated by some authorities that by 1980 the punched card will be totally obsolete and that keypunch machines and the punched card will no longer be found in data processing installations. Do you think this is going to happen? Why?

2. In addition to the methods of inputting data to a computer system which were mentioned in the chapter, a number of people have advocated that the human voice be used as input, that is, instead of writing the input and then transcribing it to computer-readable input records, the computer merely be told what the input is. In the past year or so, there has been some promising research performed in this area. What do you think of this possibility and what are some of the problems which you foresee?

3. Some analysts maintain that when converting from a manual system to a computer system, it is mandatory that the computer system be designed around existing source documents. They point out that if new source documents are designed, the personnel must be re-trained and the probability of errors in the preparation of input data is much higher. If the system is designed around currently existing source documents, there is less chance for error in the preparation of the data and there will be less cost incurred by the company because the analyst's time will not be spent designing forms, new forms will not have to be printed, and the costly training will be avoided.

   Other analysts argue that it is the function of the systems analyst to design those source documents which will be most efficient with the new computer system and that an analyst is derelict in his duty if he uses the currently existing source documents when new documents should be designed.

   Which position do you support? Why?

CASE STUDY PROJECT — CHAPTER 5
The Ridgeway Company

INTRODUCTION

After designing the output for the billing system and the related sales analysis reports, the analyst reviewed the hardware configuration of the computer system acquired by the Ridgeway Company. The hardware consists of the following:

1 - IBM System/370 Model 125 Computer
1 - Card Reader/Punch
1 - High-speed printer
5 - Disk Drives
3 - Tape Drives
9 - IBM 129 Data Recorders
1 - Burster
1 - 4-part Decollator

He also reviewed the basic input documents which are currently used in the billing system for the Ridgeway Country Club. The name and address file of all members was kept on 3 x 5 indexed cards. An example is illustrated below.

Thomas R. Lindsdale   Member Number L004
1771 Camino Real St.
Oakmont, CA 92633

Phone No.: 420-4222

Date of Membership: 9/15/75

When new members join the club a card is typed up by Gunther Swartz's secretary and forwarded to the accounting department.

The sales slips for the restaurant, bar, and pro shop are again illustrated.

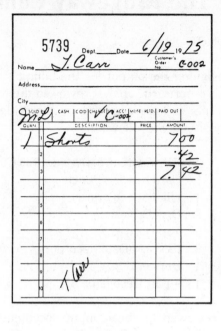

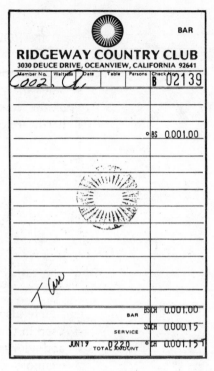

Sales Returns are handled by checking the appropriate box in the Sales form.

STUDENT ASSIGNMENT 1
DESIGN OF SYSTEM INPUT

1. Determine the data which is required to produce the required output for the billing system and prepare an input analysis form for the input data.

2. Determine the method of data entry to be used for the billing system of the Ridgeway Country Club. Any changes or additions to the hardware currently being used by the Ridgeway Company must be justified.

3. Redesign existing source documents for more efficient data entry, if required.

4. Design any new source documents which may be required for the billing system.

5. Design the input records which will be prepared from the source documents and will be input to the billing system. The formats and characteristics of each input record should be documented on a multiple card layout form or similar form.

DESIGN OF
SYSTEM FILES

CHAPTER 6

DESIGN OF SYSTEM FILES

6

INTRODUCTION

After the output which is required from a system has been determined and after the input which is necessary to create the desired output has been designed, it remains for the analyst to design the method in which the input is to be processed to produce the required output. Although the specification of the desired output may satisfy every user and the input designed contains all of the information which is required to create this output, without an efficient and practical marriage of these two elements into a smooth processing system it is likely that the system will never be functional. It is, therefore, very important that a great deal of time and design work be allocated to the specifications required for the processing of the input to create the output.

Prior to developing the detailed steps required to produce the required output, the analyst must define the characteristics of the "files" that are to be processed. A FILE is defined as a group of related records. In defining the characteristics of the system files the following steps must be undertaken by the analyst:

1. Define the method of file storage

2. Define the type of file organization

3. Design the files

Files fall into two basic categories—**MASTER FILES** and **TRANSACTION FILES**. The heart of any system are the master files. Master files contain information which reflects the current status of a system. For example, in a payroll system an Employee Master File would contain the employee number, employee name, and other information about the employee including pay records such as year-to-date earnings, year-to-date taxes, etc. Master files are periodically updated by transaction files.

Transaction Files contain the individual records which reflect the day-to-day activities of a business. For example, in a payroll system, the employee master file which contains year-to-date earnings will be updated by the transaction file which reflects the current earnings for the week. The result of the processing of the transaction file against the master file will be a new updated master file with new year-to-date earnings, new year-to-date taxes, etc.

The concept of a group of related records comprising a file is extremely important in the systems design phase of the systems project as the design of a computer system will normally involve the processing of one or more files. One of the first decisions that the analyst must make is the type of file storage devices or media that are to be used to store the files that comprise the system.

METHODS OF FILE STORAGE

Perhaps the ideal computer would be one containing unlimited primary storage within the central processing unit so that all records and files of a company could be directly and immediately accessible through the primary storage unit of the computer system. Unfortunately, primary storage is too expensive for the permanent storage of records and files; therefore, other types of hardware and storage methods must be considered by the systems analyst. Three basic storage media or methods are commonly used to store files accessible to computer systems. These include:

1. Punched Card Files

2. Magnetic Tape Files

3. Magnetic Disk Files

It should be noted that in addition to the above types of storage methods there are other devices including magnetic drum storage, magnetic card storage, and magnetic cartridge storage. These methods, however, are normally used for specialized applications in large systems.

The advantages and disadvantages of the methods of file storage should be thoroughly understood by the systems analyst prior to undertaking the file design of the systems project. These methods of file storage are discussed on the following pages.

PUNCHED CARD FILES

Historically, one of the earliest and most commonly used methods for the storage of files to be accessed by a computer system was punched cards. In a punched card-oriented system the group of individual records in any given application comprise the file that is processed by the computer. The diagram in Figure 6-1 illustrates a series of name and address cards combined together to form a Name and Address File.

Name and Address File

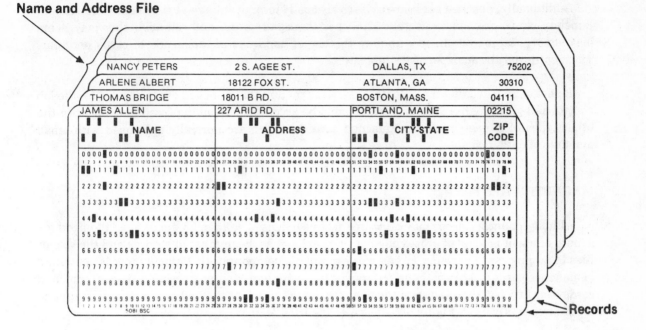

Figure 6-1 Example of Punched Card File

It should be noted that when dealing with punched card files the records are normally organized sequentially on the basis of some control field. In the example above, the cards are organized in an ascending order based upon the ZIP code for each record.

Advantages and Disadvantages of Punched Card Files

When dealing with small volumes of data, punched card files may provide an economical and efficient form of file storage. One primary advantage of punched card files is that the records of the file can be handled and viewed visually, which tends to reduce the complexity of locating and correcting errors that may occur within a file. In addition, information can be written on punched cards for manual reference as well as computer reference and this is not possible with other forms of file storage.

A significant disadvantage of punched card files is the limited storage capacity on a single card. Standard punched cards provide for the recording of 80 or 96 numbers, letters of the alphabet, and/or special characters on a single card. Thus, if it is required that a record contain 85 characters and 80 column cards were used, two cards would be required for each record. A file of 10,000 records would therefore require 20,000 cards. As can be seen, for large volumes of data, punched cards provide a very bulky form of storage.

The processing of punched cards on the computer is also slow in comparison to other means for storing data. For example, a typical card reader can read punched card files at the rate of 1000 cards per minute (80,000 characters per minute) and produce punched card output at the rate of 300 cards per minute (24,000 characters per minute). Although this may seem to be a fast form of input and output, card readers and punches are extremely slow as compared to other forms of storage devices; and this is a significant disadvantage of punched cards.

Additionally, punched cards are a relatively costly form of storage as they cannot be reused. Punched cards cost approximately $2.50 per thousand cards, and although this may seem inexpensive, large systems may require the use of hundreds of thousands of cards per year. Thus, cards may be an expensive supply item.

It should be noted that most "batch" systems require that punched cards be used for the job control and other control functions which must be accomplished on the computer; but with the other types of storage available, punched card data files are normally only used with small amounts of data, if at all.

MAGNETIC TAPE FILES

Although punched cards may offer an effective means of storage for small files, the need for a more efficient method of processing becomes evident as the number of records and the size of files to be processed becomes larger. For example, a company with 100,000 records contained on punched cards is faced with the task of sorting, handling, and processing a stack of punched cards over fifty feet high. To meet the need for a more efficient method of file storage magnetic tape is frequently used. Figure 6-2 illustrates the IBM 3420 Magnetic Tape Unit.

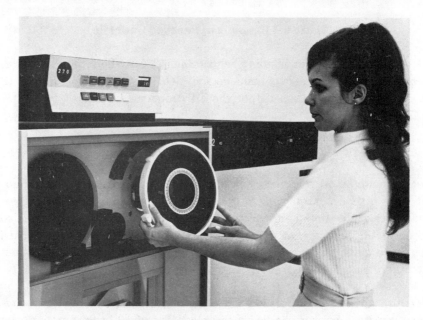

Figure 6-2 IBM 3420 Magnetic Tape Unit
(Photo Courtesy of IBM)

The data recorded on magnetic tape may include numbers, letters of the alphabet, and/or special characters. There are no restrictions on the length of physical or logical records when using magnetic tape as there are with punched cards.

An important advantage of the use of magnetic tape is the density at which the data is recorded, that is, the number of characters which may be recorded on an inch of tape. Although the density at which data is recorded varies with the type of unit being used, most magnetic tape units today read or record data at a density of 800, 1600, or 3200 characters per inch. Thus, a single punched card, with 80 characters, could be stored on 1/20 of an inch on magnetic tape if the density is 1600 characters per inch.

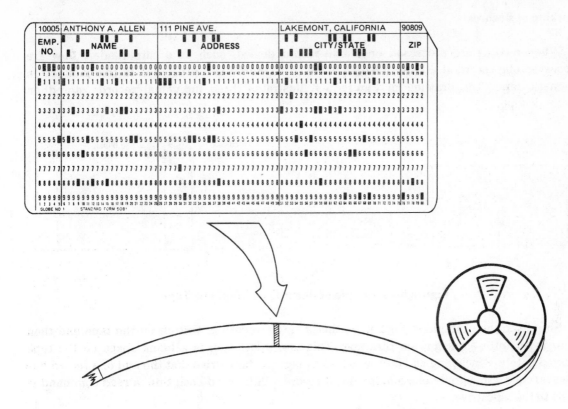

Figure 6-3 Recording Density of Magnetic Tape

As with punched cards, records stored on magnetic tape are normally stored in some predetermined sequence, such as employee number in the example above. Magnetic tape records are always stored and processed in a sequential mode, that is, records are written and read one record after another in a manner similar to reading punched cards.

Tape Transport Speed

The speed at which data on magnetic tape is processed is dependent upon the density of the data stored on the tape and the tape transport speed of the magnetic tape unit utilized. The tape transport speed of magnetic tape units varies from approximately 18.75 inches per second to 200 inches per second. The tape transport speed refers to the speed at which the tape is transported past the reading and writing heads of the tape unit.

Because of the density of magnetic tape and the speed at which the tape is transported past the read-write head, extremely fast input/output speeds are possible. To obtain the "effective" data transfer rate, that is, the speed at which data may be transferred to the central processing unit from magnetic tape, the tape transport speed is multiplied by the tape density. For example, a magnetic tape unit operating at 200 inches per second using magnetic tape with a density of 1,600 characters per inch has an effective data transfer rate of 320,000 characters per second! It is interesting to note by comparison that a card reader operating at a 1,000 card per minute rate has a data transfer rate of approximately 1,330 characters per second. Thus, magnetic tape offers a much faster input/output capability than punched cards.

Blocking of Records

When records are processed on a unit-record device, such as a card reader, they are processed one record at a time. Records can also be stored and processed in this manner on magnetic tape. The drawing in Figure 6-4 illustrates three individual records stored on magnetic tape.

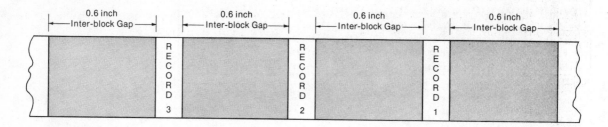

Figure 6-4 Example of Records on Magnetic Tape

Note from the example in Figure 6-4 that a single record is written on the tape and then there is an Inter-block gap on the tape. This inter-block gap is a blank space on the tape approximately .6 inch long and indicates to the magnetic tape drive that the end of a record has been reached. Thus, only one single logical record will be read each time a read command is given to the tape drive.

It can be seen from the example in Figure 6-4 that there is a great deal of wasted space on the tape because for each logical record there is an inter-block gap approximately .6 inch long. Thus, the majority of the tape is used for inter-block gaps, not for data. In addition, the fact that only one logical record is read each time a read command is given will slow down the effective processing speed considerably.

To overcome this inefficiency in utilizing the magnetic tape and to increase the speed of reading and writing, it is many times advantageous to block the logical records. Blocking refers to the process in which two or more individual records (logical records) are grouped together and written on a magnetic tape creating a "physical record" or "block." This is illustrated in Figure 6-5.

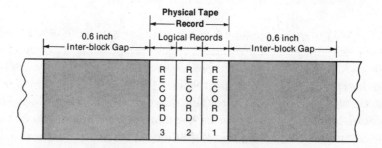

Figure 6-5 Example of Blocked Records

As will be noted from Figure 6-5 three logical records make up the physical record, or block. There is no inter-block gap separating these records so the magnetic tape is used more efficiently. In addition to the major advantage of more efficient usage of the magnetic tape, another major advantage of blocking records on magnetic tape is that records can be read faster because two or more records can be read before the read operation is stopped by the inter-block gap.

The limiting factor in blocking records is the amount of main storage available for input/output operations, as there must be enough room to store the complete block of data to be processed. Thus, the larger the block of records, the more main storage that must be allocated for storing the block. For example, if fifty 80 character records comprise the physical record, then 4,000 positions of main storage are required when the physical record is transferred from magnetic tape to main storage or from main storage to magnetic tape.

The analyst must make the decision concerning how many records will be contained in the physical record which is to be written on the tape files. This number of logical records comprising the physical record, called the **Blocking Factor**, is important because it can significantly affect the efficiency of the file processing within the system. In general, as large a blocking factor as possible should be used, limited only by the amount of main storage which is available for processing the input or output records.

Record Formats

In addition to determining the contents of the file which is to be processed in the system, the type of medium to be used, the organization of the file, and the blocking factors, the analyst must also consider the format of the records which are to be stored on the file. There are three basic formats for records—Fixed Length, Variable Length, and Undefined Length.

Fixed Length Records

A fixed length record is one which always contains the same number of positions. Thus, when a file is defined as having fixed length records and the record length is 120 positions, all records on the file will have 120 positions. Fixed length records can be blocked or unblocked. When they are unblocked, each physical record will be 120 characters long. When the fixed length records are blocked, the physical record contains more than one logical record. Thus, if a blocking factor of five is used, that is, there are five logical records for each block or physical record, then the block length would be 600 characters (120 characters/rec x 5 recs/block).

Variable Length Records

A variable length record is one which may contain a variable number of positions in each logical record, that is, each record may contain the same or a different number of characters. Variable length records are used when different amounts of data may be available for each record.

The example below illustrates the use of variable length records.

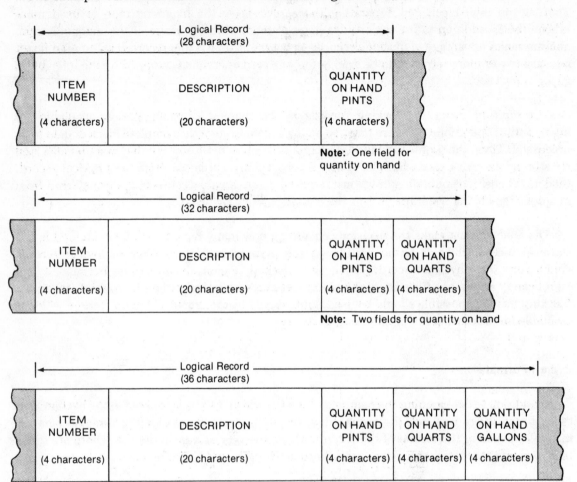

Figure 6-6 Example of Variable Length Records

In the example above the data record consists of the Item Number, the Description, and a series of Quantity On Hand fields. Some items are available in pints, some are available in pints and quarts, and some are available in pints, quarts, and gallons. It can be seen that if fixed length records were used, all records would have to contain three quantity on hand fields even though many of the records did not utilize all three fields; this is a waste of a large amount of storage space on the tape.

Thus, as illustrated in Figure 6-6, variable length records are used where one record contains only a field for pints, one record contains fields for pints and quarts, and one record contains fields for pints, quarts, and gallons. It should be noted that variable length records can be stored either as blocked or unblocked records, depending upon the application.

Undefined Length Records

An undefined length record is a record which may be any length desired by the programmer. This type of record is used whenever the length of the record is unknown. This may occur when long strings of data of unknown length are being processed, such as when analog numeric data is converted to digital data and the digital data is stored on tape. This record format has more application in scientific and engineering applications than in business applications.

Undefined records differ from variable length records in that there is a maximum length for variable length records and there is no maximum length specified for undefined length records. Undefined records cannot be specified as blocked in most systems and must be stored as single logical records.

Summary - Blocking and Record Formats

When determining the record format to be used for a file, the analyst must primarily analyze the contents of the record in terms of what data will always be present in the record and what data may be optional. If there are fields which are optional and which may or may not appear, then it may be more efficient in terms of space and processing speeds to use variable length records. Otherwise, fixed length records are normally used. Undefined length records are not normally used except in unique circumstances. It will be found that fixed length records are used in the majority of applications.

It should be noted that the blocking and deblocking of records to be placed in a file and the reading and writing of fixed or variable length records is not normally a concern of the programmer. These functions are handled by the Input/Output Control System (IOCS) which is normally available as a part of the software with the computer system which is being used. Thus, like the organization of a file, the various possibilities of blocking and record formats need not concern the analyst in terms of the programming which must take place. It is, however, quite important for the analyst to specify if the records are to be blocked, what the blocking factor is, and what the record format is. The analyst must always make these decisions regardless of whether he is concerned with the actual programming which must take place to physically process the files.

ADVANTAGES AND DISADVANTAGES OF MAGNETIC TAPE

As pointed out, the primary advantage of magnetic tape as compared to punched cards is the recording density of data stored on tape and the increased input/output speed available when processing. The increased density substantially reduces the storage required to store the file, which can be a significant factor where large files are involved. A single reel of magnetic tape which can easily be handled by the computer operator can store the equivalent of many thousands of punched cards.

A disadvantage of magnetic tape may be the increased cost of the computer hardware for processing the magnetic tape. A single magnetic tape unit has a monthly rental from typically $600.00 to over $1,000.00 per month, and many systems require two or more tape units and related interface hardware when a system is designed using magnetic tape input/output files. However, this increased cost may be offset by the increased speed of processing where large volumes of data are concerned.

Another limiting factor when using magnetic tape is that the data must be stored and processed in a sequential manner, that is, records must be stored and read or written one after the other in the same manner as punched card records. Although many applications do not require anything other than sequential processing, there are an increasingly larger number of applications which require data to be accessed in a very short period of time, that is, within seconds. Magnetic tape files cannot be used for these types of applications.

MAGNETIC DISK SYSTEMS

As noted, there is an increasingly greater need in business applications for a means of obtaining Direct Access to data, such as occurs in airline and hotel reservation systems where inquiries must be made into files and responses must be obtained in seconds. Magnetic Disk storage devices, commonly called Direct-Access Storage Devices (DASD), are used in applications that require inquiry into files or where random or non-sequential retrieval of data is desirable. An IBM 3330 magnetic disk storage facility is illustrated in Figure 6-7.

Figure 6-7 IBM 3330 Disk Storage Facility

The IBM 3330 disk storage facility allows the mounting of removable disk packs on a maximum of eight different magnetic disk drives. All eight of these drives are on-line to the computer at the same time. Each of the removable disk packs consists of eleven magnetic disks which contain 19 surfaces on which data can be recorded.

Data is recorded in the form of magnetic spots along a series of concentric circular recording positions on each disk recording surface. For the IBM 3330 disk packs, there are 411 circular recording positions per surface, with 404 of the recording positions available to the user and 7 alternate positions to be used in case one of the other recording positions is defective. Each circular recording position is capable of storing a maximum of 13,030 characters. Figure 6-8 illustrates a schematic of the recording positions on the surface of each disk. The first recording position is referenced by the number zero and the inner recording position by the number 410.

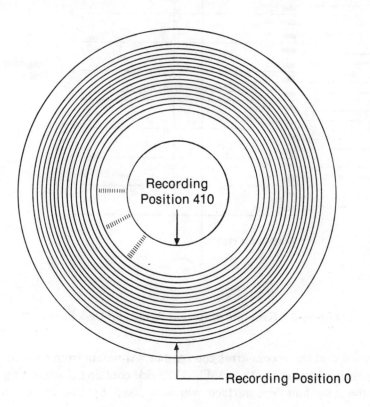

Figure 6-8 Schematic of Recording Surface

As noted, the IBM 3336 disk pack which is used with the IBM 3330 storage device consists of 11 magnetic disks such as illustrated in Figure 6-8. In order to access the data on these disks, a series of access arms move in and out between the individual surfaces. This is illustrated in Figure 6-9.

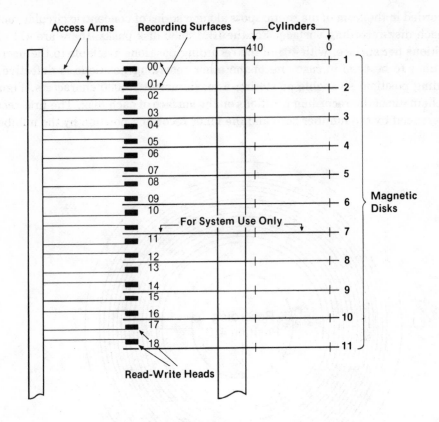

Figure 6-9 Example of Access Arms and Disk Pack

Note in Figure 6-9 that the access arms can read or write data from each of the surfaces on the disk pack. The disk pack used for the IBM 3330 device contains 19 recording surfaces which are available to the user and one surface which is used by the system to store control information.

It should be noted also that with one positioning of the access arm, all 19 recording surfaces can be written or read from. The amount of data which can be accessed with one positioning of the access arm is said to be stored on a CYLINDER of data. It will be recalled from Figure 6-8 that there are 411 concentric circles on each recording surface. Therefore, there are 411 cylinders on an IBM 3336 disk pack. Each recording surface is called a TRACK. Thus, from Figure 6-9 it can be seen that there are 19 tracks per cylinder which are available to store user data.

Each track on an IBM 3336 disk pack is capable of storing 13,030 characters. Since there are nineteen tracks per cylinder, a total of 247,570 characters can be stored on a cylinder of data on the disk pack and can be referenced by the access arm in one position. This is illustrated in Figure 6-10.

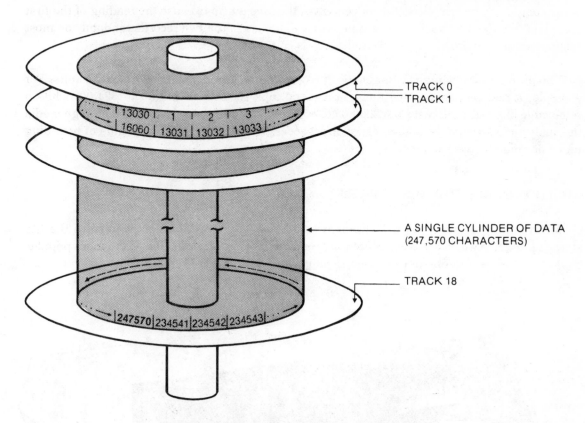

Figure 6-10 Example of the Cylinder Concept

Note in Figure 6-10 that a single circular recording position or track can store a possible 13,030 characters. Since there are 19 tracks per cylinder, 247,570 characters can be stored on a single cylinder. The disk pack consists of 404 cylinders available for storing user data. Therefore, the IBM 3336 disk pack can store a total of 100,018,280 characters (404 cylinders X 247,570 characters/cylinder). As there are eight drives on an IBM 3330 device and therefore eight disk packs which can be mounted at one time, a single 3330 unit may reference 800,146,240 characters directly on-line to the computer.

ADVANTAGES AND DISADVANTAGES OF MAGNETIC DISK UNITS

Although records may be stored sequentially on magnetic disk units, one of the primary advantages of magnetic disk is that it is also possible to have direct or random access to data. For example, it is possible to read the first record on cylinder 1, track 0, and then under control of the computer program direct the movement of the access arm to cause the reading of the first record on cylinder 199, track 17. The time required to access the two records randomly on most systems would require less than 100 milliseconds.

The primary disadvantage of magnetic disk stems from the increased costs of the units. For example, a reel containing 2400 feet of magnetic tape costs from $12.00 to $24.00, while a removable disk pack will cost from $300.00 to $600.00. Thus, in the selection of a storage media the analyst must weigh the added cost of direct access devices relative to the value of providing direct or random access to files.

ADDITIONAL FILE STORAGE DEVICES

Many additional file storage devices have come into existence to attempt to solve the file storage problems which are encountered in business data processing. One of the more popular devices is the tape cassette device, which is illustrated in Figure 6-11.

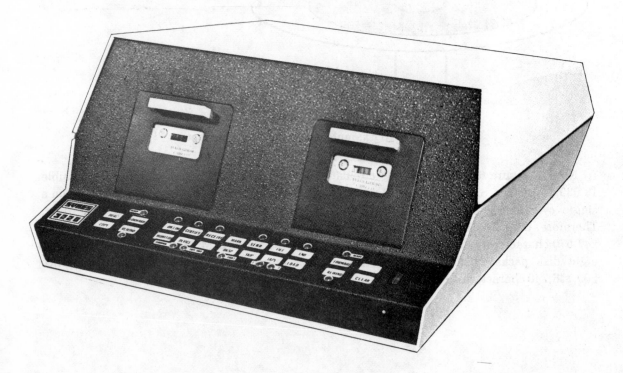

Figure 6-11 Tape Cassette Drive

Note in Figure 6-11 that two tape cassettes, which are almost identical to those used in audio cassettes, are mounted in the drive. Data is recorded on the tape cassettes in the same manner as data is recorded on the magnetic tape drives mentioned previously. Although the tape cassettes have neither the storage capacity nor the speed of the larger tape drives, in some applications they have proven to be quite suitable.

Several disk cassette drives have also been developed for applications where massive storage is not required but where the direct-access capabilities of magnetic disk are required. As with the tape cassettes, these drives may not be applicable to all applications but in some situations they are a very cost-effective method for storing files. Floppy disks, as discussed in Chapter 5, can also be used for file storage.

There have been other specialized file storage devices designed which may, in certain applications, be quite feasible. In addition, many exotic storage devices, such as laser beam memory and "bubble" memory, are in the development stages. It is not unreasonable to expect within a few years storage devices which can store billions of bytes of data which are accessible at the same speeds found in main storage today.

Summary - File Storage Methods

In determining the file storage method to be used within the system, the analyst must consider such factors as file size, processing speed required, and whether there is a need for random or direct access to the data stored in the files.

For small systems, punched cards may provide the least expensive method of file storage. Tape cassettes, disk cassettes, or floppy disks may also be quite useful in small- or medium-sized systems. Where larger numbers of records are to be processed magnetic tape offers a compact and fast method of storage and processing. In systems requiring random or direct access to data, magnetic disk storage must be used.

In addition to reviewing the types of physical storage devices or media for a system, the analyst must also be concerned with the type of file organization utilized within the system. The most commonly used types of file organization are discussed on the following pages.

FILE ORGANIZATION AND RETRIEVAL METHODS

In some cases the type of file organization and processing best suited for a particular system is immediately evident and is dictated by the size of the system to be designed. In all but the smallest of systems, however, the analyst must give serious consideration to the file organization and retrieval methods to be used. There are three types of widely used file organization methods. These include:

1. Sequential File Organization
2. Direct File Organization
3. Indexed Sequential File Organization

The analyst should be familiar with each of the basic types of file organization methods and the advantages and disadvantages of each.

SEQUENTIAL FILE ORGANIZATION

Perhaps the most common type of organization of a file of data is the sequential organization. In sequential file organization, records are stored on a storage medium one after another, usually in some type of sequence based upon a key within each individual record. Sequential files can be stored on various types of storage mediums, including punched cards, paper tape, magnetic tape, and direct-access devices. The following is an example of a sequential file which is stored on tape and is in employee number sequence.

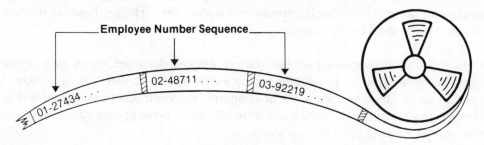

Figure 6-12 Example of Sequential File Organization

Note from Figure 6-12 that the records are stored one after another and they are in employee number sequence. Records which are stored sequentially, such as illustrated above, are normally processed sequentially also, which means that each record is read in the sequence in which it is stored in the file.

It should be noted that all card files which are processed on the computer and all magnetic tape files must be organized and processed sequentially. This is true because each record must be stored or accessed following the processing of the record previous to it. Files which are stored sequentially on a direct-access device may be processed sequentially or randomly, but in the large majority of cases they are processed sequentially.

SEQUENTIAL FILE UPDATING

As noted previously, files normally fall into two categories—Master Files and Transaction Files. Master files contain the current information relating the status of a given system while transaction files contain information which relates the day-to-day activities of a business. Since master files must contain current information, one of the important activities which must take place within the data processing system is the updating of the master files to contain the most current information. Typically, the transaction files contain the day-to-day information which is used to update the master file.

There are three basic activities which take place when processing master files. These activities include:

1. ADDITIONS - An Addition takes place when a new record is added to a master file. For example, in a payroll system, when a new employee is hired, a record for the new employee must be added to the Employee Master File.

2. DELETIONS - A Deletion becomes necessary when a record currently stored on the master file must be removed. For example, in a payroll system, when an employee is terminated, the record for that employee must be removed from the Employee Master File.

3. CHANGES - A Change must be made to the master file whenever the data on the master file no longer contains accurate, up-to-date information. For example, in a payroll system, when the payrate of an employee changes, the master record for that employee must be changed to reflect the new rate.

In most applications, transaction files will contain the data which is to be used to update the master file data. When updating master files stored in a sequential mode, the transaction file and the master file to be updated are read by the computer program and a new, updated master file is written. This is illustrated in Figure 6-13.

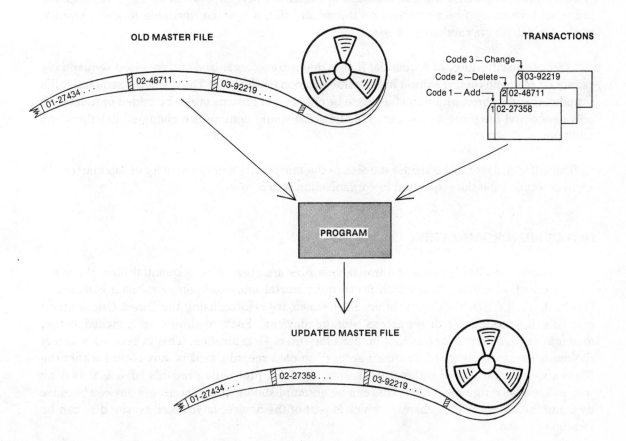

Figure 6-13 Sequential File Updating

Note from Figure 6-13 that the sequential transaction file which is stored on punched cards is processed against the sequential master file which is stored on magnetic tape. The transaction file which is stored on punched cards will contain the additions, deletions, and changes which are to occur to the master file. Note in the example that record 02-27358 has been added to the updated master file and record 02-48711 has been deleted from the master file. Record 03-92219, which is specified as being a change record, would cause some change to occur in the original master record such as a name change, a change in year-to-date earnings in a payroll update, etc.

A sequential file is most applicable in a situation where the master file is frequently updated and there are many transaction records to be processed against the master file. When this is the case, sequential file organization is an efficient form for storing and processing data.

In some applications, however, there are several disadvantages to sequential files. One disadvantage is that it is not possible on sequential files to process, for example, the tenth record on the file and then process the first record on the file. Because of this, whenever a sequential file is to be updated or otherwise processed, all of the records within the master file must be read in order to process the data.

For example, if there are 10,000 master records and only 100 transaction records to be processed against the master file, all 10,000 master records would have to be read and processed just to process the 100 transaction records. Therefore, when there is little update processing which is to be performed on the master file, it may be desirable to use a storage method which allows random processing.

Another disadvantage of sequential files is that records organized and accessed sequentially cannot be read and then rewritten in the same position on the tape. Thus, when a sequential file is updated and records within the file are to be changed or records are to be added or deleted, a new sequential file must be created. This new file would contain the changes, deletions and additions.

Thus, the analyst must consider the size of the master file and the activity of the master file when determining if the sequential file organization is to be used.

DIRECT FILE ORGANIZATION

Although many master files and transaction files are created as sequential files, there are other methods of storing data which prove quite useful under certain processing conditions. One of these is the Direct Organization. Files which are created using the Direct Organization can only be created on direct-access storage devices. Such mediums as punched cards, magnetic tape, or paper tape cannot be used for Direct Organization. This is because a Direct file requires the capability of "random access" to data records, that is, any record within the file is able to be processed without processing all of the preceeding records such as it is done with sequentially organized files. This can be accomplished with direct-access devices because by using the access arm mechanism which is part of the device, any record on the disk can be located and read.

With Direct Organization, an "address" on the direct-access device is derived from the key of the record which is stored in the file. This address on the direct-access device is unique and allows the program which is to access this record to specify the address of the record and the record willl be retrieved immediately without having to process any other records in the file. The methods of computing an address from the key of the record vary dependent upon the key to be used, the nature of the records, the size of the file, and the type of direct-access device to be used.

For example, in a simplified example of direct file organization, assume that is is necessary to store a master file of employees on a disk using direct file organization. If the employee numbers ranged from 501-999, a possible method would be to store employee 511 on cylinder 5, track 1, as the first record; employee 555 on cylinder 5, track 5, as the fifth record; employee 934 on cylinder 9, track 3 as the fourth record; etc. This is illustrated in Figure 6-14.

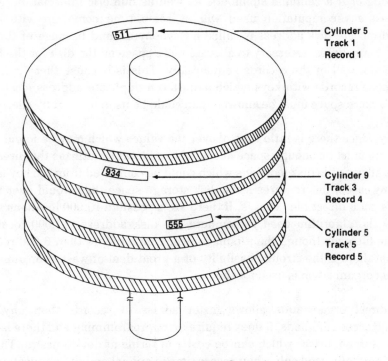

Figure 6-14 Direct File Organization

The primary advantage of direct file organization is that any record in a file can be referenced without examining all preceding records in the file. For example, if it were necessary to retrieve record 555 from the file illustrated in Figure 6-14, a command in the computer program could be given to read the record which was stored on cylinder 5, track 5 as the fifth record and the record would be available for processing in less than 100 milliseconds. Thus, direct file organization offers one of the fastest forms of retrieval. Other advantages of direct file organization are that the records do not have to be in sequence in order to load or update the file, and updating does not require the creation of an entirely new master file such as with sequential file updating.

There are, however, several distinct disadvantages with a file which has direct organization. The first is in the difficulty of computing the disk address from a key within the record. In the simple example illustrated in Figure 6-14, the disk address used was taken directly from the key, that is, the first digit was the cylinder number, the second digit was the track number, and the third digit was the record number. In most applications, this type of key-to-address conversion will not work properly or efficiently. For example, it may turn out to be very inefficient usage of disk storage to place only 10 records on each track of the disk; or cylinders 5-9 may not be available for storing the file as this portion of the disk is used for another file. When these types of problems arise, it is necessary to perform some arithmetic calculations upon the key of the record in order to generate a usable direct-access address for each record within the file.

Although various techniques exist for the computation, it may be difficult to do if the key of the record is long or if it contains alphabetic as well as numeric information. As part of this problem, almost any computation used will at some time come up with duplicate disk addresses. A duplicate disk address is called a "synonym" and because of the possibility of synonyms, it is many times necessary to allocate more space on the disk for the file than would be required to store all of the records sequentially. This is because there must be alternate locations for those records with keys which generate a duplicate address location. A general rule is that 20% more space must be allowed than could be used by a "full" file.

Additionally, since there is little control over the values which may be found in the keys for the records, there must be enough space allocated to a file which utilizes the direct organization method so that all possible disk locations which could be computed from the key are available to the file. In many cases, this requires more disk storage space than would ever be allocated to the file if it was not a direct file. That is, if there are a possible 10,000 locations which could be determined by the algorithm used to compute the disk address, 10,000 locations must be reserved for the file even though the file may never contain more than 6,000 records. Thus, it can be seen that there is the strong possibility of a great deal of wasted or unused space on a disk when direct organization is used for the file.

Although direct organization allows faster access to records than any of the other organization and access methods, it does require extra programming and there is the possibility of unused disk storage space, which can be costly in terms of device usage. Therefore, direct organization is normally used only when access time is critical and the application requires that very fast record retrieval be possible. This type of activity may be required where the file is to be referenced continuously in an on-line data communications environment which makes it mandatory that the records be available for immediate reference.

INDEXED SEQUENTIAL FILE ORGANIZATION

A file organization method which was developed by IBM and has since been implemented on computer systems manufactured by other companies is the Indexed Sequential File Organization. The indexed sequential file organization provides for both sequential and random processing through the use of indexes which are used to point to records which are stored on the file.

The records of an indexed sequential file are organized on the basis of a collating sequence determined by a specific control field or ''key'' within the record. An indexed sequential file exists in space allocated on the disk called the PRIME DATA area, the OVERFLOW area, and the INDEX areas.

When an indexed sequential file is initially established on the disk, all data records are loaded into an area called the PRIME DATA AREA. The data in this area is available to be processed by both sequential and random access methods. For example, a master employee file containing employee numbers and related information about the employees of an organization could be stored on a disk as an indexed sequential file. In the example in Figure 6-15 these records are stored in the Prime Data Area beginning at Cylinder 50, Track 1.

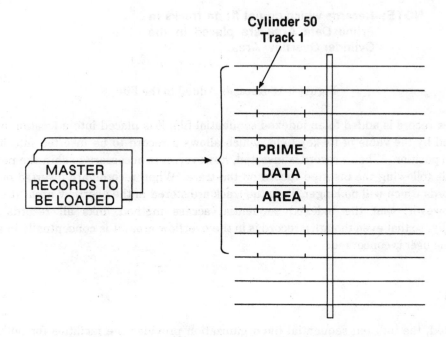

Figure 6-15 Prime Data Area

After the file is established, the user can ADD records without reorganizing the entire file as in sequential file organization. A new record added to an indexed sequential file is placed into a location on a track determined by the value of the "key" in each record. To handle additions, an OVERFLOW AREA exists as illustrated is Figure 6-16.

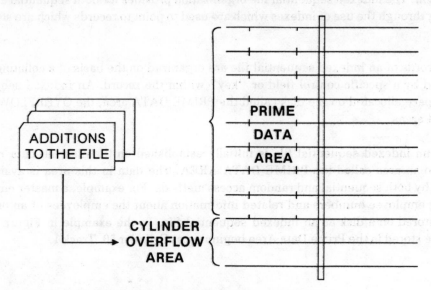

NOTE: Records which do not fit on tracks in Prime Data Area are placed in the Cylinder Overflow Area.

Figure 6-16 Records Added to the File

When a record is added to an indexed sequential file, it is placed into a location on a track determined by the value of its key field, which allows a record to be inserted into its proper sequential position. When a record is inserted, however, it is necessary to shift the positions of the records following the inserted record on the track. When records are inserted on a track, those records which will no longer fit on the track are stored in the overflow area. It should be noted, however, that the indexed sequential access method links all records together sequentially, so that even though a record is in the overflow area, it is conceptually in sequence so far as the user is concerned.

INDEXES

As noted, the indexed sequential file organization provides the facilities for both random and sequential processing of the records stored in the file. This ability is provided through the use of the key which is associated with each record in the file and Indexes, which are established when the file is initially "loaded" and which are updated as records are added to the file.

An Index is a pointer which is used by the access method to point to the disk location of a record within the file. By assigning a particular cylinder, track, and record location to a record and associating the key of the record with the address and placing this information in an index, any record for which the key is known can be located and processed.

In actual practice, a series of indexes are established so that the process of locating a record in the file involves a stepping process wherein a high level index is examined which in turn points to a lower level index which is examined, and so on until the record is actually located.

It should be noted that indexed sequential files can be stored only on direct-access devices because of the use of random retrieval techniques. Random retrieval with an indexed sequential file takes place by locating the key of the desired record in the indexes and then retrieving the record whose address is related to the key in the index.

Advantages and Disadvantages

As noted, the indexed sequential access method has the advantage of allowing both sequential and random retrieval because of the indexes. There are several disadvantages, however, which at times makes it more efficient to use either sequential file organization or direct file organization. The biggest disadvantage of indexed sequential files is that the processing of the files is relatively inefficient, compared to both sequential and direct files. Since the record locations are recorded in the indexes and the indexes are located on the direct-access device, every time a record is to be read by a program, one or more indexes must also be read in order to locate the record. These multiple reads of direct-access devices take time and this leads to slower processing.

In addition, as more records are added to the file and placed in the overflow areas, the processing becomes even more inefficient because all of the pointers which locate records in the overflow areas must be read and processed. Therefore, in a highly active file, where many records are added to a file, the indexed sequential file quickly becomes highly inefficient and the time it takes to process the file becomes prohibitive.

On the other hand, if the file will have a lot of retrieval processing but is relatively stable concerning the contents of the file itself, then the indexed sequential access method may be quite useful because it does give the capabilities of random processing while at the same time allowing sequential processing which lends itself to creating reports and such processing as this.

Although there are more file organizations which are specifically designed for a given purpose, the sequential organization, the direct organization, and the indexed sequential organization are the three which are most widely used in business data processing and which will satisy the needs of most business systems.

FILE DESIGN CRITERIA

When determining the type of files to be designed within a system the analyst is faced with two basic problems. These problems are:

1. What type of hardware or media should be used—punched cards, magnetic tape, magnetic disk, or some type of specialized hardware?

2. What type of file organization should be used—sequential, direct, indexed sequential, or a specialized file organization?

The advantages and disadvantages of the various forms of hardware and file organization have been discussed previously. In deciding upon the type of hardware and file organization methods to be used there are a number of factors which must be considered. These factors include:

1. File Accessibility
2. File Activity
3. File Capacity
4. File Processing Speed
5. File Cost

File Accessibility

An important consideration when designing a system is the accessibility to the data contained within the files required by the users of the system. Basically there are two types of processing systems in use which will determine the accessibility needs. These are Batch Processing and On-Line Processing.

In a Batch Processing System "batches" of data, such as transactions to be processed against a master file, are accumulated and processed at specific times during a given period of time, such as daily, weekly, or monthly. Certain applications, such as a payroll system which is processed once a week, are perfectly suited for batch processing. In most instances, batch processing is the most inexpensive form of processing from both a hardware and systems design point of view and this is the major advantage of batch processing.

On the other hand, it has the disadvantage that the data in the system, such as the data in a master file, is not current since there is a time lag from the time an event occurs until the time it is reflected in the master file. For example, in an inventory control system using batch processing, sales orders representing a reduction in inventory would be accumulated perhaps weekly and processed against the master inventory file at the end of the week. Thus, at any given time during the week the records of the master file would not be accurate as they do not reflect the actual amount of inventory for sale.

On-Line Processing refers to the processing of individual records or transactions at the time the transaction occurs. For example, in an on-line inventory control system, as a sale is made it would be processed through some type of point-of-sale recorder. The quantity of merchandise sold would be immediately subtracted from the quantity on hand in the proper record in the master file. Thus, as the transaction occurred, the new quantity on hand of the item sold would be reflected in the master inventory file.

On-line systems may also be designed to provide for file inquiry in which it is possible to query the status of a system by examining the contents of the master file. For example, in the inventory control system it would be possible to determine the inventory status of any item within the inventory at any time. These inquiries are normally made by means of a remote terminal and an immediate response is given to the inquiry.

On-line systems are much more complex to design than batch processing systems and normally require additional hardware and more complex programming. Therefore, although on-line systems are very effective in many applications, the analyst should give serious consideration to the cost justification of such systems.

As can be seen from the nature of the two types of processing systems, different file attributes are required. In a batch system, there is no need for fast access to the data—therefore, sequential files will many times be used in batch processing systems. In on-line systems, on the other hand, access time is critical since the on-line update to a master file must be made quite rapidly in order to be able to process the many updates which would occur each day and, in addition, on-line inquiries require that the data be accessed rapidly so that the response to the inquiry can be returned quickly. In this type of environment, files with direct organization or indexed sequential organization would be more appropriate. Thus, the type of processing which is to take place within a system can be quite important in determining the file organization to be used in a system.

File Activity

Another important factor related to file organization and design is "file activity." File activity refers to the number of transactions which are to be processed against a master file in a specific period of time. For example, assume in an inventory system there are 10,000 parts in stock. If these 10,000 parts comprise the master file but only a small percentage of the parts are regularly sold and processed against the master file, it is said that there is little file activity.

In a batch processing system, a file with high activity will quite likely be a sequential file because, as noted previously, with many transactions to be processed against a master file, the sequential organization can prove quite efficient. With a low activity file, however, the requirement of reading and processing every record in a master file in order to process a small number of transaction records will make the sequential file inefficient and the indexed or direct file organizations should be considered.

In an on-line environment, however, the file requirements change. For example, if there is a high activity file in an on-line environment, the necessity of rapid access and fast processing is one of the most important considerations of the entire system. Therefore, with a high activity file in an on-line environment, it is quite likely that a file with direct organization, which offers the fastest access times, will be required. With lower activity, an indexed sequential file can be considered. It is unlikely that any file which is to be accessed in an on-line environment will be a sequential file, regardless of the file activity.

File Capacity

File capacity refers to the maximum size of the files that are to exist within the system, that is, how many individual records are contained within the file. Although all methods of file storage with data processing systems have inherently large storage capacity, there are certain limitations which can influence the decision concerning the method of storing a data file.

As discussed previously, small systems with a limited amount of data may justify only the use of punched card files. As the size of files becomes larger, punched cards become inefficient and consideration must be given to magnetic tape or magnetic disk.

There are no physical limits on the size of files which are stored either on magnetic tape or magnetic disk. In addition, the files on disk can be stored in any of the three file organizations previously mentioned. It should be noted, however, that if the files are organized either as indexed files or direct files, the entire file must be available for processing at the same time. For example, if a file requires three disk packs for storage, all three disk packs must be mounted on disk drives and be available to the CPU for processing even if only the information on the first of the three packs is required for processing.

With sequential files, on the other hand, when stored on either tape or disk, only the tape reel or disk pack containing the data being processed must be mounted on a drive. Thus, if a sequential file were stored on three tape reels, the first reel could be mounted on a drive and the entire reel would be read. When the first reel was completed, the second reel could be mounted on the same drive and processed, and so on. Therefore, with very large files, sequential organization has the advantage of requiring fewer drives than files organized either as direct or indexed files. In an installation which has a limited number of drives, this could be a very important factor in determining the file organization and processing.

File Processing Speed

File processing speed refers to the speed at which data can be transferred into and out of computer memory. Punched cards offer the slowest form of file processing while direct-access devices normally offer the fastest file processing speed. Factors to be included in the analysis of file processing speed are not only the speed of the input/output units, such as card readers, magnetic tape and magnetic disk units, but also the type of file organization method used, as has been discussed previously.

File processing speed must also be considered in terms of the type of processing which is to take place—batch or on-line. As noted previously, the requirements of speed are normally determined by the file activity which will be found within either batch or on-line processing. Thus, the analyst must not only consider each factor by itself but must relate each of the factors to one another when finally determining the medium and file organization to be used in the system.

File Costs

In any systems project the analyst must be able to cost-justify any decision made. One of the most important decisions in the systems project relative to cost is the recommendation made concerning file design and storage media. This is because there are numerous factors related to cost, such as the cost of the unit, the cost of the storage media (cards, tape, disk packs), the cost of processing the data stored on these media (slower processing costs more in computer time), the cost of storing the cards, tapes, or disks in a "library," the costs in reruns due to the unreliability of different media, and a number of other factors.

Thus, in addition to considering the needs of the system in terms of processing, it is always necessary to consider costs at the same time. In some applications, it may be necessary to change the specifications for the system because the file capabilities needed for the system are just too expensive, regardless of the benefits to be received from the system. Thus, the decision made by the analyst at this stage of systems design will have a significant influence on the cost and effectiveness of the entire system for the life of the system.

DESIGNING THE FILES

After the analyst has arrived at a decision as to the type of file storage media and the type of file organization to be used he is now faced with the task of the actual design of the files.

When files are to be designed using punched cards, the multiple card layout form is normally used to assist in designing the individual records within the file. When files are to be contained on magnetic tape or magnetic disks, the Record Layout Worksheet is used. A Record Layout Worksheet is illustrated in Figure 6-17.

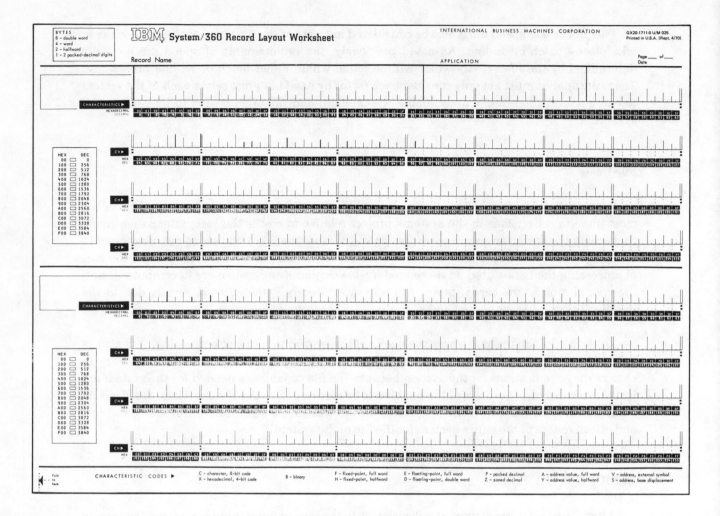

Figure 6-17 Record Layout Worksheet

Note from the example in Figure 6-17 that the Record Layout Worksheet contains space at the top of the form for the Record Name, the Application, and the Date and Page Number. On each form there is space provided to record the format of two records which could be stored on either magnetic tape or magnetic disk. There is space for a maximum of 256 characters per record, with each line containing room for 64 characters. Note that below each position within the record there is both a hexadecimal and decimal count.

The characteristics field beneath position is used to indicate the format of the data in the field. The codes which can be used are contained at the bottom of the form. Note that the data in disk and tape records can be stored in various formats and are not restricted to the character format found in punched cards.

Storing Numeric Data

As noted, numeric data contained in master files stored on disk or tape can be stored in either a packed-decimal format or a binary format if the computer utilizes the Extended Binary Coded Decimal Interchange Code (EBCDIC). Storing data in the packed decimal or binary format can result in substantial savings in tape and disk storage. The following example illustrates a card being read into computer storage in the zoned decimal or character format and then being converted to a packed decimal format. Note that the number 598 requires three bytes of computer storage when stored in the zoned decimal format but only requires two bytes when stored in the packed decimal format.

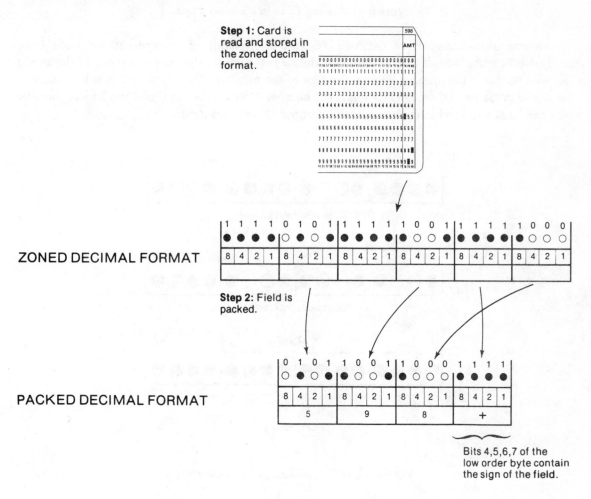

Figure 6-18 Example of Packed Decimal Format

If numeric data is stored in the packed decimal format on disk or tape, considerable storage can be saved. For example, if the field illustrated in Figure 6-18 was stored in the packed decimal format rather than the zoned decimal format and 100,000 records were to be stored in a file, then 100,000 bytes of tape or disk storage would be saved for this field alone. If all numeric fields were packed, considerably more tape or disk storage could be saved.

Data may also be stored in a binary format on magnetic tape and magnetic disk. When data is stored in a binary format, it is normally stored as either a halfword, fullword, or double word. A halfword is 16 bits, a fullword is 32 bits, and a double word is 64 bits. The following example illustrates the number 598 recorded in a binary form in a halfword.

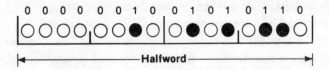

Figure 6-19 Storing Data in a Binary Form

An interesting comparison between the various methods of storage can be made from analysis of Figure 6-20. It can be seen that the largest number that can be stored in the zoned decimal format is the number 99 plus the sign of the number. The maximum number that can be stored in the packed decimal format is the number 999 plus the sign, and the largest number that can be stored in a binary format is the number 32,767 plus a sign.

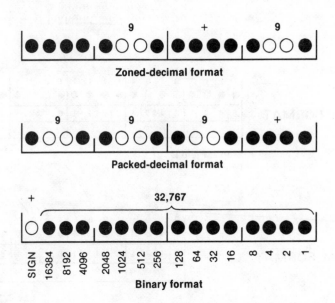

Figure 6-20 Comparison of Numeric Data

As can be seen, storing data in a binary format is more efficient than storing data in either the packed decimal format or the zoned decimal format and storing data in the packed decimal format is more efficient than storing data in the zoned decimal format. The following chart summarizes the savings that can be obtained by storing data in the various formats.

NUMBER VALUE	ZONED DECIMAL		PACKED DECIMAL		BINARY	
	Hexadecimal Representation	Bytes	Hexadecimal Rep.	Bytes	Binary Representation	Bytes
9	F9	1	9F	1	0000 1001	1
99	F9 F9	2	09 9F	2	0110 0011	1
999	F9 F9 F9	3	99 9F	2	0000 0011 1110 0111	2
9999	F9 F9 F9 F9	4	09 99 9F	3	0010 0111 0000 1111	2
99999	F9 F9 F9 F9 F9	5	99 99 9F	3	0000 0001 1000 0110 1001 1111	3
999999	F9 F9 F9 F9 F9 F9	6	09 99 99 9F	4	0000 1111 0100 0010 0011 1111	3
9999999	F9 F9 F9 F9 F9 F9 F9	7	99 99 99 9F	4	1001 1000 1001 0110 0111 1111	3

Figure 6-21 Chart of Numeric Formats

From Figure 6-21 it can be seen that it is possible to save considerable space on a disk or magnetic tape by storing numeric data in either the packed decimal format or the binary format, rather than the character or zoned decimal format. For example, with a seven digit number, seven digits are required in the character format, four digits in the packed decimal format, and three digits in the binary format. Thus, when designing any record which is to be stored on a direct-access device or on magnetic tape, consideration should be given to storing numeric values in either the binary or packed decimal formats.

Field Sizes

As has been noted previously, the analyst must be aware of the sizes of fields which must be used in the input and the output of the system. The same is true of the file records which are to be used in the system. In most instances, the analyst can determine the sizes of the fields from the known sizes in the input or the output records and reports. For example, if the Name field in the input record contains 25 characters, the Name field in the record for the file must also contain 25 characters.

There are some instances, however, where the input field and the field in the record will not be the same size. This normally occurs when the field in the file record is used to accumulate totals. For example, the input to the program which processes the file may contain the weekly pay for the employee but the master file may contain the year-to-date pay. Thus, the analyst must carefully determine the function of each field within the master record to ensure proper size allocation.

Blocking Considerations

As has been noted, one of the decisions which must be made by the analyst when designing a file to be stored on a magnetic tape or direct-access device is the size of the block or physical record which is to be used on the file. One of the factors to be considered when deciding block size is the amount of computer storage which is available for input/output areas. There must be sufficient space in computer storage to store the entire physical record or block.

Another consideration when deciding the block size is the device on which the file is to be stored. When utilizing magnetic tape, the block size can be as large as desired. When utilizing direct-access devices, however, there is a maximum number of bytes which can be stored on a single track of a direct-access device and a block should normally be kept equal to or less than this size. In addition, if sufficient storage is not available to store the maximum size block on the track, then consideration must be given to the next best blocking size. The table in Figure 6-22 can be used to determine the most efficient block sizes.

Maximum Bytes per Record Formatted without keys				Records per Track	Maximum Bytes per Record Formatted with keys			
2311	2314 2319	2321	3330		2311	2314 2319	2321	3330
3625	7294	2000	13030	1	3605	7249	1984	12974
1740	3520	935	6447	2	1720	3476	920	6391
1131	2298	592	4253	3	1111	2254	576	4197
830	1693	422	3156	4	811	1849	406	3100
651	1332	320	2498	5	632	1288	305	2442
532	1092	253	2059	6	512	1049	238	2003
447	921	205	1745	7	428	877	190	1689
384	793	169	1510	8	364	750	154	1454
334	694	142	1327	9	315	650	126	1271
295	615	119	1181	10	275	571	103	1125
263	550	101	1061	11	244	506	85	1005
236	496	86	962	12	217	452	70	906
213	450	73	877	13	194	407	58	821
193	411	62	805	14	174	368	47	749
177	377	53	742	15	158	333	38	686
162	347	44	687	16	143	304	29	631
149	321	37	639	17	130	277	21	583
138	298	30	596	18	119	254	15	540
127	276	24	557	19	108	233	9	501
118	258	20	523	20	99	215		467
109	241	15	491	21	90	198		435
102	226	10	463	22	82	183		407
95	211	6	437	23	76	168		381
88	199		413	24	69	156		357
82	187		391	25	63	144		335
77	176		371	26	58	133		315
72	166		352	27	53	123		296
67	157		335	28	48	114		279
63	148		318	29	44	105		262
59	139		303	30	40	96		247

Figure 6-22 Record Capacities on DASD

In order to use the table in Figure 6-22, first find the proper column for the device to be used to store the file. In the table, there are specifications for the 2311 disk drive, the 2314 and 2319 disk drives, the 2321 device, and the 3330 device. It must also be determined if the file is to contain keys. Normally, when a sequential file is utilized, there will not be keys associated with the file, but when indexed sequential files and direct files are used there are keys.

The records per track column indicates the number of physical records, or blocks, which can be stored with the maximum length as specified in each of the columns for each of the devices. For example, if the length of a physical record or block is 3520 bytes, then two of these records will be stored on a single track of a 2314 disk drive. If the physical record is 3600 bytes in length, only one record will be stored on the track because this is greater than the maximum for two records.

The following examples will illustrate the use of the table when determining the block lengths which should be used. The examples assume the use of a 2314 or 2319 device.

Example 1: Assume that the logical record length of records to be stored on a file is 100 bytes. The analyst must determine what the most efficient blocking factor is. If the analyst blocks the records 2 records per block, he will get a maximum of 23 physical records on the track (note 1, Figure 6-22) because each physical record or block will be 200 bytes in length (100 x 2), and 200 is greater than 199, but less than 211. Therefore, he will be able to store 46 logical records on the track (2 logical records/block x 23 blocks/track). If, however, he makes the blocking factor 72, that is, 72 logical records/block, he can store 72 logical records on a track because 7200 is greater than 3520, but less than 7294 and therefore, one physical record or block can be stored on the track (note 2, Figure 6-22).

Thus, it can be seen that by increasing the blocking factor, more logical records will be able to be stored on the track of the direct-access device.

Example 2: Assume that the logical record length of records to be stored on a file is 80 bytes, and assume further that the analyst has determined that he has enough computer storage to have a maximum block size of 3600 bytes. Therefore, a maximum block size of 3600 bytes would be the largest block length which could be used (80 x 45). Note, however, that 3600 is greater than 3520 and therefore, only one physical record or block of 3600 could be stored on a track (note 3, Figure 6-22), and the maximum number of logical records to be stored is 45.

If, however, the analyst reduced the number of records in the block (i.e. the blocking factor) to 44, he would be able to store 88 logical records on the track because 80 x 44 is equal to 3520 and this is the maximum size of a block which will allow two blocks to be stored on a single track (note 4, Figure 6-22). Thus, by reducing the blocking factor by 1 record per block, the analyst was able to store nearly twice the number of records on a single track.

By utilizing the table in Figure 6-22, the analyst should be able to determine the most efficient blocking factors to be used for a file, depending upon the size of the logical record and the type of device being used. It is quite important that efficient blocking techniques be used so that the storage space on direct-access devices is properly utilized.

Summary

As can be seen the analyst must have a thorough understanding of the characteristics of the computer and storage devices for which the system is designed; for it is only through selecting the proper file storage media, the proper file organization technique, and the proper method of storage of data that the most efficient system can be designed.

CASE STUDY - JAMES TOOL COMPANY

In analyzing the file processing needs of the James Tool Company Payroll System, the systems analysts, Don Mard and Howard Coswell, first assessed the computer system which was currently being used by the company. James Tool Company acquired an IBM System/370 Model 125 approximately two years ago. Since then a number of applications have been implemented on the system, including an order entry system, an inventory system, an accounts payable system, an accounts receivable system, and a general ledger system. At the present time none of these systems are on-line and it is not contemplated that they will be on-line in the near future. The computer system is relatively small with one card reader, one printer, three magnetic tape drives, and three magnetic disk drives.

In light of the current computer system, the applications which are being processed and their storage requirements, and the needs of the payroll system, the analysts concluded the following:

1. The payroll system did not require on-line inquiry and processing; thus, batch processing will be utilized throughout the system.

2. The records to be stored in the system, including the employee master file and the transaction files, would require more than 80 characters per record.

3. There appeared to be no reason why random access of any records within the system was required.

4. There is to be high activity of the records stored in the master file because each week when the payroll is processed every record in the employee master file will be read and updated.

5. There will be a moderate level of updating when adding employees, deleting employees and making changes to various fields within the master records.

6. File capacity may become a problem in the future because of the systems which are already being processed and which use magnetic disk storage.

On the basis of the above factors, the analysts determined that the files would be organized sequentially since there was no need for random access and the file activity in the batched environment was quite high. In addition, since there was the possibility of file capacity problems at a later date and since the tape drives were available, the sequential files would be stored on magnetic tape.

Case Study - File Design - Employee Master Record

The first file to be designed is the Employee Master File, which is to contain the master records which relate the status of each employee relative to payroll information. As was noted previously, it is necessary to examine both the source documents and the input records which will be used to construct the master record in order to determine which fields will appear and the sizes of these fields. The source document and input records which will be used to construct the master record are illustrated in Figure 6-23 and Figure 6-24.

JAMES TOOL COMPANY
EMPLOYEE COMPENSATION RECORD

EMPLOYEE NAME _Adams, Anthony David_   MARITAL STATUS _Married_

EMPLOYEE ADDRESS _7148 Groove Rd., Manchester, Calif. 92101_   NUMBER OF EXEMPTIONS _3_

SOCIAL SECURITY NUMBER _261-24-9021_   DATE EMPLOYED _01/16_

UNION AFFILIATION _Machinists (1)_   DATE TERMINATED _____

WEEKLY PAYROLL / YEAR-TO-DATE

Week Ending	Regular Earnings	Overtime Earnings	Total Earnings	Federal Tax	FICA	State Tax	Credit Union	Union Dues	Net Amount	Check Number	Regular Earnings	Overtime Earnings	Total Earnings	Federal Tax	FICA	State Tax	Credit Union	Union Dues	Net Amount
1/24	176.00		176.00	21.50	10.30	3.52		3.00	137.68	02015	176.00		176.00	21.50	10.30	3.52		3.00	137.68
1/31	176.00	4.95	180.95	23.30	10.59	3.80		3.00	140.26	02232	352.00	4.95	356.95	44.80	20.89	7.32		6.00	277.94

TO EMPLOYEE MASTER

Figure 6-23 Employee Compensation Input Record

NEW EMPLOYEE FORM

DATE ___/___/___

FIRST CARD

FUNCTION (Col 1)

[1] New Employee

EMPLOYEE NUMBER (Col 2-8)

[☐ ☐ ☐ ☐ ☐]

NAME (Col 9-38)

[☐☐☐☐☐☐☐☐☐☐☐☐☐☐☐☐☐☐☐☐☐☐☐☐☐☐☐☐]

ADDRESS (Col 39-57)

[☐☐☐☐☐☐☐☐☐☐☐☐☐☐☐☐☐]

CITY/STATE (Col 58-73)   ZIP CODE (Col 74-78)   SEQUENCE (Col 80)

[☐☐☐☐☐☐☐☐☐☐☐☐]   [☐☐☐☐]   [1]

SECOND CARD

FUNCTION (Col 1)

[1] New Employee

EMPLOYEE NUMBER (Col 2-8)

[☐ ☐ ☐ ☐ ☐]

SOCIAL SECURITY NO. (Col 9-17)   JOB CLASSIFICATION (Col 18-19)   STARTING DATE (Col 20-25)

[☐ ☐ ☐ ☐ ☐]   [☐]   [☐ ☐ ☐]

PAY RATE (Col 26-29)   UNION (Col 30-31)   EXEMPTIONS (Col 32-33)

[☐☐☐]   [☐]   [☐]

MARITAL STATUS (Col 34)   CREDIT UNION (Col 35-39)   SEQUENCE (Col 80)
(1) ☐ Married   (3) ☐ Separated
(2) ☐ Single   (4) ☐ Divorced   [☐☐☐]   [2]

MULTIPLE-CARD LAYOUT FORM

Company   JAMES TOOL COMPANY
Application PAYROLL -NEW EMPLOYEE   by   MARD   Date 12/3   Job No. _____   Sheet No 1 OF 1

Function '1'	Employee Number	Employee Name (Last, First, M.I.)	Employee Address	City/State	ZIP Code	Sequence '1'
9	9 9 9 9 9 9 9	9 9 9 9 9 9 9 9 9 9 9 9 9 9 9 9 9 9 9 9 9 9 9 9 9 9 9 9 9 9	9 9 9 9 9 9 9 9 9 9 9 9 9 9 9 9 9 9 9	9 9 9 9 9 9 9 9 9 9 9 9 9 9 9 9	9 9 9 9 9	9
1	2 3 4 5 6 7 8	9 10 11 12 13 14 15 16 17 18 19 20 21 22 23 24 25 26 27 28 29 30 31 32 33 34 35 36 37 38	39 40 41 42 43 44 45 46 47 48 49 50 51 52 53 54 55 56 57	58 59 60 61 62 63 64 65 66 67 68 69 70 71 72 73	74 75 76 77 78	80

Function '1'	Employee Number	Soc. Security Number	Job Cl.	Starting Date	Pay Rate	Union	Exemp.	Marital	Credit Union		Sequence '2'
9	9 9 9 9 9 9 9	9 9 9 9 9 9 9 9 9	9	9 9 9 9 9 9	9 9 9 9	9 9	9 9	9	9 9 9 9 9	9 9 9 9 9 9 9 9 9 9 9 9 9 9 9 9 9 9 9 9 9 9 9 9 9 9 9 9 9 9 9 9 9 9 9 9 9 9 9 9 9	9
1	2 3 4 5 6 7 8	9 10 11 12 13 14 15 16 17	18 19	20 21 22 23 24 25	26 27 28 29	30 31	32 33	34	35 36 37 38 39	40 41 42 43 44 45 46 47 48 49 50 51 52 53 54 55 56 57 58 59 60 61 62 63 64 65 66 67 68 69 70 71 72 73 74 75 76 77 78 79	80

TO EMPLOYEE MASTER

Figure 6-24 New Employee Input Record

Note from Figure 6-23 and Figure 6-24 that there are two sources of information for the Employee Master File—the Employee Compensation Record and the New Employee Form. It will be recalled from Chapter 5 that it is necessary for the analyst to consider the processing which will be necessary to initially load or build the master file as well as considering the processing which must take place once the file is built. In the payroll system, the information from both the Employee Compensation Record and the New Employee Form is necessary to build the file because the master file must contain both personal information about the employee such as name, address, social security number, etc., and also the pay record for the employee.

When an employee is added to the file after it has been built, only the information from the New Employee Form will be used since the employee has not worked for the company and therefore has no pay record. Thus, for new employees to the company, only the New Employee Form is used.

After the source of information has been determined for master records, it is normally necessary to compile an analysis of the fields to be found in the master record. The master record analysis which was made for the Employee Master Records is illustrated in Figure 6-25.

S Y S T E M D O C U M E N T A T I O N				

NAME OF SYSTEM	DATE	PAGE 1 OF 1
Payroll	January 20	

ANALYSTS	PURPOSE OF DOCUMENTATION EMPLOYEE
D. Mard H. Coswell	MASTER FILE - Record Format Analysis

FIELD	DESCRIPTION	NO. OF CHAR.	SOURCE	COMMENTS
Employee Number	Numeric-Packed	7	New Employee Form	4 bytes-packed
Employee Name	Alphabetic	30	New Employee Form	Last, First M.I.
Employee Address	Alphanumeric	19	New Employee Form	
City/State	Alphanumeric	16	New Employee Form	
ZIP Code	Numeric-Packed	5	New Employee Form	3 bytes-packed
Social Security No.	Numeric-Packed	9	New Employee Form	5 bytes-packed
Job Classification	Alphanumeric	2	New Employee Form	
Starting Date	Numeric-Packed	6	New Employee Form	4 bytes-packed
Pay Rate	Numeric-Packed	4	New Employee Form	3 bytes-packed
Union Code	Alphanumeric	2	New Employee Form	
Exemptions	Numeric-Packed	2	New Employee Form	2 bytes-packed
Marital Status	Alphanumeric	1	New Employee Form	
Credit Union Deduction	Numeric-Packed	5	New Employee Form	3 bytes-packed
Regular Earnings-YTD	Numeric-Packed	7	*Emp. Comp. Record or Calculated	4 bytes-packed
Overtime Earnings-YTD	Numeric-Packed	7	*Emp. Comp. Record or Calculated	4 bytes-packed
Total Earnings-YTD	Numeric-Packed	7	*Emp. Comp. Record or Calculated	4 bytes-packed
Federal Withholding-YTD	Numeric-Packed	7	*Emp. Comp. Record or Calculated	4 bytes-packed
F.I.C.A.-YTD	Numeric-Packed	5	*Emp. Comp. Record or Calculated	3 bytes-packed
State Withholding-YTD	Numeric-Packed	6	*Emp. Comp. Record or Calculated	4 bytes-packed
Credit Union Deduction-YTD	Numeric-Packed	6	*Emp. Comp. Record or Calculated	4 bytes-packed
Union Dues-YTD	Numeric-Packed	5	*Emp. Comp. Record or Calculated	3 bytes-packed
Net Amount-YTD	Numeric-Packed	7	*Emp. Comp. Record or Calculated	4 bytes-packed
Union Dues-This Period	Numeric-Packed	4	Calculated	3 bytes-packed

*These fields are taken from the Employee Compensation Record when the file is initially loaded. After initial loading, the data is calculated. When a new employee is added to the file the fields are initially set to zero and then the values for the fields are calculated weekly.

Figure 6-25 Record Format Analysis - Employee Master File

Note from Figure 6-25 that the field name, the description of the field, the number of characters in the field, the source of the data to be placed in the field, and a comments column are included in the record format analysis. It will be noted that the numeric fields in the master record are all to be stored in the packed decimal format. The data in the master record will come from three sources—the New Employee Form, the Employee Compensation Form, and some data will be calculated. The year-to-date pay information will come from the Employee Compensation Record when the file is initially loaded; after that, it will be calculated weekly. For those employees who are added to the file after the file is initially built, the initial values will be zero and then the values will be calculated weekly. It is the task of the analysts to ensure that all information which will be required to prepare the payroll will be available in the files which are designed for the system. Therefore, it is critical that the file contents be carefully reviewed for completeness.

After the record format analysis has been completed, the analysts should prepare the record layout worksheet. The worksheet for the Employee Master File is illustrated in Figure 6-26.

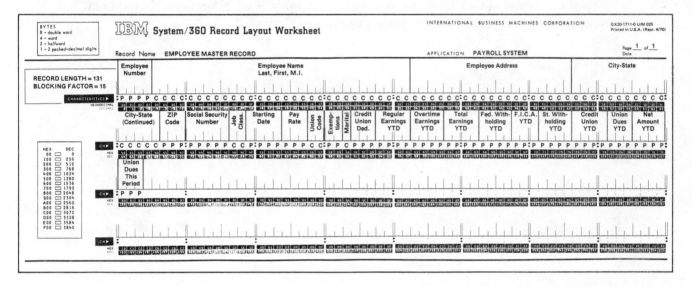

Figure 6-26 Employee Master Record

Note from Figure 6-26 that the format of the record is recorded on the record layout worksheet. The characteristics code under each of the positions in the record indicates the format of the data. When designing the record; the analysts must realize that packed decimal numeric fields do not require the same number of positions in the record as they would if they were stored in the character format. Therefore; care must be taken to ensure that the correct number of positions are reserved for each of the fields.

When designing the record it is also necessary to consider the blocking factor which will be used for the file. As can be seen, the record length is 131 positions. Since the records are to be stored on magnetic tape, the primary consideration when determining the blocking factor is the amount of computer storage which can be used to store the physical record when it is read or written. The analysts, after determining the partition size to be used in computer storage, decided that there was room for the block size to be a maximum of 2000 characters in length. In order to determine the blocking factor, the size of computer storage available (2000 bytes) is divided by the record length (131 bytes). The quotient is the number of records which will fit in the computer storage area, i.e. the blocking factor. The result of the division is 15, with a remainder of 35. Therefore, the blocking factor will be 15 and the block size will be 1965 bytes.

Case Study - File Design - Payroll Report File

The Employee Master File contains the fixed information concerning each employee within the company. In most systems, however, the master files are not used directly as input to create reports. Instead, a file must be created which contains all of the information required to create reports within the system. This information would be retrieved not only from the master file but also from the transaction files which are input to the system and calculations which may be performed within programs.

In order to determine the contents of the file containing the information to create the reports in the system, it is necessary to analyze the contents of the reports which were designed in the output design step of the systems project. It should be noted that one single file will not always suffice to contain all of the information needed for the system reports. This is because different processing steps may create different information and each of these processing steps may create information which must be used in the various reports. The first step, however, is to review the report output from the system to determine what information is required. The reports created in the payroll system are illustrated in Figure 6-27.

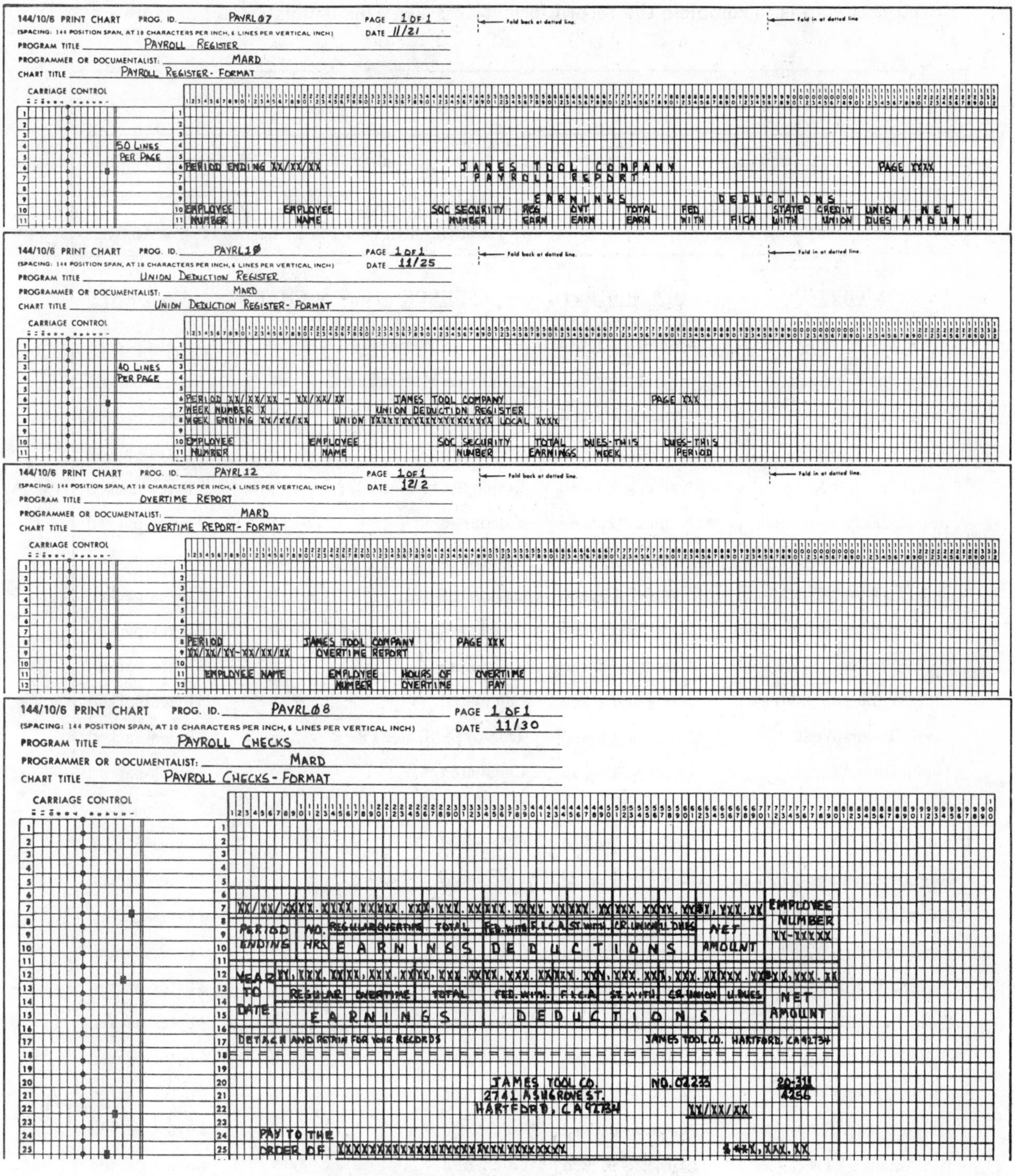

Figure 6-27 Reports for Payroll System

Note from Figure 6-27 that the four reports which are created out of the payroll system all have basically the same information on them. Therefore, if the information can be compiled at one time, such as out of an update program, then only one report file will be required. Since the time cards are to be processed against the employee master file, all of the information required for the reports should be available from the program which processes the time cards and the employee master file. Therefore, the analysts decided to design one report file containing all of the information required to produce the reports.

The first step is to complete the record format analysis. This is illustrated in Figure 6-28.

SYSTEM DOCUMENTATION			
NAME OF SYSTEM Payroll	**DATE** January 22		**PAGE** 1 OF 1
ANALYSTS D. Mard H. Coswell	**PURPOSE OF DOCUMENTATION** PAYROLL REPORT FILE - Record Format Analysis		

FIELD	DESCRIPTION	SOURCE	NO. OF CHAR.	COMMENTS
Employee Number	Numeric-Packed	Employee Master File	7	4 bytes-packed
Employee Name	Alphabetic	Employee Master File	30	Last, First M.I.
Hours Worked	Numeric-Packed	Time Card	3	2 bytes-packed
Week Ending Date	Numeric-Packed	Time Card	4	3 bytes-packed
Social Security No.	Numeric-Packed	Employee Master File	9	5 bytes-packed
Regular Earnings-Week	Numeric-Packed	Calculated	5	3 bytes-packed
Overtime Earnings-Week	Numeric-Packed	Calculated	5	3 bytes-packed
Total Earnings-Week	Numeric-Packed	Calculated	6	4 bytes-packed
Federal Withholding-Week	Numeric-Packed	Calculated	5	3 bytes-packed
F.I.C.A.-Week	Numeric-Packed	Calculated	4	3 bytes-packed
State Withholding-Week	Numeric-Packed	Calculated	5	3 bytes-packed
Credit Union-Week	Numeric-Packed	Employee Master File	5	3 bytes-packed
Union Dues-Week	Numeric-Packed	Calculated	4	3 bytes-packed
Union Code	Alphanumeric	Employee Master File	2	
Net Amount-Week	Numeric-Packed	Calculated	6	4 bytes-packed
Regular Earnings-YTD	Numeric-Packed	*Calculated	7	4 bytes-packed
Overtime Earnings-YTD	Numeric-Packed	*Calculated	7	4 bytes-packed
Total Earnings-YTD	Numeric-Packed	*Calculated	7	4 bytes-packed
Federal Withholding-YTD	Numeric-Packed	*Calculated	7	4 bytes-packed
F.I.C.A.-YTD	Numeric-Packed	*Calculated	5	3 bytes-packed
State Withholding-YTD	Numeric-Packed	*Calculated	6	4 bytes-packed
Credit Union-YTD	Numeric-Packed	*Calculated	6	4 bytes-packed
Union Dues-YTD	Numeric-Packed	*Calculated	5	3 bytes-packed
Net Amount-YTD	Numeric-Packed	*Calculated	7	4 bytes-packed
Union Dues-This Period	Numeric-Packed	*Calculated	4	3 bytes-packed

*The new Year-To-Date Figures are calculated by adding the weekly figures, which are calculated, to the old Year-To-Date totals, which are stored on the Employee Master File.

Figure 6-28 Record Format Analysis-Payroll Report File

Once the record format analysis has been completed, the analysts must check closely to ensure that all necessary information has been included. When it has, the format of the record can then be established on the Record Layout Worksheet. The worksheet for the Payroll Report File is illustrated in Figure 6-29.

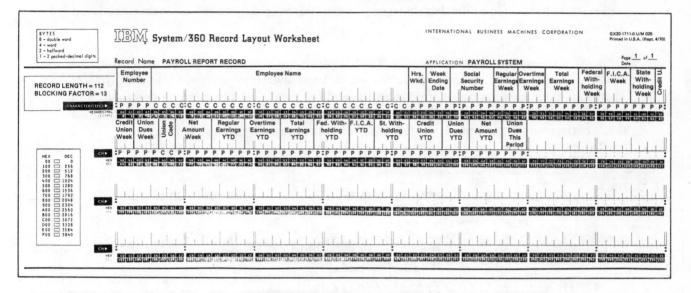

Figure 6-29 Payroll Report Record

As was noted previously, this form is used to record the format of the record to be used in the file. Again, it is important that the analysts carefully consider the contents of the file since all data which must be available for the reports generated in the payroll system should be contained in the file.

As with all files, the blocking factor must be determined. Since the files are stored on magnetic tape, the most important consideration is the amount of computer storage available to store the records when they are processed. The analysts determined that 1500 storage locations in computer storage were available. Since the record length is 112 characters, the division of the storage locations by the number of characters in the record showed that 13 logical records could be stored in 1500 bytes. Thus, the blocking factor will be 13 and the actual block size will be 1456 (13 X 112).

Case Study - File Design - Terminated Employees File and Time Card File

Two other files must be considered by the analysts for the payroll system. As will be recalled, when an employee is terminated or quits the company, his payroll records must be retained until the end of the year even though he will not be paid each week. Therefore, it is necessary to have a file containing those employees' pay records who are no longer employed by the company.

It will also be noted that the time input cards which are prepared from the time cards and which are input to the system will be stored on tape after they are read into the system. Therefore, both the terminated employees file and the time card file must be designed. The analysts would go through the procedure described previously for analyzing the file needs. As a result of the analysis performed, the following formats were designed for the files.

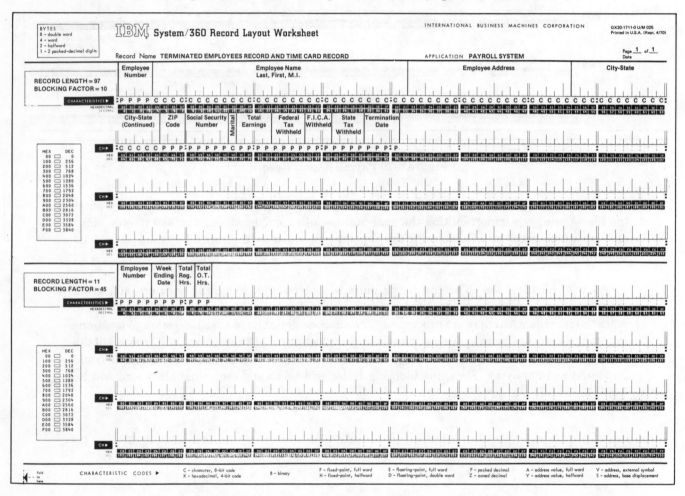

Figure 6-30 Format of Terminated Employees File and Time Card File

Note from Figure 6-30 that the record layouts and blocking factors have been decided in the same way as was used for the previous files. Again, it is quite important that close attention be paid to the contents of the files so that all the data required for the application will be included.

SUMMARY

The file design portion of the systems project is one of the most important since the files contain the bulk of the information which is to be processed within the system. The determination of the media on which the files will be stored, the organization of the records within the files, and the data within the records will have significant impact upon the effectiveness of the system when it is implemented.

STUDENT ACTIVITIES—CHAPTER 6

1. List the steps which must be undertaken by the systems analyst in defining the characteristics of the system files.

    1)

    2)

    3)

2. Explain the terms "master file" and "transaction file."

3. What are the three most common types of file storage?

4. Explain the advantages and disadvantages of the use of punched cards as a method of file storage.

5. Explain the advantages of magnetic tape as a form of file storage as compared to punched cards.

6. Explain the disadvantages of magnetic tape as a form of file storage as compared to punched cards.

7. What is meant by the term ''variable length record''?

8. What is the primary advantage of magnetic disk storage as compared to magnetic tape storage?

9. What is the disadvantage of magnetic disk storage as compared to magnetic tape?

10. List three common types of file organization methods.

    1)
    2)
    3)

11. What are the three basic activities that take place when updating master files?

12. Briefly explain the concept of ''direct file organization.''

13. Briefly explain the structure and operation of "indexed sequential files."

14. What are the advantages and disadvantages of indexed sequential file organization?

15. List five factors which should be considered when determining the type of hardware and file organization to be used in a system.

   1)
   2)
   3)
   4)
   5)

DISCUSSION QUESTIONS

1. "Magnetic tape is obsolete. With the advent of direct-access devices with data storage capabilities greater than magnetic tape, there is no reason for magnetic tape devices, with their limited processing capabilities, to be used in a modern data processing installation."

   Do you agree with this statement? Why?

2. One analyst pointed out that "we never store numeric data in a binary format in our files. The routines in our programs to process binary data consume more computer storage than if the data is stored in another format; and, since computer storage is more costly than secondary storage, it is better to use less computer storage."

   Another analyst says "secondary storage space, especially on direct-access devices, is critical. For this reason, data should be stored in files in the most compact manner available, including the binary format for all numeric data."

   What do you think?

3. With increasingly greater frequency, data processing systems are being designed using On-Line Processing. It has been said that Batch Systems should no longer be designed since an On-Line System can do everything a Batch System can.

   Do you support the position that Batch Systems are obsolete? Why?

CASE STUDY PROJECT—CHAPTER 6
The Ridgeway Company

INTRODUCTION

After the output and input have been designed for the billing system of the Ridgeway Country Club, it is necessary to design the files which will be used to process the data within the system.

STUDENT ASSIGNMENT 1
FILE DESIGN

1. Design the transaction and master files to be used in the billing system for the Ridgeway Country Club. Include the following elements for each file:

   1) Storage Medium
   2) File Organization
   3) File Retrieval Method
   4) Record Design, including a Record Format Analysis
   5) Blocking Factor

2. Document the justification for each file storage medium and the methods of organization and retrieval based upon File Accessibility, File Activity, File Capacity, File Processing Speed, and File Cost.

DESIGN OF
SYSTEMS PROCESSING

CHAPTER 7

DESIGN OF SYSTEMS PROCESSING

7

INTRODUCTION

After the input records, the output records, the reports, and the files have been designed, the analyst is faced with the task of determining how the system will operate in order to produce the results required from the given input. He must consider all of the processing from the creation of the original master and transaction files to the preparation of the last report.

The basic tool used to accomplish the design of the processing within the system is the systems flowchart. The systems flowchart, upon completion, should show all runs and their interrelationships, identify each file and report, indicate the number of machine runs to be programmed, and provide management with a means of reviewing the overall plans for the system.

FILE PROCESSING

The processing that occurs in most business data processing systems can be categorized into four basic steps. These steps are:

1. Editing
2. Sorting
3. Updating
4. Reporting

Although the type and complexity of processing varies greatly with each business application and with the type of processing to be performed (batch or on-line), most systems will incorporate some form of editing, sorting, updating, and reporting within each systems processing cycle.

Editing - Batch Systems

The processing associated with the "Edit" portion of a system consists of reading input data into the system and performing checking routines to ensure that the input records contain valid data, that is, data which is "accurate and authorized." The data normally comes from punched cards, magnetic tape, or magnetic disk in a batch system or from remote terminals in an on-line system.

In a batch system, the usual procedure is to have an "edit run" in which an edit program reads the input data and creates an output file, on tape or on a direct-access device, which contains the valid data. The edit program also creates an Exception Report which lists and identifies those records which contain errors so that the errors can be corrected prior to further processing of the records. The example in Figure 7-1 illustrates the concept of an edit run in which Employee Status Change records serve as input to an edit program.

In the example in Figure 7-1, the Employee Status Change records should contain an Employee Name, a Social Security Number, a Pay Rate and a Code of 1, 2, or 3. The Employee Status Change records will be used to update an Employee Master File. A code of "1" indicates an employee is to be added to the master file. A "2" code indicates that an employee is to be deleted from the master file and a code of "3" indicates that a change is to be made to an employee's pay rate. The criteria established by the analysts for valid data is that there must be a social security number, that is, the field in columns 31-39 cannot contain blanks, and the code field must contain the value 1 or 2 or 3. If the Social Security number field is blank or the Code field does not contain a 1, 2, or 3, the input record will be written on the Exception Report as an error record. Valid records, that is, records which do not contain errors detected by the edit program, are written onto magnetic tape for further processing.

Procedures vary concerning whether all of the data must be correct before any further processing is performed. When operating in a batch environment, with some systems the valid records will be processed immediately and those records which appear on the exception report will be corrected at some later time and processed in a subsequent run. This will normally occur when time is not a critical factor or where partial processing will suffice in the system.

In some systems, however, **all** records must be correct before any further processing is undertaken. For example, in a payroll system normally all records which are required to prepare the payroll must be correct prior to processing the payroll. In batch systems where all of the input data must be correct before any further processing takes place, such as in a weekly payroll, several methods can be used to re-enter corrected data into the system. One method is to make the corrections and then read and process the entire input file back through the edit program. Although this is a relatively easy way to make the corrections, it has the major disadvantage that the entire transaction input file must be read and edited again after already being processed one time. With small transaction files this may not be a significant problem but with input files of any size, excessive processing time would be used to process the entire file.

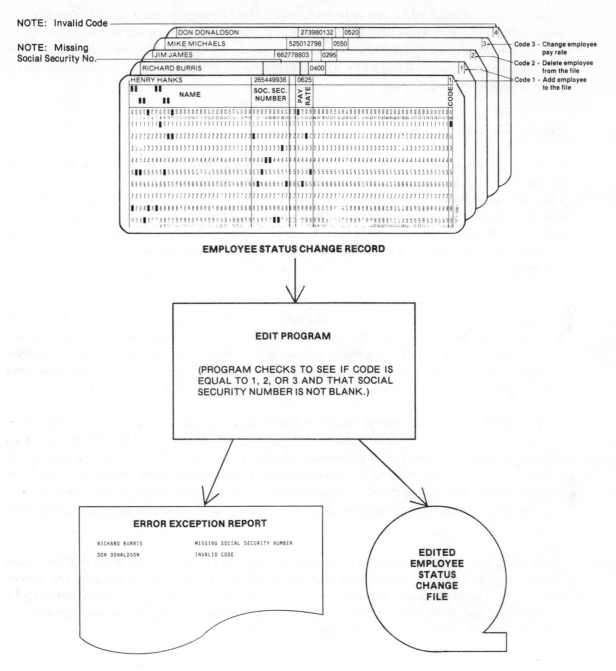

Figure 7-1 Example of Edit Run

A more common means of re-entering corrected data in a batch system is to process only the corrected data through the edit program. The corrected data would be processed in the same manner as the original transaction records, that is, the input records with no errors would be placed in an output file for further processing and those records with errors would be written on an Exception Report. It would be hoped that the records in the second edit run would be correct and that no entries would be made on the exception report. Although this is normally the case, the possibility remains that a third or fourth edit run may be necessary in order to have all input records contain valid data.

It will be noted that when two or more runs of the edit program must be performed two or more output files with valid data will be produced. Since these files each contain information which must be processed in subsequent programs within the system, they must be merged into one file at some point in the system. This will normally take place in the sorting step which occurs after the editing step.

Editing - On-Line Systems

In an on-line system where data is received from a terminal or other device located at a remote site, editing is normally required for the data being received. The checking of the data will many times be the same as if the data were being entered into a batch system. In addition, however, the editing requirements may be more stringent in an on-line system and there may be additional verification data which must be entered in an on-line system which is not required in a batch system. This is because correcting the transaction in error may be more difficult than in a batch system.

For example, in a savings and loan application, if the teller enters the wrong account number for a customer, two errors would occur: the deposit would not be credited to the account of the customer making the deposit; and it would be credited to the account whose number was entered. When the mistake was noticed by the customer, the procedure to isolate the transaction in error and make the correction would be a difficult and time-consuming process since it is likely that the mistake would not be caught until some later time.

Therefore, most banks require additional verification before a balance is changed in an account. For example, one bank requires that not only the account number but the previous balance be entered at the time a deposit or withdrawal is made. The computer program will compare the balance entered with the balance contained in the account whose number has been entered by the teller. If they match, it is assumed that the teller has entered the correct account number. If the balances do not match, the teller is informed of the discrepancy and corrections must be made.

In an on-line system, processing and updating normally take place one record at a time, that is, as an operator enters a record from the remote terminal all of the processing for that record takes place. Therefore, if an error is found in the record, the terminal operator will normally be informed of the error immediately. The correction would be made by the operator by re-entering the transaction into the system.

SORTING

After the input data has been edited, it is many times necessary to place the input data in some type of sequence before any updating can take place. Therefore, the next processing step after editing is commonly a sort program which will place the data in the desired sequence. It should be noted that the analyst need only define the files to be sorted and specify the sequence in which the data is to be sorted since utility sort programs are normally available with the software of most computer systems. Figure 7-2 illustrates the systems flowchart and the processing which takes place with a sort program.

SYSTEMS FLOWCHART   PROCESSING

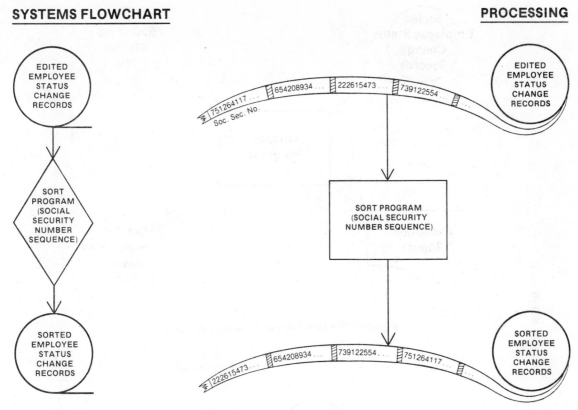

Figure 7-2 Example of Sort Processing

In the example in Figure 7-2 the edited employee status change records serve as input to the sort program. The sort program causes the records to be sorted in social security number sequence and to be placed in the output file. The output of the sort program will then be used as input to an update program.

Updating - Introduction

After the transaction data is sorted into the proper sequence to be processed against the master file, the update processing takes place. As has been noted, updating normally consists of making additions to a master file, making deletions from a master file, and making changes to records stored in the master file.

Although the basic processing which takes place in an update, that is, add, delete, and change, is fairly standard the methods used to accomplish the updating can vary considerably depending upon the type of processing (batch or on-line) and the file organization.

Updating - Sequential Batched

When updating sequential files in a batched environment, the usual technique is that the update program reads a sorted sequential transaction file such as discussed previously and matches the transaction records with the records in the sequential master file. The master records and the transaction records must be arranged in the same sequence in order for the matching of the records to take place. This update technique is illustrated in Figure 7-3 and Figure 7-4.

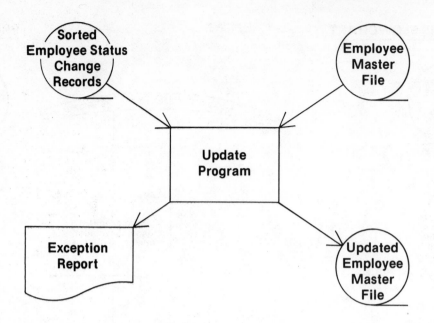

Figure 7-3 Flowchart of Sequential Update

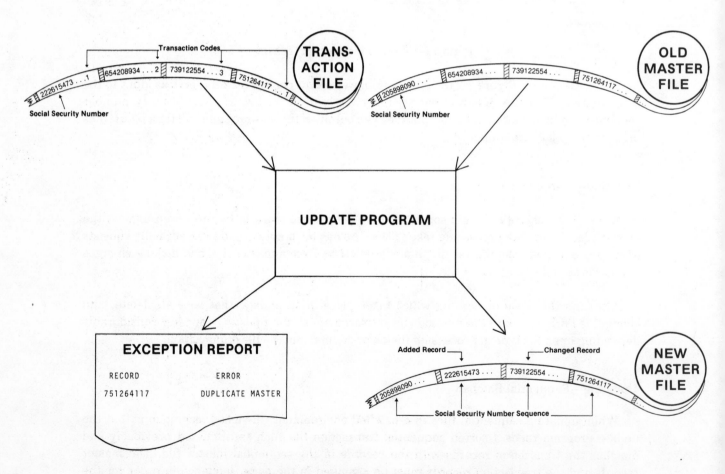

Figure 7-4 Example of Sequential Update Processing

Note in the examples in Figure 7-3 and Figure 7-4 that the transaction records and the master records are arranged in sequence by social security number. In addition, it will be noted that a "transaction code" is contained in the transaction record. This code is used to indicate the processing which is to take place. In the example, a code of "1" indicates an addition to the master file, a code of "2" indicates a deletion, and a code of "3" indicates a change to a record in the master file.

As a result of the update processing, an updated master file is created which contains master records which were changed, master records for which no changes or deletions were indicated, and any records which were added to the file. Records for which a deletion transaction was processed will not be contained on the new file.

When updating files, two basic types of errors can occur: 1) A transaction record can indicate that a record is to be added to the file and a record with the same control field, such as social security number, is already on the master file; 2) A transaction record can indicate that a master record is to be changed or deleted and there is no record to be changed or deleted on the master file. In the example in Figure 7-4 it can be noted that the record identified by the social security number 751264117 is listed on the exception report as a "duplicate record." This is because this record was to be added to the master file (indicated by the code of "1") but there already existed a record with a social security number of 751264117 on the master file.

Updating - Random Batched

As has been noted previously, files are normally organized either in a sequential, indexed sequential, or direct organization. Those files which are organized as indexed sequential or direct can both be updated randomly, that is, a transaction record can specify any record in the file to be updated, regardless of the position of the record within the master file.

The type of updating, that is, additions, deletions, and changes, are much the same for random updates as for sequential updates. There are, however, several significant differences:

1. With Random Updating a new master file is not created each update run.

2. With Random Updating the transaction records do not have to be sorted prior to the updating run.

When a random update takes place, a new master file is not normally created. Instead, the record to be updated is read into computer storage from the master file, changed, and then rewritten back onto the master file in the same location from which it was read. This is illustrated in Figure 7-5.

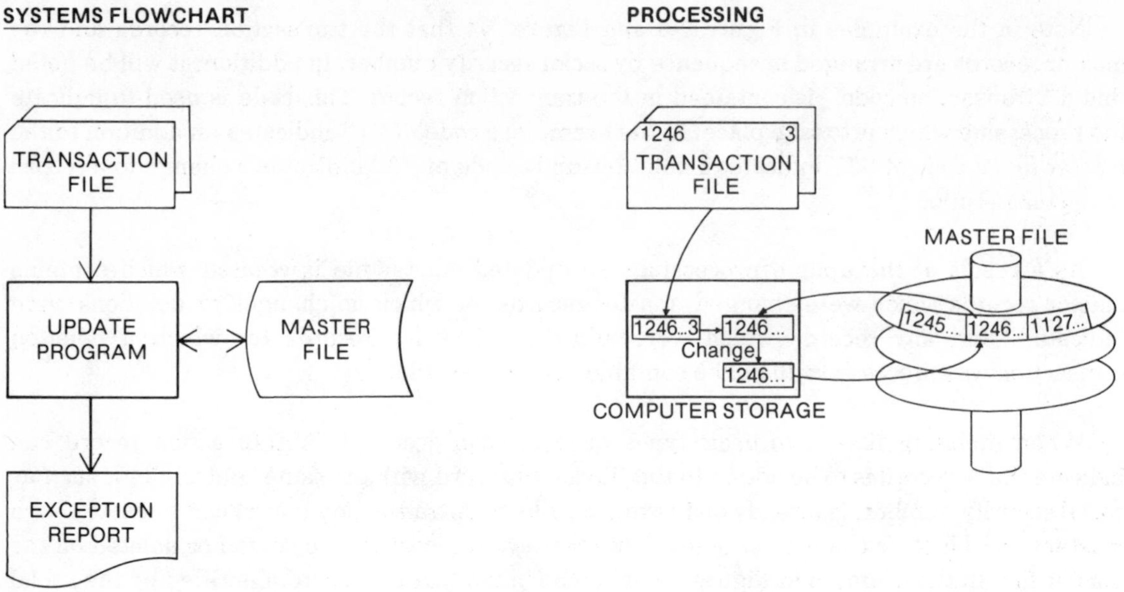

SYSTEMS FLOWCHART

PROCESSING

Figure 7-5 Example of Random Update

Note in the example in Figure 7-5 that the transaction record and the record from the master file to be updated are read into computer storage. The master record is then changed and rewritten back into the master file in the same location from which it was retrieved. A new master file is not created as when a sequential update is performed.

Since a new master file is not normally created when random updating takes place, it should be noted that a record in the master file cannot be physically deleted from the file, that is, the record is stored in the file and it cannot be removed without completely rewriting the file on another disk. Therefore, instead of physically deleting the record from the file, an indicator must be set within the record to show that the record is "logically" deleted from the file, that is, when the "flagged" record is read in future update or report programs, it will be treated as if it were not a part of the master file. For example, a single character field could be established to indicate if the record is "active" or if the record is to be considered "deleted." An "A" in the field could be used to indicate an active record and a "D" in the field could be used to indicate a deleted record. Thus, when designing a file which will be updated randomly the analyst must include a field for the indicator.

Another difference which will be noted is that when a master file is randomly updated there is no need to have the transaction records in a given sequence, that is, since any master record in the file is available for processing, the transaction records can be read in any sequence and the master record can still be updated. When randomly updating an indexed sequential file, however, there may be some advantage to having the transaction data in the same sequence as the data is stored in the master file. The reason is that even though indexed sequential files allow random processing, the records are stored in the file in a sequentially ascending order, that is, the lowest key is the first record in the file and the highest key is the last record. Thus, if the transaction records are in an ascending sequence when they are read, there may be less arm movement in retrieving the records from the indexed sequential master file and therefore less time required for the update processing. The analyst must decide whether sorting the transaction records will save sufficient time in the update processing to justify the time required to sort the transaction records.

Updating - Random On-Line

The types of updating that take place in an on-line system are, for the most part, quite similar to those in a batched environment, that is, there are additions, deletions, and changes. Virtually all files which are updated in an on-line system require random updating, for two primary reasons:

1. Speed is normally a requirement in an on-line system.

2. Transactions cannot be batched and sorted as required for sequential updating.

As has been noted previously, in an on-line system an update takes place as the transaction is sent from the remote terminal and it is normally required that the update occur quite rapidly. Therefore, random updating is preferable to other types of updating because it is the fastest means of updating a file.

In addition, when transactions are entered by the terminal operator there is no way of knowing the sequence in which transactions will be entered by the terminal operator. For example, an operator may enter a transaction to update the last record in the file, then the first record, and then a record in the middle of the file. Since each transaction normally is entered one at a time, there is no possibility of batching and sorting the transactions to allow for sequential updating.

For these reasons, most master files which are to be updated by transactions entered into the system from remote terminals will be files which allow random updating.

Complex Updating

As has been noted, updating normally consists of the three functions add, delete, and change. Updating, however, can take many different forms depending upon the application. The previous examples have illustrated very basic updating problems. In actual practice, updates of master files will usually require more complex processing.

The complexity introduced into file updating can take several forms. First, the changes made to a master file may consist of many different types. For example, the code field within the transaction record may contain many different values to indicate that different fields within the master record are to be changed. The following are some of the changes which can take place in a master payroll record:

CODE	CODE
1 - Add a New Employee	5 - Change Pay Rate
2 - Change the Name	6 - Change Exemptions
3 - Change Classification	7 - Change Marital Status
4 - Employee Termination	8 - Change Credit Union

In addition, when transactions are entered by the terminal operator there is no way of knowing the sequence in which transactions will be entered. For example, an operator may enter a transaction to update the last record in the file, then the first record, and then a record in the middle of the file. Since each transaction normally is entered one at a time, there is no possibility of batching and sorting the transactions to allow for sequential updating.

There are countless combinations which may occur when changing fields in a master file and the analyst must be aware of the possibilities. In general, it can be said that any changes which are required in order to have the master file contain valid data for the system can be made, regardless of the complexity.

Updating More than One File

An update does not always consist of just one transaction file and one master file. For example, a single transaction file may consist of records which are to update several master files. This processing is illustrated in Figure 7-6.

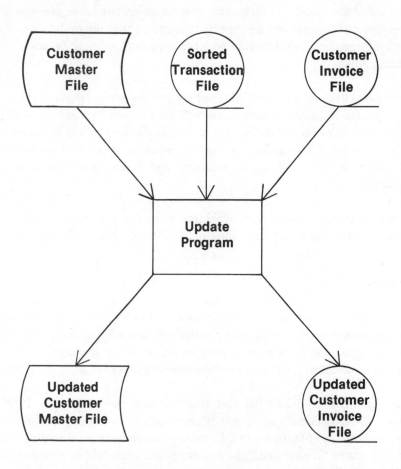

Figure 7-6 Example of Multiple File Updating

Note in the example in Figure 7-6 that the input to the update program consists of a sorted transaction file, the customer invoice file, and the customer master file. The customer invoice file contains all invoices and related charges to the customer. The customer master file contains all permanent data for the customer, such as name and address, credit status, and year-to-date payments. The transaction records will contain data which can update both the invoice file and the customer file. For example, when an invoice is paid by the customer, the invoice will be removed from the invoice file. In addition, since a payment was received from the customer, the year-to-date payments field must be updated in the customer master file. Thus, it can be seen that through one update transaction, more than one file will be changed and updated. This type of procesing is not unusual and should be considered by the analyst when designing the processing within the system.

Random and Sequential Updating

In the previous example both the customer invoice file and the customer master file were sequential files. In some cases, one file to be updated may be a sequential file and the other may be processed randomly. For example, a Personnel Master File could be organized sequentially, based upon employee number, and a Payroll Master File could be an indexed sequential file, organized on social security number. A Name and Address Change file is to update both the Personnel Master File and the Payroll Master File. The Name and Address file is a sequential file organized on employee number. Therefore, the Personnel Master file can be updated sequentially while the Payroll Master file must be updated randomly. This is illustrated in Figure 7-7.

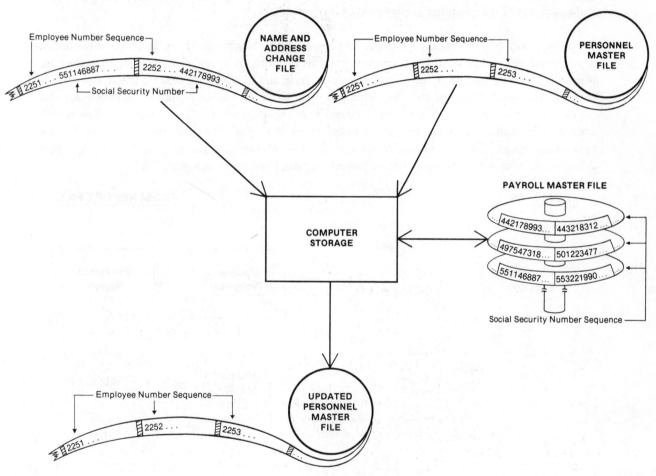

Figure 7-7 Example of Sequential and Random Updating

Note in the example in Figure 7-7 that the Name and Address Change file and the Personnel Master file are in employee number sequence and the Payroll Master file is in social security number sequence. Therefore, the Personnel Master File can be updated sequentially since it and the Name and Address Change file are in the same sequence. The Payroll Master file, however, cannot be sequentially updated because it would not match the Name and Address Change file sequence. Thus, the social security number which is found in the change record would be used as the key to randomly access the Payroll Master file. The update program could change any record in the Payroll Master file by randomly retrieving the record, changing it, and rewriting it back into the same location on the disk.

The previous examples of updating are merely a few of the many types of processing which may take place when files are to be updated. The analyst should be aware of these general types of processing when designing the update step within the system.

Reporting - Batch Systems

The next general step following the edit, sort, and update processing is the report producing step. Needless to say, reports generated from a computer system will vary greatly. Each different type of system requires its own types of reports and in some instances, even the same type of system, such as a payroll system, will have different reports depending upon the requirements of the company implementing the system.

In addition to the types of reports generated from a system there may be differences concerning the source of the information for the reports. In some systems, all that may be required is the extraction of information directly from the master file, that is, the master file is input to the print program and the print program extracts the information required for the report. In other systems, it may be necessary to create "report files," that is, files which contain information extracted from both master files and other files within the system. In this case, the report files are normally created in one of the update programs. The systems flowcharts for these two types of report processing are illustrated in Figure 7-8.

FROM MASTER FILE   **FROM REPORT FILE**

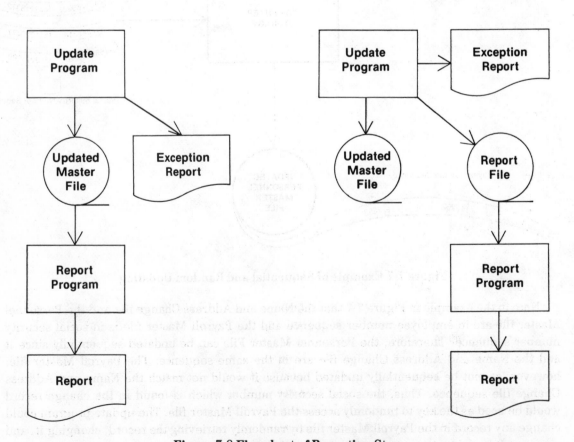

Figure 7-8 Flowchart of Reporting Step

Note from the flowcharts in Figure 7-8 that an Exception Report is prepared in the update program as the master file is being updated. Thus, the required reports can be prepared either from the master file after it has been updated or from a report file which is generated as a result of the processing which takes place in the update program. Typical of reports which will require a report file are those which contain some information which is not contained in the master file and other information which can only come from the data in the master file. For example, a payroll report may contain both the hours worked for an employee in one week and the year-to-date earnings of that employee. The hours worked for each employee is not contained in the employee master file but the year-to-date earnings are. Therefore, a report file must be prepared which contains the hours worked from another source and the year-to-date earnings from the master file.

Reporting - On-Line

In most batch systems the reports which are generated will be printed on a printer for disbursement throughout the company. When dealing with on-line systems, the analyst must determine the needs of the user and base his design of the report processing on these needs.

There are several methods for reporting in an on-line system. The first type is when specific inquiry functions are incorporated into the system, that is, an operator at a remote terminal may ask for specific information, such as the balance in a savings account. In this case, the reporting consists of making an inquiry to the master file and sending the answer back to the terminal operator. In general, this is the most common type of reporting which will be performed in an on-line system.

A second type of reporting will occur when the operator at a remote terminal transmits an update to a master file; for example, a transaction is sent which indicates that $100.00 has been added to a savings account. In this case, the master file must be changed to reflect the deposit. In addition to the update, however, it may be desirable to acknowledge to the terminal operator that the update has been made to the master file. Thus, after the deposit is made, the new balance in the account will be sent back to the terminal operator so that the customer will know his new balance. In this case, the analyst must design the system so that this acknowledgement will be made properly.

A third type of reporting which can occur within an on-line system is one in which the accumulated processing which takes place over a period of time is reported to the remote location. For example, a remote terminal may send numerous update transactions to the main computer over the period of a day. At the end of the day, it would be necessary to report these update transactions. This report would normally be prepared as a printed report. The printing may be done either by sending the information over telephone lines to the remote site which would have a printer or by printing the report at the main computer and then mailing or sending the printed report to the remote site.

In determining the type of reporting which is to occur in an on-line system, the analyst must take into consideration the type of hardware which is located at the remote site, the amount of time within which the information must be known, and other factors which are dictated by the particular system under design.

BACKUP CONSIDERATIONS

In virtually every data processing system which is developed, at some time errors may occur that are not detectable by the data editing that takes place in the edit and update programs. These errors normally are caused by improper programming, by hardware failures of the computer, or by operator errors which are made when processing the data in the system.

Regardless of the source of the error, the analyst must expect these errors and must plan some means of recovering from the errors. As has been noted, comprehensive checking of the input data by the computer programs within the system will alleviate many of the errors which may occur in the processing of the data, but for those errors which result in the deletion, loss, or destruction of the master or transaction files, other means must be used to provide for continued processing within the system. The two factors which must be considered are: 1) There can be no loss of data from either the master files or the transaction files; 2) There must be a means of restarting the processing within the system after the errors have been corrected.

Backup - Transaction Files

It is vital to ensure the integrity of the data within a system. Therefore, the analyst must design procedures which are taken when errors occur. In general, when there is an accidental loss or destruction of current transaction files, the files may be recreated by merely reprocessing the input data; thus allowing the files to be duplicated. This, of course, assumes that the input data is still available to be processed.

In most batch systems there is no difficulty but in some on-line systems, recreating the transaction data may present difficulties unless separate files of the transactions have been kept. One method which can be used is to create a transaction file whenever a message is received from a remote terminal. This is illustrated in Figure 7-9.

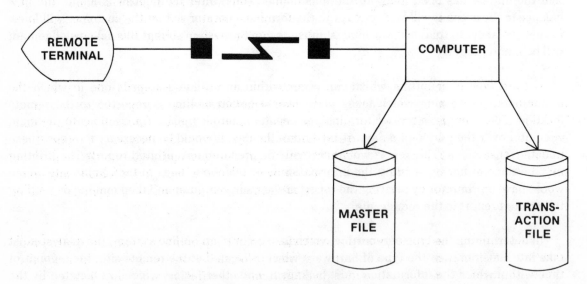

Figure 7-9 Example of Remote Processing

Note in Figure 7-9 that an update can be sent from a remote terminal to the main computer. As has been discussed previously, when an update is processed in an on-line environment, the update takes place as the transaction is received and processed at the main computer. In the example, a transaction file is created at the same time the master file is updated. Thus, if at a later time the master file is unusable, the transactions which took place would be recorded and accessible. If this transaction file was not created, the task of reprocessing the transactions against the master file would be much more difficult because they would all have to be resent from the terminal.

Backup - Master Files

Although transaction files are an important consideration, the portion of the system which is most susceptible to major errors are the master files which are used in the system. If a master file has been updated a number of times prior to becoming unusable, it may be extremely difficult to restore it to its current status unless special consideration is given to the problem by the systems analyst when designing the system processing.

The most common method used to ensure the security and integrity of a master file is the use of BACKUP files. A backup file is either an old master file which can be updated to a current status or a copy of the current master file. If it is a copy, this copy is not used in processing and will only be used when, for some reason, the contents of the master file are destroyed or otherwise become unusable. If an error occurs during the processing cycle and the master file is accidentally destroyed, the backup file can be used to recreate the master file. If there has been update processing to the current master file between the time the backup file is created and the time the error which destroys the master file occurs, this update processing would have to be rerun to bring the master file up to date.

It should be noted that when sequential file updating occurs, a new master file is created when the updating takes place. Therefore, if the old master file and the transactions which updated it are retained, there is effective backup since the update processing can be rerun to create the current master file.

Backup - Indexed Sequential and Direct Files

If the master file is organized as an indexed sequential or direct file, however, the backup problem is more involved because the file is updated without creating a new master file, that is, the updated records are rewritten back on the disk in the same location as before they were updated. Typically, the backup file is created after update processing, as illustrated in Figure 7-10.

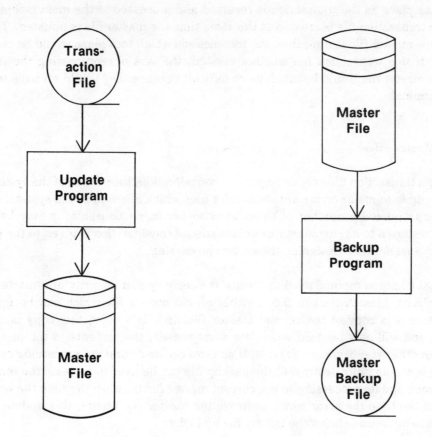

Figure 7-10 Example of Creating Backup File

Note in the example in Figure 7-10 that the update processing takes place first and then the backup file is created from the information in the updated master file. Most backup files are kept on magnetic tape because tape can be easily stored and is not as expensive as a disk pack. It should be noted that the file copying required to create a backup file may many times be accomplished by utility programs which are available with the operating system of the computer. If this is not possible, then the analyst must design the programs which will be used to create the backup file.

Frequency of Backup

In addition to planning for the creation of a backup file, the analyst must also determine how frequently a backup file is to be created, that is, is it necessary to create a backup file each update cycle or can several cycles be processed before a backup file is created. In making this determination, there are several factors which must be analyzed:

1. Is the system a batch system where all updates are processed at a given time or is it an on-line system where continual updating takes place?

2. How big is the master file to be copied? How much computer time is required for the copying process?

3. How difficult is it to bring an old master file up to current status?

When a master file is updated in a batch system, all of the update processing will normally take place at one time. Thus, at the conclusion of the update processing, a backup file can be created, if desired.

In an on-line system, updating will be a continuing process. Thus, the analyst must designate a time during the on-line processing when the system will not be processing updates so that a backup file can be created.

One of the primary considerations in determining how frequently a backup file will be created is the size of the file to be copied. For small files, the creation of backup files is no problem because little time will be required for the processing. With the creation of backup files for master files containing hundreds of thousands of records, however, the processing can be a considerable expense since much computer time will be required to copy the master file. Thus, the analyst must balance the time required to create the backup of the master file against the probability of error. There are no firm rules regarding what the final decision should be and it is up to the analyst to determine the best solution for the particular system he is designing.

The third consideration can be perhaps the most important in that if a master file cannot be recreated by any means other than a current backup file, then the analyst has little choice, regardless of the time required or the type of system in which the file is to be used, except to create a backup file at the conclusion of each update processing cycle. It should be noted, however, that with the proper planning and saving of transaction data which is used to update the master file, this situation will not occur very often and the other two considerations will carry more weight.

FILE RETENTION

Another area of concern for the analyst when designing a system is FILE RETENTION. File retention refers to the process of determining how long a file should be retained after processing before it is "scratched," that is, reused for the storing of different files.

The file rentention period can depend upon a number of factors, including:

1. The function of the file, that is, whether the file is a transaction file or a master file.

2. The file organization, that is, whether the file has sequential, indexed sequential, or direct organization.

3. Legal requirements.

File Retention - Transaction Files

In general, transaction files will normally be scratched after some predetermined period of time has elapsed so that the tape or disk on which they are stored can be freed to be used in other systems. A transaction file which is stored on tape or disk will normally be scratched when it can be assured that all the system processing has been completed accurately and that updated files are backed up in case they are destroyed. As has been noted previously, a backup master file is not always created each time an update is performed. If a current backup of a master file is not available, then the backup file which is available must be processed together with the transaction records which are required to bring the backup file up to current status. Therefore, transaction files must be retained long enough to ensure that this processing can take place if necessary and this period of time normally depends upon when master backup files are created.

File Retention - Master Files

Master files, as noted previously, can be organized as sequential files, indexed sequential files, or direct files. Whenever an update takes place on a sequential master file, a new master file is created. Since a new file is created, a point in time will be reached when the older files can be scratched. Many companies retain old sequential master files for three generations, that is, for three updating cycles. This concept is called the "grandfather," "father," "son" concept. The "grandfather file" represents the oldest master file retained, the "father file" represents the second oldest file, and the "son file" represents the current master file. By retaining these three files and the transactions which have been used to update the files, there is adequate backup since the current file can be recreated by rerunning the processing which changed the "father file" to the "son file."

When a master file is stored on magnetic disk as an indexed sequential file or a direct file, however, this concept changes because a new master file is not created when update processing takes place. Instead, after being updated, records are rewritten back onto the disk in the same location from which they were retrieved. Therefore, the master file stored on the disk in the indexed sequential or direct organization should not be scratched after three update cycles since it always contains the most up-to-date information, unlike the "grandfather file" when dealing with sequential files. Thus, the retention period on an indexed sequential or direct file will normally be such that the file cannot be scratched.

The method used for indicating retention periods is normally the use of file labels which are actually written as the first record or records of the file when the file is to be recorded on tape or in a special label area on direct-access devices. This label, in addition to indicating the name of the file and other information concerning the file, contains the date on which the file can be scratched. The example in Figure 7-11 illustrates the contents of a part of a tape header label which contains the expiration date of the file.

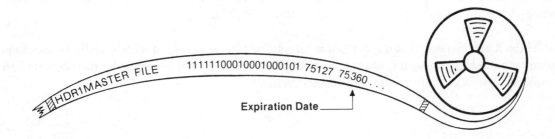

Figure 7-11 Example of Header Label with Expiration Date

Note in the portion of the tape header label illustrated in Figure 7-11 that the expiration date is specified as "75360." This value is known as a Julian Date; the first two digits are the year and the last three digits are the day within the year. Thus, the tape can be scratched on the 360th day of 1975, which is December 26, 1975. The information stored in the label is recorded as magnetic inpulses in the same manner that other data is stored on tape.

If an attempt is made to **write** over the file prior to the expiration date specified in the label, the computer operator is normally notified that the file has not reached its expiration date and that the tape should not be used for another output file. A typical message (IBM Disk Operating System) notifying the operator that a tape file is mounted and ready for processing with an unexpired date is illustrated in Figure 7-12.

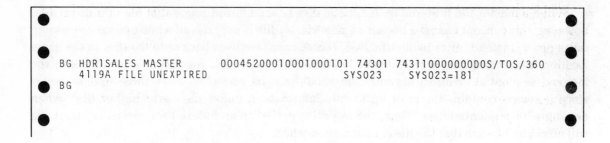

```
BG  HDR1SALES MASTER            00045200010001000101 74301 743110000000DOS/TOS/360
    4119A FILE UNEXPIRED                        SYS023     SYS023=181
BG
```

Figure 7-12 Example of Operator Message

Note in the example above that the operator is notified that the tape has an unexpired date in the label. The operator has several decisions that can be made relative to processing the job with the tape that is mounted: 1) The operator can cancel the job so that the tape file is not destroyed; 2) The operator can dismount the tape with the unexpired date and mount a tape which can be written upon; 3) The operator can type a response such as "DELETE" in which case the new data will be written on the tape even though there is an unexpired date in the tape label. This action would destroy the contents of the old files.

Thus, it can be seen that the computer operator must be provided with specific instructions relative to the processing of files that are unexpired. An incorrect response by the operator can readily destroy a valuable master or transaction file!

Legal Requirements

In some systems external requirements may exist which dictate how long a file should be retained. The most common occurrence is a legal requirement which indicates that certain data must be retained for a certain length of time. For example, in an accounts receivable application there may be a requirement that all "receipts" be kept for a certain length of time by a company. If the system is designed so that all records of payments received are kept on a master file, then this master file would normally be retained until the legal time period had elapsed. Although this may not be a common restriction on scratching files, it is one which the analyst must consider when specifying retention periods.

Summary - File Retention

The file retention periods which the analyst determines for each of the files within the system can be quite critical since if files are scratched too soon there may not be adequate backup to recover from unforeseen errors which may occur within the system. Therefore, considerable thought should be placed in determining file retention periods for both the transaction and master files found in a system.

RESTART CONSIDERATIONS

When an error occurs within a system, for example a hardware failure which causes the master file on a disk to be destroyed, the first event which must take place is that the failure must be corrected. This is true in the case of a hardware failure and in the case of a software (programming) error. Once the correction is made, however, the system must be restarted from the point where the error occurred. Part of the analyst's task in designing the processing within the system is specifying points within the processing where restarts can occur, depending upon where the error occurred.

In determining restart points the analyst must consider two important elements:

1. Before a program can be rerun all data and files must be exactly the same as when the first run was accomplished, that is, the analyst cannot allow any program to process data which has been altered. Thus, before any processing can be restarted, master files which have been altered must be restored to their original status through the use of backup files. The analyst must inform the operations staff what files must be restored to their original status, depending upon where the restart must take place within the system.

2. Depending upon where the error occurs, there may be the possibility that the program with the error can be temporarily bypassed and other programs within the system can be processed while the corrections are made to the program in error. For example, if a report program did not function properly, it may be bypassed and other report programs in the system executed while the print program in error is corrected. It should be noted that this technique is not usually available for programs which must be executed before any other programs can be run, such as edit programs and update programs. If a file is not properly updated, the results of subsequent programs will be invalid, but there may be some programs which can be bypassed.

Checkpoints

Another type of restart procedure which is sometimes useful, especially on very long programs which may take an hour or more to process the data, is the use of Checkpoints. A checkpoint is a point within the processing of a program where all of the important indicators which designate the current processing status of a program are saved. Such values as which records are being processed, the contents of computer storage, the contents of registers in the computer, and all other information necessary to restart the program at the point where it is checkpointed are saved, usually on a magnetic tape file. The indication that a checkpoint is to be taken is given by special instructions in the program itself. After the checkpoint is taken, the normal processing of the program resumes. After another period of time, the instructions within the program can again be executed and another checkpoint will be taken.

If an error in processing occurs, as a result of either program or hardware failure, the program will be stopped and the corrections will be made. When processing can resume, the program need not be completely rerun. Instead, the program can be restarted at the last checkpoint taken before the error occurred. In order to restart, the computer operator must enter the proper instructions to the operating system. With the proper commands, the restart will take place automatically.

Thus, if a program required three hours for processing and a checkpoint was taken every 15 minutes, the most time that would be wasted by rerunning data which had already been processed would be 15 minutes; whereas if checkpoints were not taken and the program failed after processing data for 2-1/2 hours, that entire amount of processing would have to be rerun.

As noted, checkpoints are not normally taken for programs which run a relatively short time and, as a general rule, the program must run longer than 30 minutes before it becomes economically feasible to take periodic checkpoints. If the program runs less than 30 minutes, it is usually more economical to rerun the entire program rather than take the time required to process a checkpoint.

ROLE OF THE OPERATING SYSTEM IN FILE PROCESSING

Most systems which are designed today will be processed on a computer which is under the control of an "operating system." An operating system is a series of programs which are normally supplied by the manufacturer of the computer and which essentially control the operation of the computer. The function of the operating system includes controlling the physical process of transferring data from external devices into main storage, allocating main storage for program use, and processing job control statements which are used to cause programs to execute and to define files on the various input and output devices.

Additionally, an operating system will allow special types of processing to occur, if requested. Two important areas which directly affect both the procedures followed within the system and the efficiency of the system are **Multiprogramming** and **Spooling**.

Multiprogramming

Multiprogramming refers to the process of executing more than one program in the same computer during the same period of time, that is, computer storage is partitioned and two or more programs are placed in computer storage at the same time. When one program is not executing because it is waiting for an input/output operation or for some other reason, portions of the other programs can be executed. Depending upon the operating system being used, up to sixteen different programs can be in computer storage at the same time to be executed alternately. Multiprogramming is extremely important when "on-line" programs are being processed because, generally, on-line programs do not require all of the computer resources in order to be executed. Therefore, the on-line program and other programs can be executing during the same period of time.

One of the difficulties when processing programs in a multiprogramming environment is the availability of peripheral devices such as card readers and printers. For example, many computer systems do not have more than one or two printers attached. Therefore, if there are five or six programs requiring a printer being executed in a multiprogramming environment, the output from these programs must be placed on a tape or direct-access device to be read and printed at a later time when a printer is available. The placing of output from a program into a temporary hold area on disk or tape from which it is retrieved and later written on a unit-record device such as the printer is called **Spooling**.

Spooling is another feature which is allowed by many operating systems. Its primary advantage, besides the fact that it will hold data for later output on an unavailable device, is that it is normally faster than writing directly on a printer or reading directly from a card reader. The reason is that the printer is one of the slower devices which is a part of a computer and when the processing program is slowed down to wait for the printer, a great deal of time can be wasted. It is more economical in terms of computer usage to place printer output into a temporary file by writing this data on a direct-access device or magnetic tape device when the program is actually processing.

After the program has completed processing, a small program in another partition can be used to merely transfer the data from the temporary storage area to the printer while another applications program is processing more data. Thus, spooling is used both because a device is unavailable and to save a great deal of time in a multiprogramming environment. It should be noted also that card input and output as well as printer output can be spooled. The input cards are read from the card reader and placed in a temporary file to be read by the program which is going to use the data. The card output is processed the same as printer output.

Both spooling and multiprogramming are a function of the operating system and although the user may need to specify that he wants it to occur, the actual handling of the processing which allows these features is done by the programs which constitute the control portion of the operating system. Thus, the systems analyst need not be concerned with designing programs which actually perform the spooling and multiprogramming functions. He must be aware, however, that programs in the system he is designing are going to be spooled or executed in multiple partitions and must design the processing of the system to conform to any restrictions which may be inherent in this type of processing.

Utility Programs

It has been noted previously that a utility sort program which is normally a part of the operating system may be used during the processing of the system to sort data so that it is in the proper sequence. There are, in addition to the sort program, numerous other utility programs which are normally a part of an operating system which may be of aid to the systems analyst and which can be used without requiring a special program to be written. For the most part, these utility programs are used for transferring data on a file to file basis. For example, files which are stored on disk can be transferred to magnetic tape through the use of a file-to-file utility program. Thus, utility programs can be quite useful in creating backup files. Printed reports can also be prepared using these utility programs, although the formatting of the report is somewhat limited in most cases.

SYSTEM PLANNING

When designing the system, the analyst must incorporate many of the concepts discussed previously into the actual steps which must occur within the system, that is, he must determine what data must be edited and in what sequence within the overall processing the editing will take place. Similarly, he must decide which files are to be updated, what data is to update them, and how the update processing must take place. The programs to generate the required reports must be determined and placed within the operating sequence of the system. The primary tool used to accomplish this sequencing of programs into a functional system is the systems flowchart.

An example of a systems flowchart for an accounts payable system is illustrated in Figure 7-13.

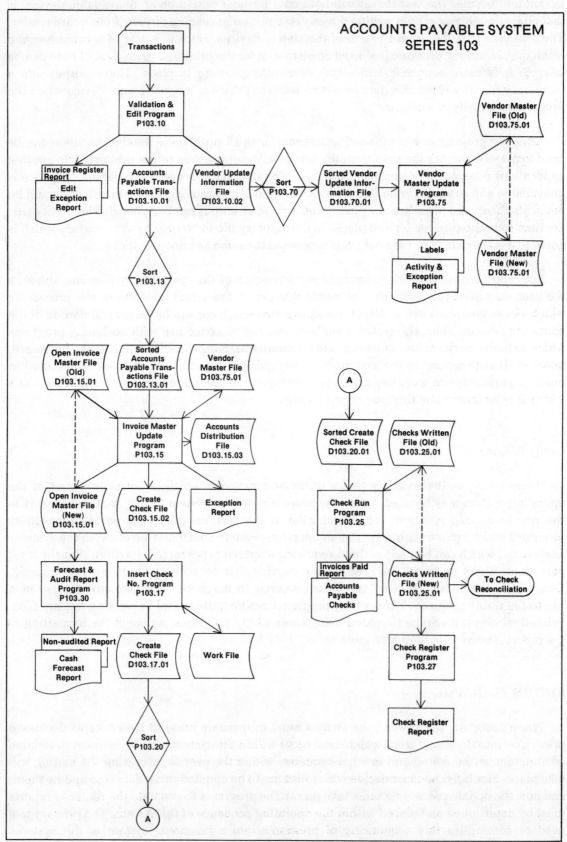

Figure 7-13 Example of Systems Flowchart

Note from the flowchart in Figure 7-13 that all the processing to take place within the Accounts Payable system is illustrated in the form of a systems flowchart. The processing follows the typical business applications sequence, that is, edit, sort, update, and report. It will be noted that there are several file update programs in the system, several sorts, and a number of reports are created, but the basic sequence is still followed.

Program and File Identification

Another feature of the flowchart illustrated in Figure 7-13 is that all of the files and programs are identified by a unique value. For example, the Validation and Edit program is given the identification "P103.10" and the Invoice Master Update Program is given the identification "P103.15." These values give each program a unique identification which can be used throughout the documentation and operation of the system. It will be noted further that the files created from the Validation and Edit program are given file numbers. The Accounts Payable Transaction File is given the indentification "D103.10.01" and the Vendor Update Information file is given the identifier "D103.10.02."

In most systems, as in this accounts payable system, there is a reason the identifiers are chosen. The prefix "P" or "D" appears on every identifier. The "P" indicates a program and a "D" indicates a disk file. If a tape file were to be used, it would be prefixed with a "T." As will be noted from the headings at the top of the page, the Accounts Payable system is designated Series 103. Therefore, all identifiers contain the value 103 for the first three numeric digits. This coding scheme could be used where there were a number of systems performing the accounting functions within a company. For example, the Accounts Receivable System could be series 101, the General Ledger System could be 102, the Accounts Payable System series 103, etc.

Following the series identification is a program identification. For the Edit program, the identification is 10, for the invoice master update program it is 15, and so on. The files are given the same program identification as the program in which they are created. In addition, they are given an unique suffix which uniquely identifies them when there is more than one file created in a program. For example, the Accounts Payable Transaction file is named "D103.10.01" and the Vendor Update Information file is named "D103.10.02."

In any good systems design effort, the files and programs must be given identifiers. The method illustrated in Figure 7-13 is merely one method which has been used successfully in a business application.

Program Specifications

In the flowchart in Figure 7-13 it is obvious that the analyst has made a determination of the flow of the data throughout the system. In addition, however, it is mandatory that the analyst, when designing this flow, know every function which must be performed within the system and which programs within the system must perform these functions. Therefore, a very important part of designing the systems flow involves the analysis of where within the system all of the processing required for the system is to take place.

There are no specific techniques which can be used to ensure that all processing will take place as required. The skill and experience of the analyst must be used to determine both that all required processing is included and that the processing takes place in a logical sequence, ensuring that the design will work once implemented on the computer.

START-UP PROCESSING

The processing which has been illustrated and discussed previously in this chapter is the processing which will take place when the system is in production, that is, when the system has been totally implemented and is producing reports, etc. in the manner in which it was designed. In many systems, however, especially those which are being changed from a manual system to a computer system, there is a certain amount of "start-up" processing which must take place so that the system can run on its day-to-day schedule. The primary requirement is that the master files which are to be used in the system be built from the currently existing data which is being used in the manual system. It is quite important that the computer system begin operation with the current information since all subsequent processing within the system will be performed on the computer. For example, in a payroll system, the year-to-date earnings must be entered properly into the computer master file so payroll checks can be printed correctly.

It should be noted that this processing is normally a "one-shot" run, that is, the only time the start-up programs need to be executed is when the files are created prior to beginning actual operation of the system. Therefore, if it is possible, it is maybe more efficient in terms of programmer time to have the programs in the start-up portion of the system be the same ones which will process the data in the normal operation of the system. Thus, some of the edit and update programs which are used to add records to master files after they have been created may be able to be used, with slight modification, to initially load the master files.

If it is not possible to do the start-up processing with programs which are a part of the normal processing then it is necessary to design programs for the specific purpose of initially loading master files. It should be noted that the analyst must determine if it is more economical to include the routine for the start-up processing in the regular processing program or to write simple start-up programs. When the start-up processing is included in the regular processing programs these programs become more complicated and it may be that less time would be required for the programming effort to write small start-up programs rather than include extra routines in the regular programs.

Regardless of the decision reached concerning where the start-up processing will be included within the programs of the system, it is critical that the analyst consider these programs in the processing design stage since they must be available to load the master files prior to the actual implementation of the system.

DOCUMENT AND INFORMATION FLOW

The processing which has been discussed previously in this chapter has concerned the activities which take place in the actual computer processing of the data within the system. In addition, the analyst must, when designing the complete system, concern himself with the processing which must be accomplished in order to prepare data for input to the computer system and disburse the information generated from the computer system after processing has been completed. The analyst must normally be concerned with the following tasks:

1. Gathering the source documents into one area for further processing.

2. Specifying control functions which must be performed on the source documents.

3. Converting data from source documents to computer input and the disposition of source documents after preparing computer input.

4. Making computer input available for processing.

5. Controlling special forms used for computer output reports.

6. Preparing system output for distribution.

7. Distributing output reports to users.

In addition to the above areas, an on-line system requires consideration of the following:

1. Designing the network, including the number of terminals and their placement at the remote sites.

2. Determining the methods of reporting for the system.

Gathering Source Documents

In most systems, source documents do not come from one single source located at one location within a company. For example, in a payroll system, there may be time cards located throughout the company and documents for updating the employee master file may come from the personnel department or the payroll department. Therefore, the analyst must determine the methods to be used for bringing all similar source documents together, that is, he must determine how the time cards are to be gathered, how the file update forms are to be handled, and so on. This is a very important concern because the most sophisticated data processing system ever designed will not function properly if data is not made available to it properly.

In order to determine the processing which must occur for the gathering of source documents, the analyst must consider three primary factors:

1. The timing required to allow the source documents to be processed on schedule.

2. The action to be taken with the source documents once they arrive at the location to be processed.

3. The procedure to be followed if the source documents arrive with invalid or incomplete data.

Since source documents are normally located at different locations throughout the company, the analyst must determine when these source documents must be shipped, mailed, or otherwise transported from the location where they are prepared to the location where they are to be further processed. For example, it may be necessary to establish a 2:00 p.m. shipping time for the personnel department and a 4:30 p.m. deadline for picking up time cards to enable the source documents to arrive at the collection point by 5:00 p.m. for further processing. Obviously, such elements as distance, availability of a means for transporting the documents and other such considerations will determine this schedule.

Once the source documents arrive at the collection point, the analyst must determine what next happens to them. Some of this decision will be based upon the nature of the collection point and some will hinge on the type of processing which may be prescribed for the documents. For example, some documents may go directly to the data entry department for conversion into a computer-readable form. Others may have to go to a control station where control totals, batch totals, etc. must be prepared prior to being sent to the data entry department.

One problem which the analyst must always consider is the arrival of incomplete source documents, documents with errors, and the late arrival of the documents. He must specify what procedures should be followed if, for example, a payment check is received from a customer but the accompanying bill is not returned; or what takes place if three copies of a source document are required but only two copies are received. The actions taken will, of course, depend upon the system and the requirements of the system but the analyst must consider these elements when determining the processing of the source documents.

Control Functions

The analyst must also be concerned with what happens to the source documents once they are received at the collection location. In general, these concerns must center around three primary areas:

1. Control of the documents themselves as they are received and processed.

2. Control of the data on the source documents.

3. The use of the source documents in other systems or for other purposes.

In some systems, the source documents, besides being used for generating information for computer processing, are important documents in their own right. For example, in an accounts receivable or billing system, a portion of the source documents may consist of checks received from customers. In this case, strict controls must be instituted by the analyst so that all checks are accounted for and processed properly.

In addition to controlling the documents themselves, controls are often necessary for the data which is found on the documents. This includes such processing as organizing the data into batches, totaling the number of documents, and totaling values contained on the documents. A detailed discussion of systems controls is contained in Chapter 8.

The analyst must also allow for the use of the source documents in more than one system or for purposes other than as input to a computer system. For example, it may be necessary to prepare deposit slips for checks received in a billing system as well as using the checks to prepare computer input records. Thus, the analyst must determine the sequence for handling these checks and any other documents which must be similarly processed.

Conversion of Data for Computer Processing

As has been noted, normally the most important function of a source document is its use as a document from which computer input data can be prepared. The analyst must determine the processing which will be followed to convert the data on the source documents to computer input data. The elements of concern include the following:

1. Scheduling of source documents from the control point to the data entry department.

2. Determining the time required to prepare the computer input.

3. Exception procedures to be followed.

4. Disposition of source documents once the input has been prepared.

Once the processing has been completed at the collection point, the source documents must be transmitted to the data entry stations for preparation. The analyst must coordinate this action so that the data entry department will be expecting the source documents and will be ready to begin preparing the computer input for them.

In addition, the analyst must be concerned with the length of time required to prepare the computer input. This time will be determined by several factors—the volume of input data to be prepared; the difficulty in preparing the input data, that is, the expected error rate of the data entry operations; the requirements for verification of the input data, that is, whether it is "sight-checked" or machine-verified; and how many different types of source documents are to be used. The analyst must also determine the procedure to be used for transporting the input data to the computer staging area.

As in most steps within a data processing system, there must be provision made for problems which arise. When dealing with the input data, problems arise when the data is not ready on schedule, when data cannot be properly prepared from the information which is on the source documents, and when corrections must be made to the input data based upon the results of edit processing on the computer. There must be detailed instructions developed to handle these and other exception conditions which may arise in the processing of input data.

After the computer input data has been prepared, the source documents must be disposed of. In some systems, this means merely storing them for a period of time depending upon the backup requirements of the system. In other systems, as has been noted, the source documents will be used for other purposes within the company and arrangements must be made for them to be handled properly.

Making Computer Input Available

Once the input data has been prepared, the analyst must be concerned with getting the data into the computer system. A number of factors are involved in this process, including:

1. The scheduling of the data into the computer operations area.

2. Procedures to follow if the input data is incorrect or incomplete.

3. The disposition of the input data after processing.

4. The retention period of the input data.

As was noted previously, the analyst must determine the time when the input data is to arrive in the staging area for input to the computer system. In addition, he must develop the procedures which ensure that other necessary information, such as balancing sheets, etc. also arrive at the proper time and place.

Decisions concerning incomplete or incorrect input data, such as whether the system can be processed even when all input data is not available, must be made so that the operations department is directed properly in its processing of the data.

After the input data has been read into the computer system, the analyst must determine how it is to be stored or otherwise processed. In most systems, it will be required that the input data be retained for a given length of time so that in case of difficulties in the computer processing the system can be restarted using the original input data. As has been noted when discussing backup, a variety of requirements may be presented by a system and the analyst must determine the requirements of the system being designed.

Controlling Special Output Forms

In Chapter 4 it was mentioned that special forms, such as payroll checks, may require special handling in order to enforce security requirements and to ensure that unauthorized use of forms and information within a system will not occur. Although consideration must be given to these requirements in the design of the output for the system, it is in the design of the processing that formal procedures must be developed. The following questions must be answered concerning the handling of special forms:

1. Where are the forms to be stored and what kinds of controls are required for them?

2. Under what circumstances will the forms be allowed to leave their storage area and how are they to be returned?

3. What elements determine authorization for having access to the forms?

4. What is the procedure if special output forms are found missing?

Preparing System Output for Distribution

As noted in Chapter 4, once the output is produced from the computer processing, the information must be disseminated to the user. Thus, when designing the processing which is to take place within the system, the following elements must be determined and specified:

1. Who gets copies of the reports generated?

2. In what format must the reports be presented? Is decollating, bursting, separating, or binding required?

3. Is any special handling of the output, such as making microfilm copies, required?

It is important to determine who receives copies of the reports so that the correct number of copies can be prepared. It is also necessary to do this so that instructions can be prepared for the operations staff in distributing the reports.

As noted previously, it is often required that reports be decollated, bursted, separated in some way, or bound in some manner prior to distribution. Again, these requirements must be specified during the detailed design of the processing so that instructions can be prepared for the operations staff.

In some cases, not only must the reports be prepared for distribution, but other special processing must occur prior to distribution. For example, it may be necessary to make microfilm copies of the reports prior to distributing them or there may be requirements that certain values must be checked prior to distribution to ensure the accuracy of the reports. Again, all these factors must be considered during the design of the system processing.

Distributing Output Reports

Once reports have been prepared for distribution, the final step is to determine the method for distribution. The following elements should be considered:

1. When do reports have to reach the user?

2. What methods are to be used for distributing the reports?

3. If reports are kept for historical purposes, when can they be destroyed?

An obvious consideration when determining the distribution of the output is the time factor, that is when do these reports have to reach the user. In some systems, the time requirement for information and reports will, to some extent, determine the entire processing schedule. Thus, the analyst must know when reports are required and must determine a method for ensuring that the reports reach their destination on time.

Methods for distribution will vary greatly depending upon the time requirements, the means available within a company, and the location of the persons or departments receiving the reports. The methods can vary from hand-carrying reports to mailing reports across the country.

In many systems it is required that copies of reports be kept for historical purposes, that is, copies are required for reference even though they are not distributed to any single individual or department. When this is a requirement, the reports will not normally be kept indefinitely. Therefore, a determination must be made concerning the length of time a report must be kept and when it can be destroyed. Typically, a report may be kept until the next update cycle when a new report is produced, but in some systems reports may be kept for months or even years.

Network Design - On-Line Systems

The elements of systems design discussed in the previous sections apply to all data processing systems. When dealing with on-line systems, however, there may be additional considerations. The most obvious is the design of the network itself. The analyst must, among others, make the following determinations:

1. The types of terminals.

2. The number of terminals.

3. The placement of terminals at the remote sites.

4. The methods of communicating with the terminals.

5. The methods of data input from the terminals.

Methods of Reporting - On-Line Systems

As noted previously, the primary method of reporting information with a batch system is the printed report. With an on-line system, a number of alternatives are available, including:

1. CRT Display

2. ''Hard copy'' on a terminal

3. Remote printers

4. Mailing printed reports from the computer center to the remote site.

The first three alternatives listed above will depend primarily upon the hardware selected at the remote site and will not involve any special handling problems. If reports are to be prepared at the computer site and mailed to the remote site, however, the analyst must determine the procedures which are to be followed both at the computer site and the remote location to ensure that the reports reach the proper recipients. The requirements of reports prepared and distributed in a batch system, such as bursting, decollating, etc., must all be met when reports are distributed as a result of on-line processing.

DOCUMENTING THE SYSTEM FLOW

As has been noted, the primary method for indicating the processing which occurs within a computer system is the systems flowchart. When indicating the document flow within a system, a management-type flowchart is normally developed. This type of flowchart indicates the flow of data and documents from one department to another department throughout the company as they are processed within the system. The flow diagram in Figure 7-14 illustrates a simplified flow diagram of the processing occurring in a retail store when an item is purchased from the catalog.

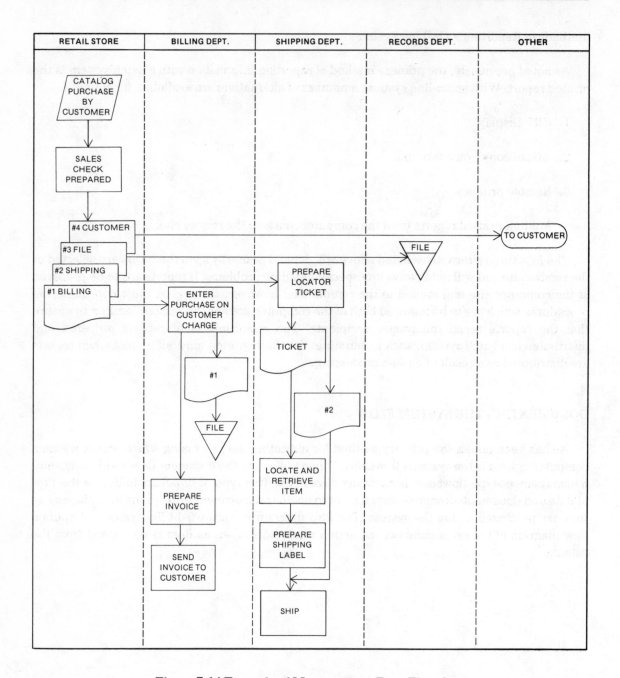

Figure 7-14 Example of Management-Type Flowchart

Note from the example in Figure 7-14 that the flowchart is segmented into various departments which are involved in the processing of the information. For the purchase of a catalog item, four departments are involved—the Retail Store, the Billing Department, the Shipping Department, and the Order Filing Department. When a catalog order is made, a four-part sales slip is prepared. The fourth copy goes to the customer when the order is made. The third copy goes to the order filing department for filing in a historical file.

The second copy goes to the shipping department. There, a locator ticket is prepared and the item ordered is located and retrieved from stock. A shipping label is made and the item, together with copy 2 of the sales slip, is shipped to the customer.

The first copy goes to the billing department. The purchase is entered on the account of the customer making the purchase and the sales slip is filed. An invoice is then prepared and sent to the customer.

Although the ''system'' illustrated in Figure 7-14 is very much simplified, it nevertheless illustrates the type of flowchart which can be prepared to document the flow of documents within a typical system. This type of flowchart is important when presenting the system to management for their approval and also for preparing the documentation and instructions necessary to properly process the documents within the system.

SUMMARY

The determinination of the processing which is to take place within a system entails many considerations, both in terms of the actual computer processing and in the data flow which must take place in order for the system to effectively operate. By determining this processing, the analyst has tied together the entire input, output, and file design into a workable unit which, with management approval, is ready for the first steps in implementation.

CASE STUDY - JAMES TOOL COMPANY

After designing the files to be used for the payroll system, the analysts Mard and Coswell were faced with determining the processing which is to take place in the system. In analyzing the system, they noted that it had already been determined that on-line capabilities were not required and the processing would take place in a batch mode. In addition, the processing required followed the typical flow found in most business systems, that is, edit, sort, update, and report. The systems flowchart for the time card processing portion of the payroll system is illustrated on the following pages.

SYSTEM DOCUMENTATION		
NAME OF SYSTEM Payroll System	**DATE** January 17	**PAGE** 1 **OF** 2
ANALYSTS D. Mard H. Coswell	**PURPOSE OF DOCUMENTATION** Systems Flowchart	

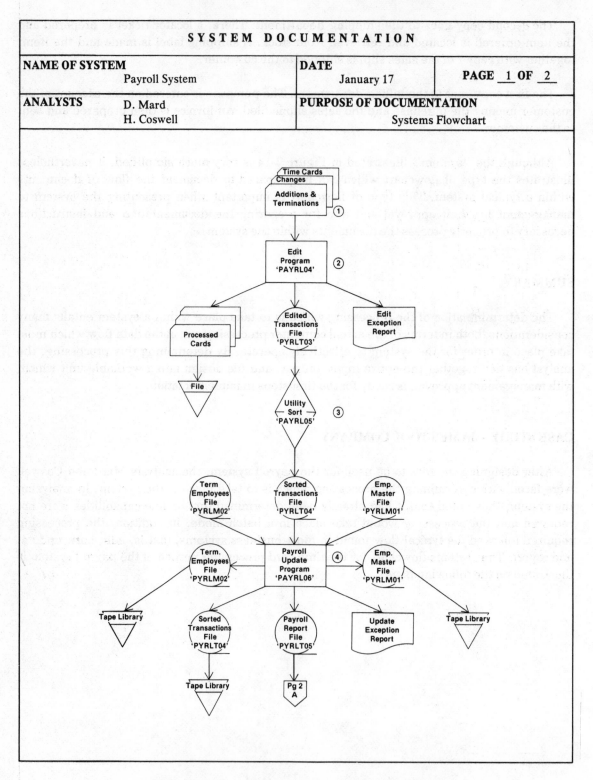

Figure 7-15 Payroll System Flowchart - Part 1 of 2

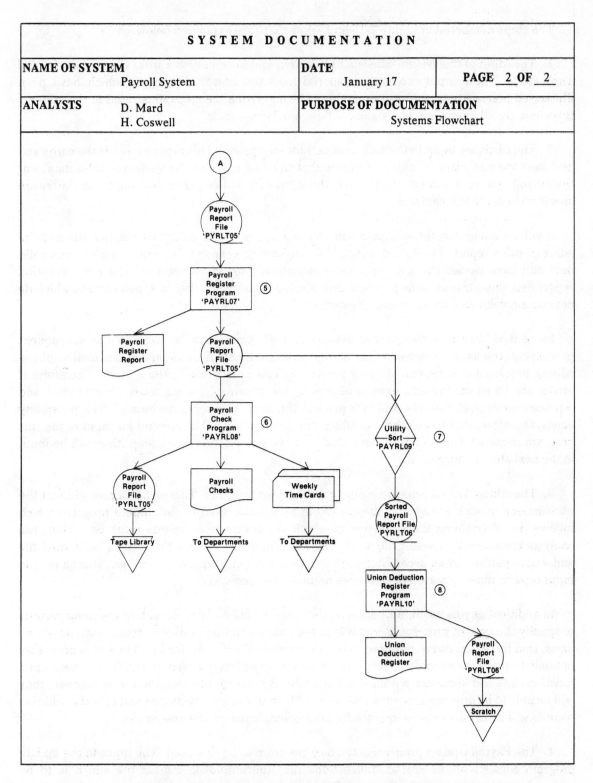

SYSTEM DOCUMENTATION		
NAME OF SYSTEM Payroll System	**DATE** January 17	**PAGE** __2__ **OF** __2__
ANALYSTS D. Mard H. Coswell	**PURPOSE OF DOCUMENTATION** Systems Flowchart	

Figure 7-16 Payroll System Flowchart - Part 2 of 2

The steps numbered in Figure 7-15 and Figure 7-16 are explained below.

1. The input to the system consists of Addition and Termination cards, Change cards, and Time cards. These input cards are prepared from the source documents which have been illustrated previously for the system. It should be noted that the steps illustrated in the systems flowchart are all represented by standard flowcharting symbols.

2. The cards are input to the Edit program for the system. This program reads the cards and performs the necessary checking to ensure that the data entering the system is valid data. For the payroll system, a number of checks on the input data will be performed and these checks are specified in detail in Chapter 8.

It will be noted that the edit program has two outputs—the edited transaction file and the edit exception report. The edited transaction file contains those transactions which have valid data and have passed the editing criteria established for input records. The edit exception report lists those transactions in which errors have been found, that is, those records which do not contain valid data for entry into the system.

Since it is critical in the payroll system that all input data be used in the subsequent processing, the analysts decided that error correction must be made at the completion of the editing process and before any further processing takes place within the system. Therefore, if errors are listed on the edit exception report, the records in error must be corrected and reprocessed through the edit program prior to the next step being performed. The possibility exists, therefore, that more than one edited transaction file will be created for input to the sort program in step 3. Once all of the transactions have passed the editing step, they will be input to the next step in the system.

3. The edited transactions are input to a utility sort program. This sort program will sort the transactions in such a way that they are able to be processed in the update program which follows. In determining the sequence in which the transaction records must be sorted, the analysts knew that the master file to be updated (Employee Master File) was a sequential file and was organized in an ascending sequence based upon Employee Number. Therefore, the input records must be sorted in employee number sequence.

In addition, as was noted in Chapter 5, there are function codes in each of the input records to specify the type of processing which is to take place; and an addition record consists of two parts, that is, two cards are required to add an employee to the master file. Thus, it is necessary not only to place the records in employee number sequence but also to sort them based upon function code and sequence within function code. By sorting the records in this manner, they will be able to be processed against the master file in the proper sequence and also the addition records will be in the correct sequence for adding employees to the master file.

4. The Payroll Update program is the next program to be executed. The input to the update program consists of the sorted transactions file, the employee master file which is to be updated, and the terminated employees file which contains the records of those employees who no longer are employed by the company. The output from the update program is an updated employee master file, an updated terminated employees file, an update exception report, and the payroll report file.

The processing which will be accomplished by the update program is as follows:

a. The Employee Master File will be updated to contain any new employees which are added to the company payroll, terminated employees will be removed from the file, changes will be made to those records which are indicated by the transaction records, and the year-to-date fields in the master record will be changed for each employee which earns a weekly pay.

b. The records for those employees which are terminated will be transferred from the employee master file to the Terminated Employees file for use in year-end processing.

c. The Payroll Report File will be created. This file will contain all of the information required to produce the reports which are needed from the system. The source of data for the payroll report file will be information contained on both the Employee Master File and the Sorted Transaction File.

d. The Update Exception Report will be created as the transactions are processed against the Employee Master File. This report will contain a listing of those records which were changed in the master file and what the changes were, any transaction records which could not be processed because of errors such as transactions for master records which did not exist and transactions which attempted to add master records already on the master file, and all employees which appear on the Employee Master File but for whom there was no time card.

The processing in the Payroll Update Program must be carried out accurately before any further processing can take place within the system because after the update program, the payroll register, payroll checks and other reports are prepared which will be distributed throughout the company. Therefore, any errors which appear on the update exception report, such as transaction records for non-existing master records or employees for which there are no time cards, must be corrected or justified prior to creating the reports from the Report File. Thus, the analysts decided that if corrections must be made to the transaction data as a result of the update processing, these corrections must be processed through the edit program; merged with the sorted transactions file by using the sort program; and then the payroll update program must be rerun using the new transaction data. In this way, the information stored in the Payroll Report File will be entirely accurate and the reports to be prepared from the system will be correct.

5. After the update processing has been certified as correct, the preparation of the reports required from the system can begin. The first report to be prepared is the Payroll Register Report. As will be recalled from the design of the output in Chapter 4, the Payroll Register consists of the weekly earnings and deductions of each employee within the company. It should be noted that the payroll register is to be printed in employee identification number sequence within department. Since the employee number consists of both the department number and the employee identification number, the records on the Payroll Report File will be in this sequence and will not require more sorting prior to being used as input to create the payroll register report.

6. After the Payroll Register is created, the payroll checks can be printed. The Payroll Check program will read the Payroll Report File as input and will both print the payroll checks and create the weekly time cards which are distributed to the departments in the company.

7. The Union Deduction Register must also be prepared from the Payroll Report File. This report, however, is not to be in employee number sequence but rather it is to be separated by each union and within each union the employees must be listed in alphabetical sequence. Therefore, the information on the Payroll Report File must be resorted into an ascending sequence based upon Union Code and by employee name within each union. A utility sort program will be used to perform this required sorting.

8. After the payroll report file is sorted, it will be input to the program which creates the Union Deduction Register. After the program is completed, the Register will be distributed to the proper unions and the input tape can become scratch, since it can be recreated with relative ease by resorting the payroll report file again.

Case Study - Exception Report Formats

It will be noted from the systems flowchart that several reports, notably the exception reports, are to be produced from the system but their formats were not determined when the output for the system was designed. The reason that they were not designed as a part of the system output is that the exact processing within the system had not yet been determined and, therefore, it was not possible to determine where in the system exception reports would be required and what information should be on them.

When the flow within the system and the programs in the system have been designed, however, it is then possible to determine what information should be on these error reports. Thus, the design of the error exception reports will not normally take place until the processing within the system has been determined.

Since the error exception reports are internal documents to be used for control purposes within the data processing department, they need only have enough information to identify the particular error and the record in error. The printer spacing chart in Figure 7-17 illustrates the exception report which is produced from the update program.

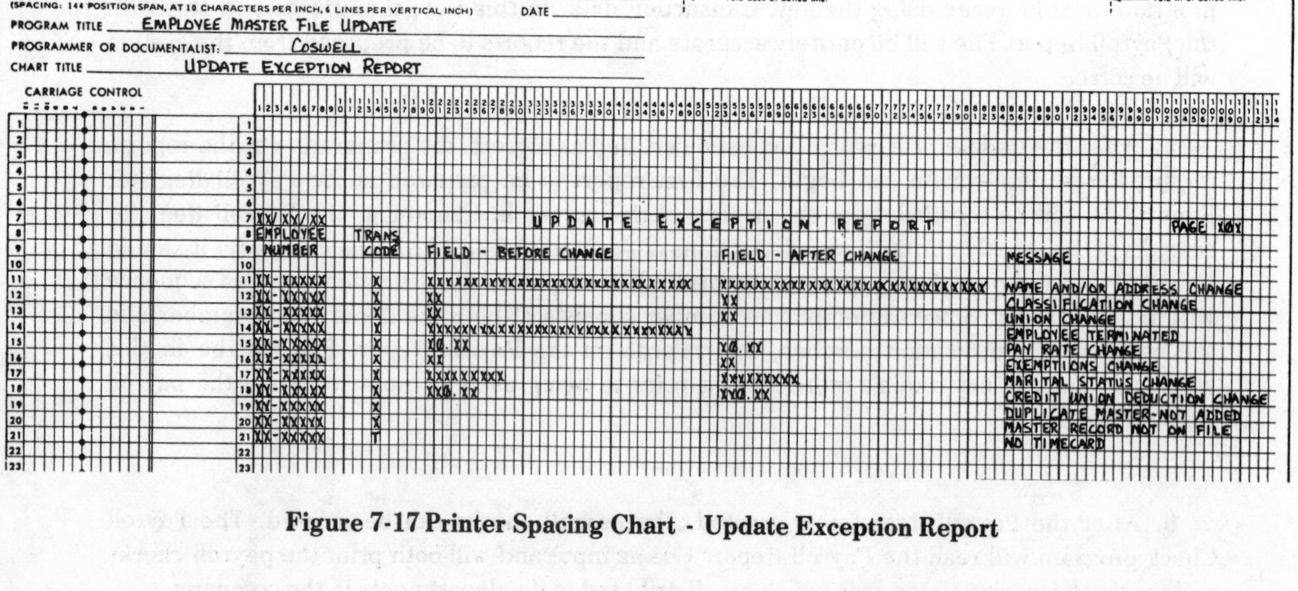

Figure 7-17 Printer Spacing Chart - Update Exception Report

Note in Figure 7-17 that the Employee Number and the Transaction Code are included on the Update Exception Report to identify the record which either contained an error or which caused a change to be made to the master file. When a transaction record causes a master record to be changed, the field which is changed is shown both with the value it had before the change and the value it has after the change. In addition, the message field contains the type of change which took place in the record.

When a transaction is in error, that is, when it cannot be processed against the master file, the error message is placed in the Message field of the report. In addition, a warning is placed on the report whenever there is no time card for an employee which is contained on the master file. Although this may be an acceptable situation, it is uncommon and therefore the report will warn that this has occurred. It many times is desirable to have warning messages such as this to point out unusual occurrences although they may be valid.

The Edit Exception Report would be designed in a similar manner to identify those records which contain invalid data and to specify the error or errors which occurred in each record.

Case Study - Program and File Identification

As was noted previously, it is mandatory that some type of program and file identification be devised by the analysts to ensure positive identification of files and programs throughout the systems documentation and for operations purposes. In the case study for James Tool Company, the analysts decided that each program would be identified with the prefix "PAYRL" followed by the program number. Thus, as can be seen from the flowchart in Figure 7-15 the Edit Program is named PAYRL04 and the Payroll Update Program is named PAYRL06.

The files contain the prefix "PYRL" followed by a "T" for transaction files or an "M" for master files. The file number follows this prefix. Thus, the Edited Transactions File is identified by the value PYRLT03 and the Employee Master File is identified by PYRLM01. These program and file names would be used in all systems documentation and also in any job control or other operations control statements which are required by the computer system.

Case Study - Start-Up Processing

As has been noted, the systems analysts must be concerned with the processing which must take place in order to "load" the files which will be used in the weekly processing of the system. In determining the needs of the James Tool Company Payroll System, the analysts Mard and Coswell decided that a separate edit program and a separate load program had to be written in order to load the master file with the information for each employee which is available from both the Addition cards and from the Compensation card which includes the year-to-date pay information for each employee (see Figure 5-37 for the format of the Compensation input card). Therefore, they developed the processing steps illustrated in the flowchart in Figure 7-18.

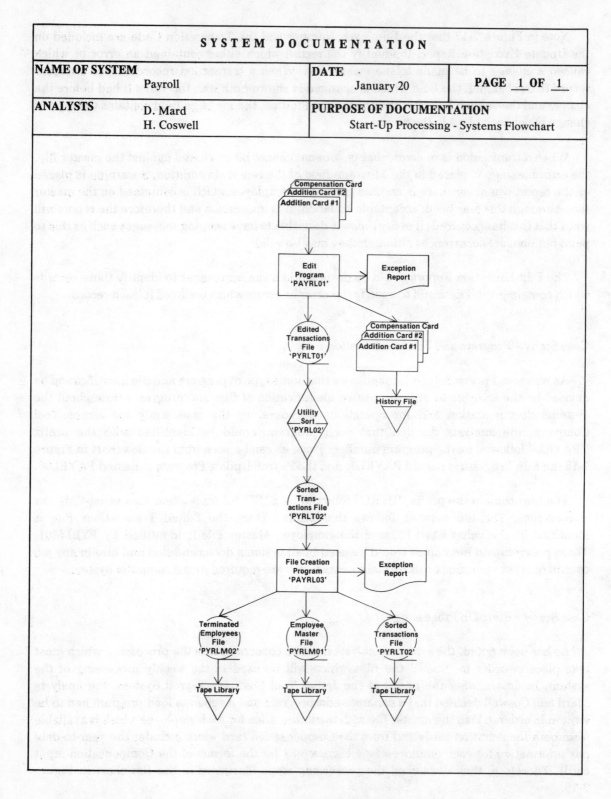

SYSTEM DOCUMENTATION

NAME OF SYSTEM	DATE	PAGE 1 OF 1
Payroll	January 20	

ANALYSTS	PURPOSE OF DOCUMENTATION
D. Mard H. Coswell	Start-Up Processing - Systems Flowchart

Compensation Card
Addition Card #2
Addition Card #1

Edit Program 'PAYRL01'

Exception Report

Edited Transactions File 'PYRLT01'

Compensation Card
Addition Card #2
Addition Card #1

History File

Utility Sort 'PYRL02'

Sorted Trans-actions File 'PYRLT02'

File Creation Program 'PAYRL03'

Exception Report

Terminated Employees File 'PYRLM02'

Employee Master File 'PYRLM01'

Sorted Transactions File 'PYRLT02'

Tape Library   Tape Library   Tape Library

Figure 7-18 File Creation Flowchart

Note from the flowchart illustrated in Figure 7-18 that the two addition cards and the compensation card are input to the edit program. In this program, the cards are checked to ensure that they contain valid data to be used in the system. The Exception Report will contain those input records which do not satisfy the editing criteria established by the analysts.

The output of the Edit program is the exception report and the Edited Transactions File (PYRLT01). This file contains all of the records which will be used as input to the File Creation Program. It should be noted that there is no predetermined sequence in which the input records to the edit program will be read. Therefore, after the edited transaction file has been created, these records must be sorted into a sequence in which the master file is to be created.

The edited transactions file is input to a utility sort program which will sort the records in the prescribed sequence. Since the Employee Master File is to be stored in employee number sequence, the primary field in the sort program will be the employee number. In addition, however, it is important that the records be in the proper sequence for processing within the file creation program. The sequence in which the records should be read is: The first addition card, the second addition card, and the compensation card. This sequence should be followed both for employees which are actively employed and for employees which have been previously terminated and whose record will be placed in the Terminated Employees File.

After the sort processing has been completed, the sorted transactions file will be read by the File Creation program (PAYRL03) and both the Employee Master File and the Terminated Employees File will be created. The Employee Master File, as has been noted previously, contains the employees who are currently employed by James Tool Company and the Terminated Employees File contains those employees who were terminated in the current year but yet must receive W-2 forms at the end of the year.

In addition to the two files, the File Creation Program will produce an Exception Report. This report will contain entries for two primary areas: First, it will contain records which are in error when the file is created. These errors would consist of employee records in which either one or more "addition" cards are missing or employees for which there is no compensation record. It will be recalled that both the addition cards and the compensation card are required for each employee to be added to the master file.

The second entries on the report will be a listing of all of the employee records which are processed and are placed either in the Employee Master File or the Terminated Employees File. This listing would be used for checking purposes to ensure that the records added to the files contained the proper data.

As with the Exception Reports, special reports such as the one listing the employee records added to the master file and the termination file will normally not be designed when the system output is designed because the detailed processing of the system is not known at that time. Instead, these reports must be designed after it is determined by the analyst that they will be needed during the course of the processing of the system.

Case Study - System Backup and File Retention

As has been noted previously, one of the more important considerations in the design of the system processing is the determination by the systems analyst concerning what backup and file retention requirements must be met. In the payroll system for the James Tool Company, there are two master files which must be backed up—the Employee Master File and the Terminated Employees File. Since both these files are sequential files stored on magnetic tape, the analysts Mard and Coswell decided that the typical file backup which can be used with sequential files, that is, the "grandfather," "father," "son" approach, would be applicable to the payroll system and adequately back up the master files.

Thus, for the Employee Master File and the Terminated Employees File, there will be two backup files plus the current file. Since the backup files must be processed with transactions in order to make them current, the analysts must also consider the backup needed for the transaction files in the system. In examining the system, the analysts determined that it was necessary to save the Sorted Transactions File (PYRLT04) in order to be able to bring the backup files up to a current status. Thus, the Sorted Transactions File required to update the "grandfather" file to create the "father" file and the Sorted Transactions File required to update the "father" file to create the "son" file must both be retained. This is illustrated in the diagram in Figure 7-19.

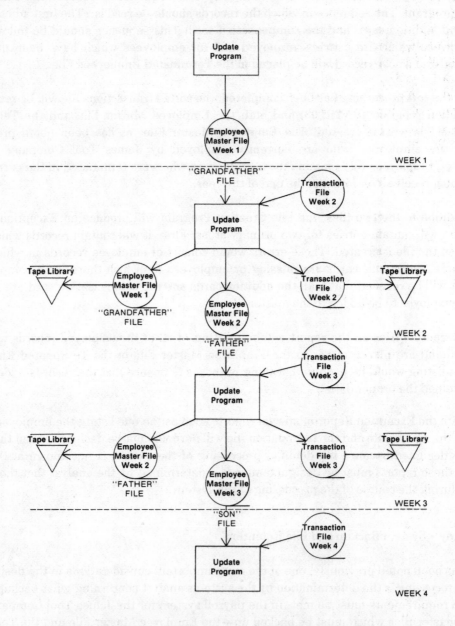

Figure 7-19 Grandfather, Father, and Son File Processing

From Figure 7-19 it can be seen that if the grandfather file is created in the week 1 processing, then the transaction file from week 2, which is used to update the grandfather file and create the father file, must be retained. Similarly, the transaction file used in week 3 to update the father file and create the son file must be saved so that, if necessary, the son file can be recreated from the father file. Thus, it will be noted that three "generations" of the master file will be kept and two "generations" of the transaction file will be kept.

As noted previously, the retention periods for the files to be kept as backup must also be determined by the analysts. Since the payroll system for the James Tool Company is to be processed weekly, a "generation" consists of one week. Therefore, the determination of the time required to keep the files will be based upon the number of files which must be kept and the "cycle" of one week. From the diagram in Figure 7-19, it can be seen that the Employee Master File and also the Terminated Employees File must be kept for three weeks so that, if needed, it can be updated twice and become the current file. Since several days of overlap time are normally included in retention times, the analysts determined that the master files should be kept for a period of 25 days. This covers the three week period plus several days.

As can be seen from Figure 7-19, the transaction file created in week 2 is required to update the master files from week 1. Therefore, each Sorted Transactions File must be saved for two weeks plus several days. Thus, the analyst determined that the retention period for the Sorted Transactions File would be 17 days.

In addition to the transaction and master files which must be considered for purposes of recovery of the master file in case of the inadvertent destruction of the file, it is also necessary to consider the other transaction files in the system as well as any report files which are created within the system processing. In the payroll system for the James Tool Company, these files consist of the input records which are on punched cards, the Edited Transactions File (PYRLT03), the Payroll Report File (PYRLT05), and the Sorted Payroll Report File (PYRLT06), as illustrated in the systems flowchart in Figure 7-15 and Figure 7-16.

The input cards, which consist of Additions and Terminations, Changes, and Time Cards, are not critical to restarting the system once the Edit program has been run. There may be occasions, however, when the payroll system will be run with no apparent discrepancies but after the payroll checks are distributed, an employee disagrees with the amount he was paid or with other information which appeared on his paycheck. In this case, the source document from which the input data was prepared and the input data itself must be examined to ensure that the correct information was entered in the system. For example, if an employee worked 47 hours but the data entry operator misread the source document and punched 41 hours, then this error would have to be corrected. Since there is the possibility of complaints arising for a period of time after the weekly payroll has been run, and because the source information is the best source to reference to determine what data actually entered the system, the analysts determined that the input data and the source documents from which the data is prepared should be retained for a six-month period.

The Edited Transactions File (PYRLT03) contains the edited transactions and serves as input to the Sort program which creates the Sorted Transactions File (PYRLT04). Thus, once the Sorted Transactions File is created, there is no need for the Edited Transactions File unless the file is destroyed in the Update program. Therefore, the analysts determined that after the Update program has been run, the tape on which the Edited Transactions File is stored can be scratched, that is, it can be used for other output files from other systems.

It will be recalled that the Payroll Report File (PYRLT05) contains all of the information required to prepare the reports for the system, that is, the Payroll Register, the Payroll Checks, and the Union Deduction Register. The file is created in the Update Program (PAYRL06). Although this file can be recreated by rerunning the Update Program, rerunning the Update Program involves several elements such as assembling the proper transaction and master files and, of course, the time on the computer. Therefore, the analysts determined that the Payroll Report File should be retained for a period of 10 days so that if any reports had to be rerun, the information would be available without rerunning the update program.

The Sorted Payroll Report File (PYRLT06) is created by sorting the Payroll Report File. Since the Payroll Report File is being retained for 10 days and since the Sort program to sort the file runs quite quickly, the analysts decided that there was no need to retain the Sorted Payroll File after the printing of the Union Deduction Register. Therefore, after the Union Deduction Register is printed, the tape on which the Sorted Payroll File is stored can be scratched.

It should be noted further that a similar type of analysis would be performed on the files which are used in the Start-up processing in order to determine their retention requirements.

Case Study - Restarting the System

In addition to determining the backup for the system, Mard and Coswell must determine where the system can be restarted in the case of a failure within a program or the destruction of a file while in the process of being read or written. In the case of the Edit program (PAYRL04) and the Update program (PAYRL06), both must be restarted from the beginning because the volume of data to be processed does not justify taking checkpoints. If the Edit program fails or the Edited Transactions File is destroyed during the processing of the edit program, no backup situation occurs because the only input to the program is the card input. Therefore, if the program fails, the corrections would have to be made and the entire program rerun. If the file is destroyed, for example if the tape drive snarls the tape, then the entire program will have to be rerun.

If the Update program fails for some reason, it will have to be rerun from the beginning using the Sorted Transactions File as input. The "son" employee master file and terminated employees file would be input the same as if the program had not been restarted. If one of the master files is destroyed, it would be necessary to rerun an update program using the "father" file in order to recreate the "son" file and then the current update processing would have to be rerun from the beginning. If the sorted transactions file is destroyed, then the sort program (PAYRL05) would be rerun to recreate the sorted transactions file and then the update program would be rerun from the beginning.

As can be seen, the primary elements in determining the restart procedures are that the files are available and that the programs work properly. The analysts would conduct a similar analysis on the remainder of the payroll system in order to determine restart points for each program within the system.

Case Study - Document and Information Flow

As noted previously, in addition to determining the computer processing which must take place within the system, the analysts must design the flow of the documents and other information which is required as input to the system and also the distribution of the information which is produced from the system. Although in the James Tool Company Payroll System there are a number of different documents which are required in order to process the system, the changes to employee master records are typical and will be examined in detail.

It will be recalled that changes are made to the master records of employees through the use of the Employee Change Form. This form is again illustrated in Figure 7-20.

EMPLOYEE CHANGE FORM

DATE___/___/___

CHECK THE APPROPRIATE FUNCTION (Col 1)

(2) ☐ Name and/or Address Change   (6) ☐ Exemptions Change
(3) ☐ Classification and/or Union Change   (7) ☐ Marital Status Change
(4) ☐ Employee Termination   (8) ☐ Credit Union Deduction Change
(5) ☐ Pay Rate Change

EMPLOYEE NUMBER (Col 2-8) All Functions

NAME (Col 9-38) Function 2

ADDRESS (Col 39-57) Function 2   CITY/STATE (Col 58-73) Function 2   ZIP CODE (Col 74-78) Function 2

JOB CLASSIFICATION (Col 18-19) Function 3   TERMINATION DATE (Col 20-25) Function 4

PAY RATE (Col 26-29) Function 5   UNION (Col 30-31) Function 3   EXEMPTIONS (Col 32-33) Function 6

MARITAL STATUS (Col 34) Function 7   CREDIT UNION (Col 35-39) Function 8
(1) ☐ Married   (3) ☐ Separated
(2) ☐ Single   (4) ☐ Divorced

Figure 7-20 Employee Change Form

The Change Form which is illustrated in Figure 7-20 is filled out in the personnel department of the James Tool Company. Thus, there are several steps which must be taken in order for the data to be input to the payroll system. These steps include:

1. The source document, that is, the Employee Change Form, must be transported from the personnel department to the data entry section of the data processing department.

2. From the Employee Change Form the data entry operators must prepare the computer input, that is, punch and verify the cards which will be input to the payroll edit program.

3. The input cards must be transported from the data entry section of the data processing department to the staging area in the computer operations section to be ready for the computer processing.

4. After the input cards are prepared, the source documents must be stored in a storage area for a period of six months, which was the time determined by the analysts that the source documents should be saved.

5. After the input cards are processed on the computer, they too must be stored for a period of six months, as discussed previously.

After determining the steps which must be taken to process the employee change forms in order to prepare the computer input data, it is necessary for the analysts to determine the schedule which must be followed in order to ensure that these steps will be performed in a manner which allow the computer processing to take place at the proper time. Prior to defining the schedule for each document within the system, however, the analysts must review the overall schedule for the complete payroll system. Since it was decided that no change would take place in the schedule for the overall computer system from the manual system, the analysts reviewed the processing of the manual system and found the following:

1. The week in the manual system, and therefore, the computer system, runs from Monday through Sunday.

2. Time cards are picked up Monday A.M. and new ones are distributed at the same time.

3. Checks are distributed to the employees on Thursday afternoon for the previous week.

4. The Union Deduction Register is due Thursday afternoon to the respective unions.

Based upon these requirements, the analysts decided that the processing of the company payroll would take place on the second shift Tuesday evening. By running the payroll on Tuesday evening, there was adequate time to have the payroll checks signed, placed in envelopes, and sent to the various departments for distribution to the employees on Thursday afternoon.

Since the computer input data must be ready for processing late Tuesday afternoon, the analysts had to determine a schedule for the Employee Change Forms which would allow time for the preparation of the punched card input files, including the transportation time from the personnel department to the data entry department, the actual preparation of the data itself, the time required to correct any errors which are found in the source documents, and the time required to transport the input records from the data entry department to the computer staging area. The analysts, therefore, prepared and documented the schedule as illustrated in Figure 7-21.

SYSTEM DOCUMENTATION		

NAME OF SYSTEM
    Payroll System
	DATE
	    January 24, 1976
		PAGE 1 **OF** 1

ANALYSTS
    D. Mard
    H. Coswell
	PURPOSE OF DOCUMENTATION
	    Schedule for Employee Change Form	

TIME	ACTION
Monday, 9:00 A.M.	Personnel Department sends the Employee Change Forms, via intracompany mail, to the Data Entry Department.
Monday, 10:00 A.M.	Employee Change Forms arrive at Data Entry Department. Preparation to begin.
Monday, 5:00 P.M.	All Employee Change Form data to be keypunched and verified.
Tuesday, 9:00 A.M.	Data Entry Department places in intracompany mail any Employee Change Forms which are illegible, contain invalid information, or are not completely filled out.
Tuesday, 10:00 A.M.	Invalid Employee Change Forms arrive in Personnel Department. Corrections begin immediately.
Tuesday, 1:00 P.M.	Corrections are completed and placed in intracompany mail.
Tuesday, 2:00 P.M.	Corrected Employee Change Forms Arrive in the Data Entry Department. Preparation of the punched cards from the corrected forms to begin immediately.
Tuesday, 4:30 P.M.	All input cards for Employee Change Forms to be completed and sent to computer staging area.
Tuesday, 5:00 P.M.	All input cards to arrive at computer staging area for processing.
Tuesday, 5:30 P.M.	Source documents to be bound and placed in storage area.

Figure 7-21 Schedule for Employee Change Form

Note from the schedule in Figure 7-21 that each step which must be performed in order to prepare the computer input cards from the Employee Change Form is specified together with the day and time the step must be completed. It should be noted that this schedule is not normally prepared by the analysts alone. The people in the personnel department who will be involved in the preparation of the Employee Change Forms and the people in the data entry department who are responsible for data preparation must be consulted to ensure that the schedule is realistic and that it can be met. Although the payroll must be processed on Tuesday evening in order to meet the Thursday deadline, there can be some flexibility in the handling of the source documents and the preparation of the input data, based upon the capabilities of the department preparing the source documents and the capabilities of the data entry department.

In addition to the schedule for the movement of documents and data through the system, it is also normally necessary to prepare a management-type flowchart to indicate the flow of the information within the various departments of the company. The flowchart prepared for the handling of the Employee Change Form is illustrated in Figure 7-22.

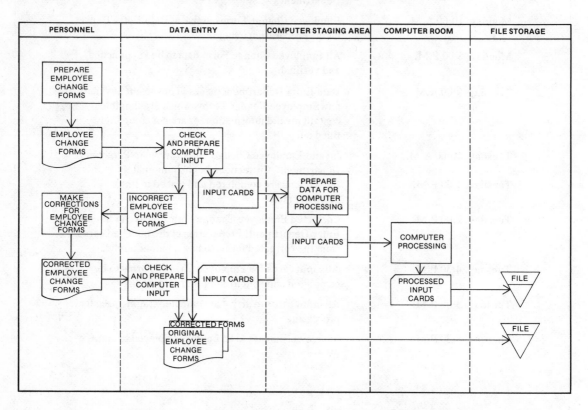

Figure 7-22 Management-Type Flowchart of Employee Change Form Processing

Note from Figure 7-22 that five different areas within the company will be involved in the handling of the Employee Change Forms. Each department is specified together with the functions which they will perform and the output which is produced as a result of their function. The Change forms are prepared in the personnel department and are sent to the data entry department where the input cards are prepared. Those change forms which are improperly filled out or which contain other errors will be sent back to the personnel department for correction. After correction, they are returned to the data entry department where the input cards are prepared. All of the input cards are then sent to the computer staging area where they are combined with the other input to the payroll system. The input data is then read by the Edit program and processed on the computer. The processed input cards are then sent to the File Storage area where they are filed. The source documents from which the computer input data was prepared are also sent to the File Storage area.

It should be noted that additional flowcharts such as illustrated in Figure 7-22 would be prepared for the other documents which are used within the system. In addition, similar flowcharts would be prepared for the processing of output documents, such as the handling of the paychecks.

SUMMARY

In the design of the processing to take place within the system, the analysts must consider many different aspects, including the computer processing and the handling of source and output documents. Without considering in detail the processing and handling of all information flowing into and out of the system, there is little chance that the system will be successfully implemented, regardless of the sophistication of the computer processing itself. Therefore, the analysts must spend considerable time analyzing the processing of the system to ensure that all parts of the system will function smoothly.

STUDENT ACTIVITIES—CHAPTER 7

REVIEW QUESTIONS

1. List the types of processing that occur in most business data processing systems.

   1)
   2)
   3)
   4)

2. Explain the term "editing" as related to file processing.

3. What are the basic types of activities that occur when updating files?

4. Explain the differences between sequential updating and random updating.

5. Explain the term "report files." Give an example of a report file.

6. What basic factors must be considered when planning for file backup?

7. Explain how transaction files are recreated when accidentally destroyed. Are backup files normally required?

8. Explain how backup files for transaction records of an on-line system are created.

9. Explain how backup files are created for sequential master files.

10. Explain how backup files are created for indexed sequential or direct files.

11. List the factors which should be considered when determining the frequency of creation of backup files.

12. List three factors which must be considered in determining the file retention period.

13. Explain the terms ''grandfather, father, and son'' files.

14. Explain the term ''restart.''

15. Discuss the factors which must be considered by the analyst in determining restart points.

16. What is meant by the term "checkpoint"?

17. Explain the term "multiprogramming."

18. Explain the term "spooling."

19. List the factors which the analyst must consider relative to preparing the input data for processing and the disbursement of the reports generated from the system.

    1)

    2)

    3)

    4)

    5)

    6)

    7)

20. What factors must be considered by the analyst relative to the control of source documents?

    1)
    2)
    3)

21. List the factors which must be considered relative to the conversion of data for computer processing.

    1)
    2)
    3)

22. After the input data has been prepared what factors must be considered in "getting the data into the computer system."?

    1)
    2)
    3)
    4)

23. Explain some of the procedures which must be followed in controlling special output forms such as blank payroll checks.

24. Discuss the steps which must be undertaken in preparing systems output for distribution.

DISCUSSION QUESTIONS

1. One data processing manager states "all of our input data is keypunched and verified...an edit run is a waste of computer time! If you have a good data entry supervisor, edit runs aren't needed!"

   Do you support this position?

2. Some authorities feel that standards must be imposed upon the data processing profession, including flowcharting standards, programming standards, and documentation standards. They argue that "Millions of dollars and thousands of hours of time are wasted in retraining whenever individuals change jobs because every installation is 'different'...Standards would eliminate this problem."

   Other authorities argue that standards "stifle creativity. We will never progress as a profession if all people in data processing are required to follow prescribed standards. Standards imply there is a 'best way' — We haven't found the 'best way' yet in data processing. Data Processing is much too 'young' as a profession to begin to impose standards."

   What is your opinion?

3. An operations manager of a large installation made the statement that "although edit runs are important and must be included in any business system, the corrections to errors create havoc when scheduling the computer. We never know how many errors will be found or how long it will take to correct the errors. So, in many cases, the computer sits idle while corrections are being made. My suggestion to analysts is that more manual checking be instituted to cut down on the number of errors in the edit run and that all systems be designed so that corrections can be processed at a later time. By doing this, once the processing of the system has begun on the computer, there will be no stopping and we can schedule the computer more efficiently, thereby saving the company a lot of money."

   What do you think of the operations manager's suggestion? Do you have any ideas how his suggestion could be implemented?

CASE STUDY PROJECT—CHAPTER 7
The Ridgeway Company

INTRODUCTION

After designing the output, input, and files for the billing system of the Ridgeway Country Club, the analyst is faced with the task of designing the systems processing.

STUDENT ASSIGNMENT 1
COMPUTER PROCESSING

1. Design the processing which must take place on the computer for the billing system and the preparation of the required management reports. This design should include, but is not limited to, the following elements:

   1) A systems flowchart illustrating the programs to be used in the system and the files which are to be processed by the programs.

   2) Documentation of the backup requirements for the transaction and master files contained within the system.

   3) Documentation of the retention periods for all files used in the system.

   4) Documentation of the Restart Procedures to be used for the system.

   5) Documentation and systems flowchart of the initial start-up procedures to be used for the billing system.

2. Using a printer spacing chart, design any Exception Reports which are required for the billing system.

STUDENT ASSIGNMENT 2
DOCUMENT AND INFORMATION FLOW

1. Determine the procedures to be followed for preparing the input to the computer billing system and disbursing the output from the system. These procedures should include at least the following:

   1) The gathering of the source documents from their areas of preparation.

   2) Control functions which must be performed on the sales slips and other source documents.

   3) Converting the data on the source documents to computer input and the procedures to follow for making the input data available to the computer operations department.

   4) Control and distribution of the output produced from the system.

2. Document the procedures which are designed, including schedules for document and output handling and document flowcharts showing the movement of the data throughout the system.

DESIGN OF
SYSTEM CONTROLS

CHAPTER 8

DESIGN OF
SYSTEM CONTROLS

8

INTRODUCTION

An important aspect of the design phase of the systems project is the establishment of a comprehensive set of system controls. System controls are defined as "a plan to ensure that only valid data is accepted and processed, completely and accurately." Controls should be built into and become an integrated part of the system when it is originally designed rather than being superimposed upon the system after it is designed. It is the responsibility of the systems analyst to design an efficient processing system with adequate controls to ensure the accuracy of reports and files generated during the processing of data.

Adequate controls must be established for two basic reasons:

1. To ensure the accuracy of the processing and the accuracy of the reports generated from the system.

2. To prevent computer-related fraud.

The necessity of ensuring the accuracy of the processing and reports is obvious; for unless correct reports are obtained the output data is useless in providing management with the information needed to direct the operations of the business. In addition, good controls provide the system with a high degree of reliability. The accuracy and orderliness which result from good controls will lead to greater processing efficiency by reducing the number of errors that enter the system and that require manual intervention and subsequent reprocessing.

Another important reason for controls, but one often overlooked by the systems analyst, is to prevent losses to the business because of computer-related fraud. Computer-related fraud refers to the process in which records of the company are purposely altered to allow the assets of the company to be extracted for personal use. Computer-related fraud can be caused either by persons within the company or by persons outside the company who gain access to data which is processed on the computer. In 1974 the total loss to companies from computer-related fraud was reported to be in excess of $200 million and some authorities feel that the potential loss to companies far exceeds this figure. The development of proper system controls can greatly reduce the danger of computer-related fraud.

TYPES OF CONTROLS

One of the basic responsibilities of management is control. Therefore, controls of all types will be a part of every business organization. As related to the design of electronic data processing systems there are four types of controls that must be considered by the systems analyst. These controls include:

1. Source Document Control

2. Input Control

3. Processing Control

4. Output Control

SOURCE DOCUMENT CONTROL

There are many sources of input data in a business organization—salesmen prepare sales orders, the payroll department originates time cards, the production department reports on materials and products manufactured, etc. If the management reports to be generated from these sources are to contain accurate, timely information, the data processing department must make sure that the input to a system is complete and accurate. Therefore, controlling the reliability of the computer system should begin at the starting point, that is, with the source documents and the data which is entered on the source documents.

Batching

One of the first steps in establishing an effective system of controls in a data processing system is the development of a method for the control of source documents which will be used as the basis for input to the system. One of the more common forms of source document control is the "batching" of source documents when they are transmitted to the data processing department.

Batching is the technique by which source documents of a common type are grouped together over a period of time and then are transmitted as a unit to the data processing department for processing. These source documents are normally transmitted to the data processing department for processing together with a batch control form. The batch control form contains information to identify the group of source documents and also contains control data, such as a document count and control totals. Figure 8-1 illustrates a batch control form for a group of sales orders which are batched by date, that is, all the orders for a given day are considered to be one batch.

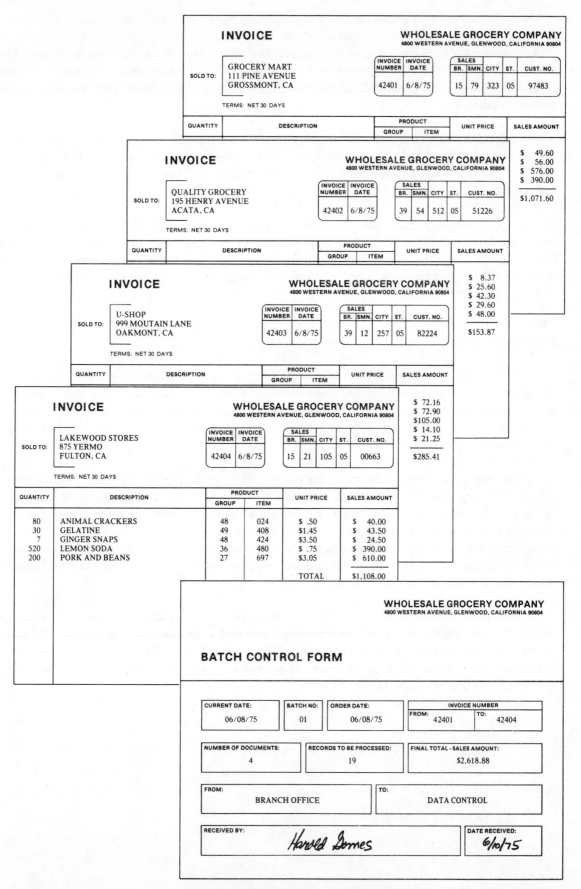

Figure 8-1 Example of "Batching"

Note from Figure 8-1 that the source document batch control form contains identifying information for the batch orders that are to be transported to the data processing department. Information such as the number of orders, and the final totals of the sales amounts would be calculated manually. These totals would then be balanced against a computer generated report to ensure that all input data has been prepared accurately from the information supplied on the source documents.

A copy of the batch control form would normally be retained in the originating department. Thus, if a record is "lost" during processing, the originating department has direct evidence that the original source documents were transmitted to the data processing department as required.

Although there are numerous other document controls which may be established, including serial numbering of input documents such as invoices and paychecks, and document registers in which each input document is recorded as it is received, batch control is one of the more common control mechanisms found in data processing installations which are handling large volumes of source documents.

INPUT CONTROLS

Input controls are established to assure the complete and accurate conversion of data from the source document to a machine processible form for entry into the computer. After the source documents have been received by the data processing department, the data contained on the source documents must be converted to some input media such as punched cards, magnetic tape or magnetic disk.

To ensure the accuracy of the recording of the data from the source documents to a machine processible form, some type of "key verification," as explained in Chapter 5, is commonly used. When "key verifying," the data on the original source document is keyed twice. The first time the data is recorded. The second manual keying of the data allows the comparing of the data originally recorded to the characters keyed the second time. This technique should normally detect over 95 percent of the errors in manually transferring the information from the source document to a machine processible form.

It should be noted, however, that key verifying will not detect all possible errors. Duplicate cards may be generated; both the recording and verifying operator may make the same mistake; the data on the source documents may be misread, or, in fact, the data on the source documents may not be correct or complete. It is important, therefore, for the analyst to include other controls throughout the system.

INPUT CONTROLS - ON-LINE SYSTEMS

With on-line systems, that is, systems incorporating computer terminals as a form of input, new problems arise concerning the methods and techniques of input control, as the concept of batching the source documents does not apply. With on-line systems transactions will normally be entered into the system when encountered.

Three basic types of input controls should be exercised when designing on-line systems. These include control over those using the terminal, and transaction checks as the data enters the system. The following is a brief explanation of these control methods:

1. With an on-line system each user of the terminal must be assigned a specific code, key, or card which identifies him as an authorized user. When the terminal is used this code should be logged together with such factors as time, type of transaction, and the actual transaction. If someone other than an authorized user attempts to use the terminal security personnel should be immediately advised so that action can be taken.

2. All transactions entered via the terminal should be checked by the program for validity, with any error causing the transaction to be rejected. Records of these rejected transactions should be maintained.

3. A listing of all transactions should be sent to the supervisor at the related terminal sites for review and approval.

On-line systems typically pose serious control problems because data is frequently entering the system at many remote sites; therefore, controls become an extremely important aspect in the design of such systems.

PROCESSING CONTROLS

Processing controls refer to those procedures that are incorporated into computer programs within the system to assure the complete and accurate processing of the data throughout the system. Although the input data to a system will normally have been visually or mechanically verified when the data is converted to punched cards, magnetic tape or magnetic disk, the largest and single most important checking problem in a data processing system normally occurs when the input data initially enters the system. At this time, those fields essential to accurate reporting should be checked for accuracy and completeness. Any record which cannot be processed by all subsequent programs in the system should normally be rejected.

There are numerous types of controls which can occur at this time, including the listing of input records and the taking of control totals which are to be balanced against the original source documents; the checking of the fields in the input record for valid data; the checking of the contents of the fields for "reasonableness," etc. Since meaningful information cannot be produced from the system until the input data is entirely accurate and since incorporating these controls into the system requires an extensive amount of programming, it is normally desirable to have a separate program for input "editing" or input "validation." An "edit run" using the "edit program" is thus used to ensure the accuracy of the input data prior to further processing.

It is extremely important to perform adequate input validating, for if a single record that is not correct enters the system, all subsequent processing using that record will be incorrect. Thus, if there are ten programs in a system that are to be used to prepare ten management reports of various types, a single input error could result in the generation of ten incorrect reports!

TYPES OF CONTROLS

There are basically two approaches to processing controls which may be utilized to ensure the accuracy of the data entering the system. These include:

1. Group Controls

2. Individual Record Checking

As the name implies, "group controls" are established to ensure that all groups of records which are to be processed are accounted for and are accurate. Individual record checking is utilized to ensure the validity and accuracy of the data within each field of the input record.

GROUP CONTROLS - List and Balance

The concept of batch control of the source documents may also be applied to the initial processing of the data in the "edit run." One method of group control involves the preparation of a computer listing of the records created from the batches of source documents transmitted to the data processing department. This computer listing would normally contain identifying information from the input records plus control information such as record counts and totals. These totals are then compared to the data on the original batch control form or to the totals on the original source documents.

If the totals do not agree, each of the records must be examined to detect any errors. Figure 8-2 illustrates a series of sales orders and the related computer listing in the form of an invoice register. The invoice register is used for control purposes and is balanced to the original batch control form and the original source documents to ensure the accuracy of the input data.

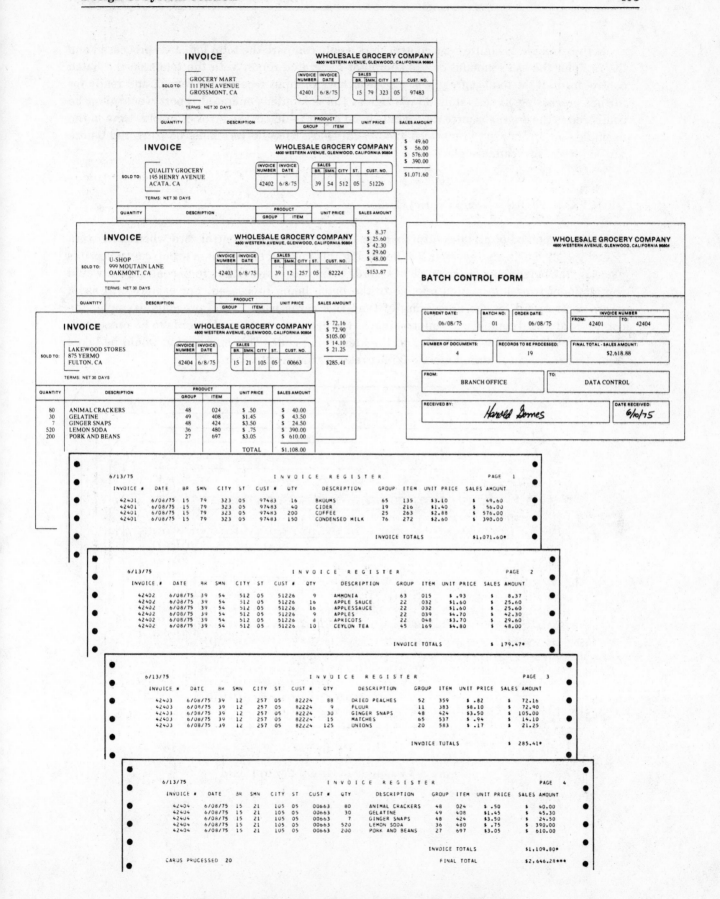

Figure 8-2 Example of List and Balance Group Controls

In the example in Figure 8-2 a control clerk would compare the total for cards processed and the total for the sales amount on the computer-generated report with the totals on the batch control form. If the totals agree, it is assumed that the input records are correct and ready for further processing. If the totals do not agree, the computer-generated report would then be compared to the original source documents in an effort to detect the error or errors. Note in the example that errors exist in invoices 42402 and 42404. These errors must be corrected before further processing can take place.

GROUP CONTROLS - Group Control Cards

Another method of group control involves the use of a group control card when processing "batches" of data in an "edit run." When using this method of control, a group control card is placed at the beginning of each batch of records to be processed. This group control card would contain identifying information relative to the batch being processed. For example, to use a group control card for the processing of the sales cards used in the previous example, the control card would contain such information as the date, the number of records to be processed, the final total of the sales amount, and the identifying code. This information would be taken from the batch control form. Figure 8-3 illustrates a group control card.

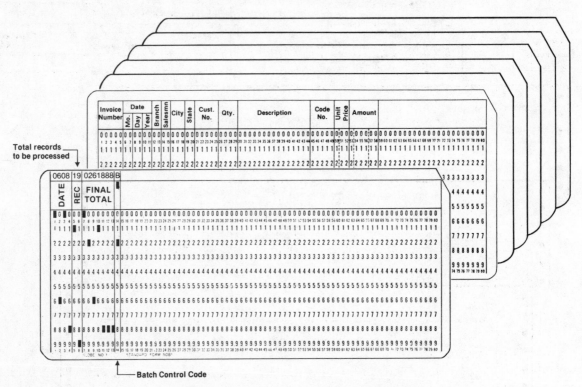

Figure 8-3 Example of Group Control Card

When the processing of the batch of records occurs, the group control card is read. In the example in Figure 8-3, the total number of records to be read and the total sales for the record in the batch would be stored in the computer. After the batch of records is processed the totals accumulated from the processing of the records would be compared, by the edit program, with the totals on the group control card. If the totals are equal it is apparent that the input records are correct. If the totals are not equal an error exists within the input records and a message would be printed on an Exception Report identifying the batch and indicating an error. A sample Exception Report is illustrated below.

```
        06/13/75      EXCEPTION REPORT        PAGE  001

        INVOICE                 CUSTOMER        SALES
        NUMBER       DATE        NUMBER         AMOUNT

        BATCH IDENTIFICATION 06/08/75

        42401    06/08/75     97483      $     49.60
        42401    06/08/75     97483      $     56.00
        42401    06/08/75     97483      $    576.00
        42401    06/08/75     97483      $    390.00

                     INVOICE TOTAL  $1,071.60*

        42402    06/08/75     51226      $      8.37
        42402    06/08/75     51226      $     25.60
        42402    06/08/75     51226      $     25.60
        42402    06/08/75     51226      $     42.30
        42402    06/08/75     51226      $     29.60
        42402    06/08/75     51226      $     48.00

                     INVOICE TOTAL  $  179.47*

        42403    06/08/75     51226      $     72.16
        42403    06/08/75     82224      $     72.16
        42403    06/08/75     82224      $     72.90
        42403    06/08/75     82224      $    105.00
        42403    06/08/75     82224      $     14.10
        42403    06/08/75     82224      $     21.25

                     INVOICE TOTAL  $  285.41*

        42404    06/08/75     00663      $     40.00
        42404    06/08/75     00663      $     45.30
        42404    06/08/75     00663      $     24.50
        42404    06/08/75     00663      $    390.00
        42404    06/08/75     00663      $    610.00

                     INVOICE TOTAL  $1,109.80

        CONTROL TOTALS  RECORDS 19  SALES  AMOUNT $2,618.88
        ACTUAL  TOTALS  RECORDS 20 SALES AMOUNT $2,646.28
        ** ERRORS **        RECORDS PROCESSED AND SALES AMOUNT **
```

Figure 8-4 Example of Exception Report

In the example above it will be noted that the batch of data for the date 06/08/75 is processed. Each of the individual records in the batch is listed and then the control totals found on the group control card are printed together with the totals which were calculated from the input data. It will be noted that there are errors both in the number of input records which were read and the total for the sales amount. In this case, the report and the input data would normally be returned to a control clerk where the errors would be found and corrected prior to allowing any further processing on the data. In this manner, errors which have been committed in the preparation of the input data will normally be caught and corrected so that invalid data will not be processed within the system.

INDIVIDUAL RECORD CHECKING

In addition to group controls which can be used to validate input data, checking may also be performed on individual records and fields within the records. This is accomplished by including routines in an edit program which perform checks on the data in fields within each input record. This checking is done to ensure that the data in the fields has been recorded accurately when it was transferred from the source documents to the input media and also to ensure that values within the fields being checked are ''reasonable'' or contain data within certain prescribed limits as defined within the system.

There are numerous checks that can be performed to ensure that valid data is contained in critical fields of the input record. These tests may include checking individual characters within a field or checking the contents of the entire field. Common tests include:

1. Tests to determine if a field contains numeric data only.

2. Tests to determine if alphabetic data is contained in a field.

3. Tests to determine if a field contains all blanks.

4. Tests to determine the sign of numeric data in a field.

5. Tests for the reasonableness of data in a field.

6. Tests to determine if data falls within certain limits established in the system.

7. Tests to determine if data in a field falls within certain ranges of values.

8. Tests to determine if data within related fields is consistent.

9. Tests to determine if records are read in a prescribed sequence.

TEST FOR NUMERIC DATA

In many applications it is necessary to test fields to ensure that the fields contain strictly numeric data and do not contain interspersed letters of the alphabet or blanks. With many computer systems, such as the IBM System/360 or System/370, if a field is to be utilized in a calculation and that field contains a non-numeric character in the low order position, processing will be terminated when the computer executes the instruction to perform the required calculation.

Thus, it is essential when using systems of this type that fields involved in calculations be checked to ensure that they contain valid numeric data in all positions. Figure 8-5 illustrates several input records and the resulting error message that occurred during the processing of those records. The output illustrated is from an IBM System/360 operating under the Disk Operating System.

COBOL Statement:

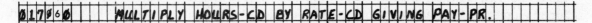

```
Ø17Ø6Ø        MULTIPLY  HOURS-CD  BY  RATE-CD  GIVING  PAY-PR.
```

Input:

NOTE: The second input record contains
a blank in the Rate field.

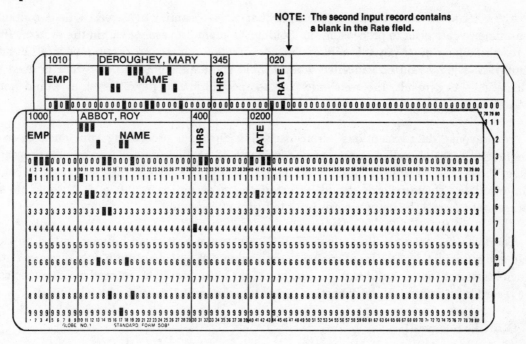

Output:

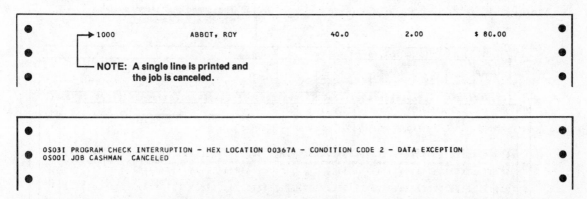

Figure 8-5 Example of Data Exception with Non-Numeric Data

In the example in Figure 8-5 a COBOL statement is illustrated which would cause the data in the Hours field in the input record to be multiplied by the data in the Rate field, giving the pay for an employee. For the first record the pay is calculated properly but since the second input record contains a blank in the Rate field, processing is terminated and an error message indicating a "data exception" is printed, together with additional information including the storage address of the instruction being executed. A data exception will occur when an attempt is made to perform a calculation on a field that does not contain valid numeric data.

Thus it can be seen from the example that it is essential to detect a field containing non-numeric data prior to processing the field in the main processing within the system. If this were not done, in a payroll application for example, after processing several hundred employees a job would be cancelled when a field containing non-numeric data is used in a calculation. As a result, the remainder of the records in the payroll system would not be processed.

To prevent "data exceptions" from occurring during a processing run, such as in the preparation of the payroll, a statement or routine could be incorporated into an edit program to detect those records which contain non-numeric data in fields which are supposed to contain numeric data. These records would be listed on an exception report and it would then be necessary to correct the input records prior to further processing.

The following example illustrates a COBOL statement that could be used to test an Hours field and a Rate field on an input record to determine if the fields contain only numeric data. If the fields do not contain valid numerc data an error routine would be entered which would cause the record to be printed on the exception report together with the error which is found.

COBOL Statement:

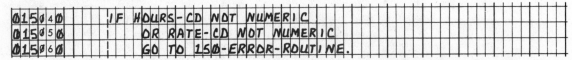

Input:

NOTE: The second input record contains a blank in the Rate field.

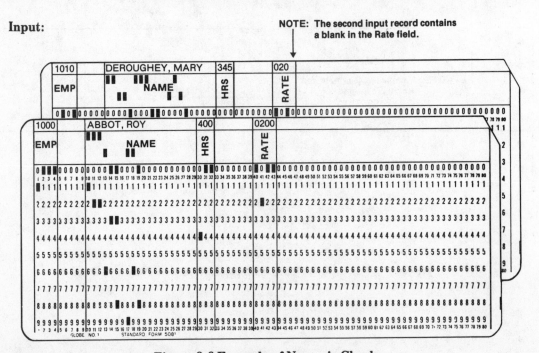

Figure 8-6 Example of Numeric Check

In the examples in Figure 8-5 and Figure 8-6 COBOL is used to illustrate the concepts of "programmed tests." It should be noted, however, that the capabilities to perform these tests exists in many programming languages and COBOL is used merely for demonstration purposes.

TESTS FOR ALPHABETIC DATA

In some applications it may be desirable to determine if a field contains only alphabetic data. For example, in a billing application using a Name and Address record, it may be desirable to assure that the State field is present in the input Record and contains alphabetic data. If the State field is blank or contains numeric data, the record cannot be utilized in subsequent processing. To check for the presence of a valid State field, an "alphabetic test" could be performed, as illustrated in Figure 8-7.

COBOL Statement:

```
021160        IF STATE-CD NOT ALPHABETIC
021170           GO TO 200-ERROR-ROUTINE.
```

Input:

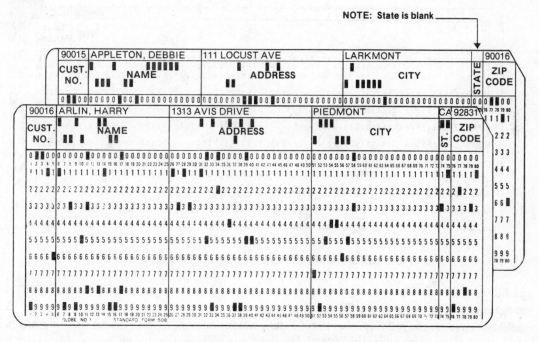

Figure 8-7 Example of Alphabetic Test

Figure 8-7 illustrates several input records and a COBOL statement to test if the State field contains alphabetic data. In the example, if the State field is blank or contains non-alphabetic data, the record would be listed on the Exception Report during the edit run and would have to be corrected before further processing could take place. If the state field is present and contains alphabetic data, the record would pass the edit "test" and be available for processing in subsequent programs within the system.

TESTS FOR BLANKS

In some applications it may be desirable to test a field to determine if the field contains blanks. For example, in a sales analysis application, one field in an input record may contain a salesman's sales and another field may contain the sales returns. When there are no sales returns, the sales returns field will be left blank rather than requiring the data entry operator to repeatedly record zeros in the field. When the field is involved in a calculation, however, the field must contain zeros. The example in Figure 8-8 illustrates a test for blanks and, if the sales return field does contain blanks, the blanks are replaced by zeros.

COBOL Statement:

```
011050    IF SALES-RETURNS-IN IS EQUAL TO SPACES
011060       MOVE ZEROES TO SALES-RETURNS-IN.
```

Input:

NOTE: Sales Return field is blank when zero

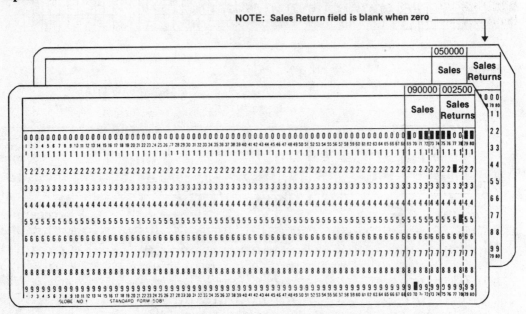

Figure 8-8 Example of Testing for Blanks

In the example in Figure 8-8, the Sales Returns field is to be involved in calculations in later programs within the system. Therefore, in the edit program, it is necessary to ensure that the field contains numeric data; otherwise, a data exception would occur when the field was used in the calculations. Since the Sales Returns field can validly contain all blanks but must contain numeric data for subsequent programs, the blanks in the field will be replaced by zeros in the edit program.

It should be noted that these editing functions are not normally applied independently of each other, that is, more than one test can be performed on a given field. Thus, after the sales returns field is checked for all blanks, it is quite likely that it would additionally be checked for numeric data. The COBOL coding to accomplish this testing is illustrated in Figure 8-9.

COBOL Statement:

```
011120      IF SALES-RETURNS-IN IS EQUAL TO SPACES
011130         MOVE ZEROES TO SALES-RETURNS-IN.
011140      IF SALES-RETURNS-IN NOT NUMERIC
011150         GO TO 310-ERROR-ROUTINE.
```

Figure 8-9 Example of Testing for Blanks and Numeric Data

Note in the example in Figure 8-9 that the Sales Returns field is first checked for blanks and if the field is totally blank, zeros are moved to the field. After the check for blanks, the field is checked to determine if numeric data is contained in the field. If not, an error routine is entered. This type of checking is quite common and the analyst, when designing the controls and checks to be incorporated into the system, must be aware that a number of different types of checks can be performed on a given field.

TESTS TO DETERMINE THE SIGN OF THE DATA IN A FIELD

It is sometimes necessary to perform tests to determine the sign of numeric data in a field, that is, is the numeric data positive or negative. In some applications it is essential that the field either be positive or be negative for proper processing to occur. For example, a test which is commonly performed on payroll checks is a test to ensure that the net amount of the check is not negative. A negative amount is not valid on a payroll check.

This situation could occur, for example, if an employee who normally works 40 hours per week, earns $3.00 per hour and deducts $40.00 a week for savings in the credit union, works only 10 hours in a week. When his wages are calculated, a negative amount would result for the net pay amount, as he would be earning $30.00 and deducting $40.00. As a negative pay amount is not valid, it would be necessary to flag this transaction as an error and take some corrective action, either manually or within the program. The example in Figure 8-10 illustrates the COBOL coding necessary to check for a negative pay amount.

COBOL Statement:

```
023170      IF NET-PAY IS NEGATIVE
023180         GO TO 440-PAY-ERROR.
```

Figure 8-10 Example of Test for Negative Value

Note in the example in Figure 8-10 that if the value in the NET-PAY field is negative, an error routine would be entered. This routine would make an entry on an Error Report which would indicate that a negative Net-Pay had been calculated so that corrective action could be taken.

TESTS FOR REASONABLENESS OF DATA

Reasonableness checks may be performed on either original input data or on data that is calculated within a program. A reasonableness check is utilized to determine whether data in the field is within "normal" or "accepted" boundaries. For example, in a payroll application it may be desirable to check to ensure that the hours worked by an employee do not exceed 70 hours per week or that the pay rate does not exceed $19.99 per hour. This decision is based upon the fact that no one in the company is normally authorized to work in excess of 70 hours per week or normally earns over $19.99 per hour. If figures in excess of these amounts are found in the input records, an entry on an Exception Report would be made indicating that the data was not within reasonable boundaries. The example in Figure 8-11 illustrates the COBOL statements which would be used to check the reasonableness of data in the hours and pay rate fields and the input records which are checked.

COBOL Statement:

```
017090        IF HOURS-IN IS GREATER THAN 70.0
017100           OR RATE-IN IS GREATER THAN 19.99
017110           GO TO 200-ERROR-ROUTINE.
```

Input:

Figure 8-11 Example of Test for Reasonableness of Data

Reasonableness checks may also be taken upon calculated values. For example, it may be determined that no employee normally receives more than $999.99 per week; therefore, a reasonableness check could be incorporated into the program so that if an individual's weekly pay exceeds $999.99, a check would not be prepared but rather the employee's name would appear on an exception report for investigation.

The effectiveness of reasonableness checks can save many embarassing moments for those in the data processing department, for even today one reads of checks being issued for $5,000,000.00 or water bills being sent to a new home owner for $100,000.00. The common comment that "the computer made a mistake" in many cases means the analyst did not consider certain checks, such as reasonableness checks, in the original design of the system. Many errors "by the computer" can be easily avoided by including reasonableness checks within the processing programs.

LIMIT TESTS

Somewhat related to the reasonableness test is the limit test. Whereas a reasonableness check is used to flag unusual occurrences, a limit test is used to place upper or lower boundaries on values found in fields and if these boundaries are exceeded, then the record cannot be processed.

For example, it may be determined that the maximum weight of a shipment which can be made by airfreight is 500 pounds. If an input record indicates that airfreight is to be used for a given shipment, but after the weight of the shipment is calculated by the program it exceeds 500 pounds, then the record will be rejected and will have to be corrected to show some other type of shipping method. The COBOL statement to perform this limit test is illustrated in Figure 8-12.

COBOL Statement:

```
024020        IF SHIP-TYPE = AIR-FREIGHT
024030           AND SHIP-WEIGHT > 500
024040           GO TO 720-REJECT-AIR.
```

Figure 8-12 Example of Limit Test

RANGE TESTS

Another common technique for checking the data within a field to determine if it is correct is the Range Test. A range test is done when some predetermined limits or standards are compared to values which are contained within a field in the input record. If the value in the field does not fall within the range of the predetermined values then the input data is considered in error.

For example, in a payroll system it may be determined that there are only certain departments within the company. Assume that there are departments 05 through 14. Any department number 05 through 14 would be considered valid but any others would be considered invalid. Therefore, if an input record was read which contained the value 22, it would be rejected as an error and placed on an exception report. The example in Figure 8-13 illustrates a COBOL statement from a payroll program to perform a range test.

COBOL Statement:

```
018070        IF DEPT-IN IS LESS THAN 05
018080           OR DEPT-IN IS GREATER THAN 14
018090           GO TO 273-DEPT-ERROR.
```

Figure 8-13 Example of Range Test

CONSISTENCY TESTS

It is possible that there can be even more detailed checks dependent upon the particular values within an input record. A check for consistency is utilized when two or more fields within a record are considered in relation to one another for proper processing to take place. For example, in an inventory application it has been found over a period of time that the number of parts sold in a given time rarely exceeds 50 percent of the number of parts found in stock. Therefore, although it is never known what the correct number of parts in stock and the correct number of parts sold is when the edit program processes the input data, a consistency check can be performed to determine if the number of parts sold exceeds 50 percent of the number of parts in stock. If it does, then it is likely that an error has been made in entering the values in the input record. This check is illustrated in Figure 8-14.

COBOL Statement:

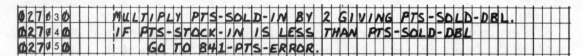

```
027030        MULTIPLY PTS-SOLD-IN BY 2 GIVING PTS-SOLD-DBL.
027040        IF PTS-STOCK-IN IS LESS THAN PTS-SOLD-DBL
027050            GO TO B41-PTS-ERROR.
```

Figure 8-14 Example of Consistency Check

SEQUENCE CHECKS

A sequence check is performed to assure that incoming data is properly arranged in either an ascending or a decending order, as required by the application. If applicable; this type of check can also be performed to include a check to determine if multiple records making up a transaction are present and in the correct order.

For example, in a billing application a complete transaction may consist of a name and address record, a balance due record, a payments record, and a purchases record. In addition, these records must be present in the order stated. In such applications a sequence check should be performed to assure that all required records are present and that the records are in the proper sequence. Figure 8-15 illustrates this concept.

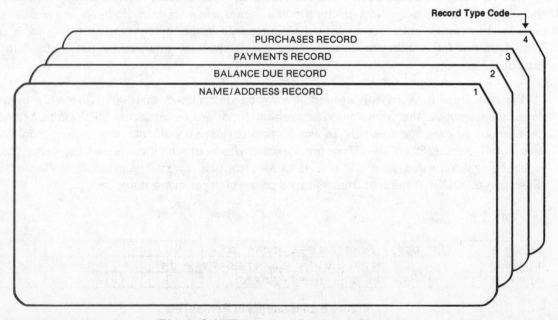

Figure 8-15 Example of Sequence Check

In the example in Figure 8-15 it can be seen that a billing transaction for a given customer consists of the four records and they must be in sequence by record type code. It should be noted that when records are read by an edit program, they are not normally sequenced, that is, no sorting has taken place prior to being read by an edit program. Therefore, if sequence checking is required, it commonly takes place in a program following the sorting step in the system processing, usually in the update program.

CHECK DIGITS

It will be noted that the previous tests are primarily concerned with determining that the values contained within various fields to be checked are valid and correct in terms of content. There are also checks which may be applied to fields such as part numbers or employee numbers to ensure, as much as possible, that there are no errors made by the data entry operator when the data is transcribed from the source document to the input document. The errors by data entry operators are generally classed as transcription errors, where there is an error in copying a digit of a number, or transposition errors, where there is a "swapping" of numeric values. For example, a transcription error would occur where the number 2865 is copied on the input record as 2365. There are several reasons this may occur; for example, the writing on the source document may be unclear or the operator may misread a document. Transposition errors occur when two numbers are switched, for example, the number 2457 is copied as 2547.

The primary means used in data processing systems to find these types of errors is the use of a method known as "check digits." A check digit is a number which is appended to the normal number used for a particular function which will allow the program to test and ensure that the normal identification number is properly recorded. For example, assume a part number consists of six digits, as illustrated in Figure 8-16.

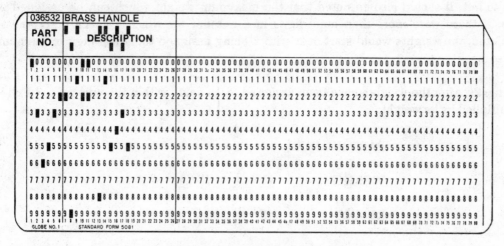

Figure 8-16 Example of Inventory Card

Note from Figure 8-16 that the part number is 036532, which is six digits. Using a check digit, a seventh digit would be added to the number, as is illustrated below.

OLD PART NUMBER = 036532

NEW PART NUMBER
WITH CHECK DIGIT = 0365327

— Check Digit

The seventh digit is the check digit and is used to ensure that all of the numbers in the part number are properly recorded. This is quite important for the part number because the part number could act as the key for the inventory record and it is critical that there be no errors in the recording of the part number from the source document to the input record.

There are a number of ways in which a check digit can be generated. The most accurate and widely used is known as the Modulus 11 check digit. The process of using a check digit includes the calculating of a unique digit from the original number and then placing this unique digit as the last character of the original number. In order to illustrate the use of a Modulus 11 check digit the following example will be used.

EXAMPLE

Step 1: Assume that the old part number for a part in inventory was 036532. In order to calculate the modulus 11 check digit and, therefore, the new part number, the first step is to assign a "weight" to each of the digits in the old part number. When using the modulus 11 technique, "weights" are assigned to each of the digits of the part number. The "weights" run from 2 through a possible maximum of 10, beginning with the low order position in the field. Therefore, the following weights would be assigned to the old part number.

PART NUMBER: 0 3 6 5 3 2

WEIGHTS: 7 6 5 4 3 2

Note that beginning with the low-order digit a weight, starting with 2 and increasing by 1, is assigned from right to left. It must be noted that the weight assigned has no relationship to the value contained within the number to be checked; the weights are always the same moving from right to left. It should also be noted that the maximum weight which can be assigned is 10. Thus, if there were more than nine digits in a number for which a check digit was to be calculated, the weights would start over with 2 being assigned the tenth digit in the number, and so on.

Step 2: After the weights have been assigned, each value within the number to be checked is multiplied by its weight.

PART NUMBER:	0	3	6	5	3	2
WEIGHT:	x7	x6	x5	x4	x3	x2
RESULT:	0	18	30	20	9	4

Note from the example that each of the digits within the number to be checked is multiplied by its respective weight.

Step 3: The results of the multiplication are added together.

$$0 + 18 + 30 + 20 + 9 + 4 = 81$$

Note that the sum 81 is derived by adding together the products of the multiplication of the values in the number by their respective weights.

Step 4: The product of the multiplication is divided by the modulus number, in this case, the number 11.

$$81 \div 11 = 7, \text{ with a remainder 4}$$

Step 5: The remainder is then subtracted from the modulus number, giving the check digit.

$$11 - 4 = 7$$

Therefore, the modulus 11 check digit for the part number 036532 is 7. The new part number to be used in the inventory system would then be 0365327.

In order to use the check digit, all part numbers would have to include the check digit as part of the number. Thus, the part number which would actually be used throughout the inventory system would be 0365327. Since there are some calculations which must be performed to determine the check digit, a computer program will normally be written which will convert the part number without a check digit into a part number with a check digit. This program would then list all of the part numbers with correct check digit and these are the part numbers which would be used when the new system was implemented.

When a number which has a check digit is read as input to an edit program, the program will check the number to be sure that none of the data entry errors mentioned previously have taken place. The process which the edit program would perform to check the check digit is illustrated in the following example.

EXAMPLE

Step 1: The entire number is multiplied by the same weights as when the check digit was determined. The check digit itself is multiplied by one.

PART NUMBER:	0	3	6	5	3	2	7
WEIGHTS:	x7	x6	x5	x4	x3	x2	x1
	0	18	30	20	9	4	7

Step 2: The products of the multiplication are added together.

$$0 + 18 + 30 + 20 + 9 + 4 + 7 = 88$$

Step 3: The sum is divided by 11. If the remainder is equal to zero, the number is correct. If the remainder is not equal to zero, the number is incorrect.

$$88 \div 11 = 8, \text{ remainder is 0}$$

Note in the example above that the remainder after the division by 11 is zero. Therefore, the number which was checked is correct. If the remainder was not zero, then it would indicate that some type of error had occurred in the preparation of the input value. If an error is found, then the input record would not normally be processed and the record would be listed on an exception report.

Thus, it can be seen that the use of the check digit can lead to verification of the proper number being used and this can be quite important, especially when the value being checked is used as a key to a file. The check digit method will normally catch all transcription and transposition errors and about 95 percent of the other random errors which occur in the preparation of data.

OUTPUT CONTROLS

The function of output controls is to ensure that all transactions are included in the processing, that arithmetic calculations have been performed correctly, that the processing has taken place as specified, and that no unauthorized alterations to the data have occurred from the time the input data is edited until the time of the preparation of the final reports and updated files. Such alterations could occur because of the omission of records from processing, from incorrect programming or, perhaps, by intentional alteration of data by those in operations who are processing the data.

Some common output control techniques include:

1. COMPARISON OF TOTALS - The basic output control technique is based upon the comparison of totals produced on output reports with summary totals obtained manually or independently from the original source documents. This includes comparison of dollar amounts and counts of items processed.

2. CONTROL OF EXCEPTIONS - Investigations should be made of all limit violations and tests for reasonableness to determine the cause of such exceptions. Other techniques involve the comparison of totals produced this period with totals produced during a similar past period and the analysis of large deviations, and the use of statistical analysis of totals such as gross profit percentage, average hourly rate, etc. to assist in detecting deviations from the norm.

3. SAMPLING OF DATA - Another method of output control is based upon the periodic checking of representative transactions and the following of data through the system to assure that the record has been processed as required.

These output controls should be designed and applied to each system so that the integrity of the data which is processed is ensured. Without some type of output controls, there is the possibility that some type of alteration of data during the processing of the system could take place, leading to the possibility of computer-related fraud or theft within the system.

ORGANIZATIONAL CONTROLS

In addition to controls over input, processing, and output there are other controls which should be considered by the systems analyst in the design of a new system. One of the most important of these controls is ORGANIZATIONAL CONTROLS. Proper organizational controls may exist within the current system; however, if they do not, the analyst should recommend new or improved organizational controls to ensure the proper operation and control of the new system being developed.

An important form of organizational control is achieved by the separation of job duties. To achieve organizational control job duties should be divided and responsibilities delegated among employees in such a manner so that no single employee or group of employees has complete control over a transaction from its initial preparation and processing to its ultimate distribution in the form of a report. Thus, in a data processing system, no single employee or group of employees should have control over all aspects of input, processing, and output. By organizing a system so that each of the three elements is performed by different groups of people, a measure of control is acheived.

Thus, for example, in a payroll system a single payroll clerk or group of clerks should not have responsibility for recording employee pay rate changes, preparing time cards, preparing the pay checks and distributing the pay checks. Such a system could readily lead to internal fraud.

Segregation of Functions

There are five major functional responsibilities in a data processing department where segregation of duties should occur. These include:

1. Systems Design and Development
2. Programming
3. Input Preparation
4. Operations
5. Data Control

The degree that duties within each of the five areas can be segregated is directly dependent upon the size of the organization, the volume of work processed and the number of employees involved. In large organizations it is normally desirable that none of the employees in any of the five groups be permitted to participate in and control the work of another group. Where practical, employees should be restricted from having access to work areas of other groups or to the records which they maintain.

Some of the areas where segregation of duties is possible include the following:

1. Systems Design and Development
2. Program Writing
3. Program Review and Approval
4. Input Preparation
5. Input Control and Balancing to Source Documents
6. Computer Operations
7. Reviewing Exception Reports of Edit and Update Runs
8. Balancing Input Control Totals to Computer Printouts
9. Distribution of Output

As much as possible a division of job duties in the above areas provides a greater amount of control. In small data processing installations segregation of control is often more difficult because of the relative few employees, and a single individual may serve as systems analyst, programmer and even operator; nevertheless, organizational control should be an important aspect of all systems.

THE AUDIT TRAIL

Beyond the need to control the day-to-day operation of a system to ensure correct and accurate output, there is a need to design the system in such a manner that any transaction, total, or resulting output may be traced back to the original source data. This concept is referred to as an AUDIT TRAIL.

Most business applications require that audit trails be a part of every system for the following reasons:

1. The audit trail may be necessary for recreating files if they are accidentally destroyed.

2. Business and legal considerations, such as a company audit by the internal revenue service, requires audit trails.

3. The audit trail provides a means for checking and correcting errors which occur within the system.

In creating an audit trail it is necessary for the analyst to provide:

1. Documentation of all transactions to permit the association of all transactions with their original source documents.

2. A system of accounting controls which proves that all transactions have been processed and that all accounting records are in balance.

3. Documentation from which any transaction can easily be recreated and processing continued, should any transaction record be lost or destroyed at some point in the system.

In many cases it is necessary to check the accuracy of an amount or total within a report. This may be by a user department suspecting an error or governmental auditors reviewing the financial data of a company. In any case the system must be designed and documented so that by reference to the documentation, not including the program coding, it is possible to manually construct the records comprising the data. This may necessitate referring to input documents, printing intermediate files, and reviewing a variety of reports within a system, but in all cases all totals and transactions must be able to be accounted for.

Creating the Audit Trail

In establishing an audit trail the original source documents from which the transactions were generated should be sorted and arranged in some logical sequence. For example sales orders could be arranged by order number, payroll records by employee number, and inventory records by item number. If the source documents are sorted into some logical sequence, the basic problem is one of assuring that the sequence in which the records are stored is somehow reflected on the reports reflecting the transactions. In this manner any transaction can be immediately traced to its source. Where summary reports are produced which include summary totals provision must be made to allow reference to the batch of source documents from which the summary totals were derived.

In addition to arranging the source documents for easy reference it is necessary to provide some method of reviewing the status of master files after updating has taken place, particularly in those cases where no output report is produced. In this case updating is taking place and records are being changed, but there may be no visible record of these changes on a daily basis if magnetic disk or magnetic tape files are being used. The one basic method of creating an audit trail in such cases is by means of a "file dump." By dumping the updated tape on the printer or on another tape an audit trail is created.

The audit trail must provide for the tracing of all transactions and processing to satisfy all legal, accounting, and practical requirements to assure the effective operation of the system.

UNDETECTABLE ERRORS

In any system errors can occur which defy detection. These types of errors normally result from human mistakes. An example of an undetectable error is one in which the original amount is recorded incorrectly on the source document, i.e., a customer orders 10 items but the amount is recorded as 01 items ordered. Not until the customer receives the merchandise will this type of error be detected.

While it must be realized that human errors can be made that cannot be detected within the system, this should not detract from the use of a comprehensive set of checks within the system. By far the majority of error conditions are detectable and can be discovered by a complete system of controls.

SUMMARY

It should be emphasized that controls cost money. The hiring of data control clerks, the writing of edit programs, and the correction and detection of errors are expensive procedures to implement. Excessive controls can hamper processing; however, inadequate controls can make the processed data useless to management. Controls should, therefore, be used wisely. Only those which satisfy a need should be included in the system, and these controls should be simple and easy to maintain. It must be noted, however, that a system lacking in adequate controls is susceptible to both intentional and unintentional errors on the part of the persons using the system. Therefore, adequate controls for the system must be designed and implemented.

CASE STUDY - JAMES TOOL COMPANY

In the payroll system for the James Tool Company, controls are quite important because, as mentioned previously, whenever money is involved in a system it is critical that controls be firmly established. It addition, it was a lack of controls in the manual system which helped justify the computer system.

Case Study - Source Document Controls

As was noted in Chapter 7, a schedule was determined for moving the source documents from their point of origin to the data entry department for preparation as input records to the computer system. The analysts Mard and Coswell determined that some controls should be placed upon this movement. Therefore, they designed a form to be used for the time cards, the new employee forms, and the employee change forms to keep track of the number of forms transmitted and the persons who received the forms in the data processing department. This form is illustrated in Figure 8-17.

PAYROLL FORM CONTROL		
DATE:	FROM:	TO:
TYPE OF FORM: ☐ Time Cards ☐ New Employee Forms ☐ Employee Change Forms		
QUANTITY SENT:	QUANTITY RECEIVED:	
SENT BY: DATE: TIME:	RECEIVED BY: DATE: TIME:	

Figure 8-17 Example of Payroll Control Form

Case Study - Input Controls

As was noted in Chapter 5, it was determined by the analysts that the most economic means for inputting data into the computer system was for the data to be keypunched on punched cards. Since the payroll data is quite critical, the analyst determined that all data which is keypunched for the payroll system must be verified also. Therefore, all data preparation will include the verification of the punched data as well as punching the data itself.

Case Study - Processing Controls

As has been noted previously, the Edit Program is the primary focal point for the processing controls. Therefore, the controls which are to be included in the Edit Program for the James Tool Company will be examined in detail.

New Employees - Format

The two cards which are required to add an employee to the master file contain most of the information which is input to the system concerning employee data. This data must be edited quite closely. The format of these cards is illustrated in Figure 8-18.

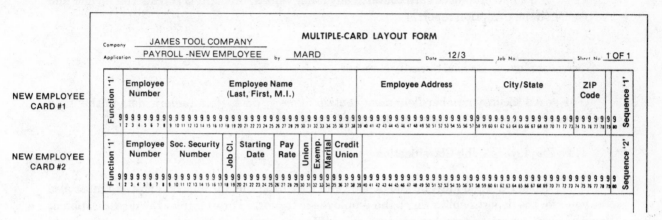

Figure 8-18 Format of New Employee Input Record

The editing which is performed in the New Employee Input Record is explained in the following paragraphs.

New Employees - Function

The function code which appears in column 1 of each input record to the Edit Program must contain one of the valid codes to be used in the system, that is, it must contain the value 1 through 9. Any other value in the function code field will make the input record invalid and will cause the record to be placed on the exception report and not used in further processing.

New Employees - Employee Number

As was noted in a previous chapter, the employee number is to consist of a two digit department number and a four digit seniority number. In addition, the analysts determined that since the employee number was the key for all of the master files and mistakes could not be allowed when dealing with the employee number, a check digit would also be used with the employee number. Thus, when the employee number is input to the system, it must consist of all numeric values and the check digit associated with the employee number must be correct. The technique for determining the correctness of the check digit is explained in detail earlier in this chapter. If the employee number is entered improperly in the input record, the record will not be processed but rather will be placed on the exception report for corrective action.

New Employees - ZIP Code

The ZIP Code for all input data must be numeric. If it is not numeric, the record is not processed but should be recorded on the exception report for corrective action.

New Employees - Sequence Number

The sequence number in column 80 of the input cards is used when new employees are being added to the master file but not when changes are being made. For new employees, the field must contain either the value 1 or the value 2, to indicate which data card is being processed. If a new employee card contains any other values, the card is considered invalid and is placed on the exception report.

New Employees - Social Security Number

The social security number field must contain numeric data. Non-numeric data in this field will cause the record to be placed on the exception report.

New Employees - Job Classification

The job classification must be one of the valid classifications for the company and must also be valid for the department in which the employee is to work. For example, the classification of payroll clerk cannot be found in an input record if the department in which the employee is to work is a machine shop. Therefore, a table would be established with the valid job classifications for each department and if the classification was incorrect, then the record would be rejected and placed on the exception report.

New Employees - Starting Date

The month portion of the starting date must contain the value 01 through 12, depending upon the month in which the employee was hired. Any other value in the month portion of the field will cause a rejection of the record. The day portion of the starting date must correspond to the number of days in the respective month. For example, a date of 0931 would be rejected because September only has 30 days.

New Employees - Pay Rate

The data in the pay rate field must be numeric and fall within the range dictated by the job classification. For example, if the job classification of ''3Y'' had pay rates of from 4.40 to 5.80 per hour, then a pay rate of 3.95 would not be valid and would cause the record to be placed on the exception report and not processed any further.

New Employees - Union

The Union Code must reflect one of the valid unions which are found in James Tool Company and must also correspond to the type of union or unions found in the department in which the employee works. For example, the designation for the machinists union would be incorrect for an employee who was to work in the personnel department. If an invalid union code is found in an input record, the entire record is rejected and placed on the exception report.

New Employees - Exemptions

The field containing the number of exemptions for the employee must contain numeric information. If the value is not numeric, the record will be rejected and placed on the exception report.

New Employees - Marital Status

This field must contain the value 1, 2, 3, or 4, as these are the only valid values for the marital status. Any other value will cause the record to be placed on the exception report and it will not be processed further.

New Employees - Credit Union Deduction

The value specified for the credit union deduction must be numeric. In addition it has been found through historical analysis by Mard and Coswell that the credit union deduction very rarely exceeds 30 percent of the gross income that an employee will earn from his regular earnings. Therefore, in addition to verifying that the field contains numeric values, the edit program will also check to ensure that the value contained in the credit union deduction field does not exceed 30 percent of the regular earnings pay that an employee will normally earn. The regular earnings pay would be calculated by multiplying the pay rate in the input record by 40. If the credit union deduction does exceed 30 percent, the record will still be processed but a warning message will be written on the exception report.

Case Study - Employee Change Form

As will be recalled, the data and the format of the employee change records is almost identical to the new employee input records. Therefore, the checking which is performed on the fields in the new employee input records will also be performed on the fields in the employee change input records.

In addition, however, when a change record is read in the edit program, only a particular change will take place, that is, when a change in pay rate is to be performed, only the pay rate must be contained in the input record. Thus, only the pay rate must be subjected to the editing criteria specified above. In addition, however, it will be noted that since only the pay rate must be contained in the change record, the other fields must not contain information. Therefore, in addition to checking the pay rate for valid numeric data, the edit program will also check to ensure that the rest of the fields do not contain data. If they do, the record will be rejected and placed on the exception report because there is the possibility that an error was made since there is additional data in the record.

Case Study - Update Programs

The update programs in the payroll system for the James Tool Company would perform many of the checks which have been discussed previously in the chapter. For example, it is known that when processing additions to the file, the input records must be in sequence by employee number and, additionally, each set of records to add an employee must be in sequence by the sequence number in column 80 of the input records (see Figure 8-18). Therefore, a sequence check will be performed in the update program to ensure that both the "1" and the "2" records are present before an employee is added to the file. If both of the records are not present, the employee will not be added and an entry will be made on the update exception report.

In addition, as has been mentioned previously, checks will be included in the update program to report those employees for whom time cards did not appear in the input stream and tests will also be performed to ensure that only one time card appears for a single employee.

Case Study - Output Controls and Editing

As the analysts developed the output reports from the payroll system, they were aware of the need for controls on the output reports and also the need for an auditing trail to be established. Therefore, the output reports such as the Payroll Register were designed to take totals when necessary so that all of the transactions were accounted for and the processing which took place in the payroll system was evident. Thus, in reviewing the output which is to be generated from the system, and in checking with the persons who will be involved in the auditing process, it was determined that the output produced from the system in the exception reports and the reports generated from the system were sufficient to satisfy the requirements of the James Tool Company.

Case Study - Organizational Controls

As noted, Mard and Coswell were the analysts in charge of the development of the payroll system. Since the job was quite a large one, neither of these analysts would be involved in the programming aspect of the system. In addition, neither the analysts nor the programmers would be involved in any of the operations of the system once it is implemented on the computer. Therefore, it was felt that there was sufficient separation of duties in the development of the system that the risk of any one person being able to "sabotage" the system was quite small.

In developing the operational methods to be used in the system, the analysts purposely kept the personnel department responsible for changes to the employee master but required that any checking of the exception reports was to be handled by personnel in the data control section of the data processing department. Therefore, again, it was felt that there was little opportunity for any one or two persons to be able to manipulate the data in the system in such a way that illegal or fraudulent actions could be taken with the data which is processed in the payroll system.

SUMMARY

The use of controls in a data processing system are quite important. First, the high speed of the computer allows more thorough checking of data than would be available in a manual system. Second, the controls must be established in a data processing system because the processing of the data takes place out of the sight of any person, that is, it is done on the computer. Therefore, if tight and adequate controls were not established, there is the definite risk of invalid information being produced from the system or even illegal practices on the part of the people who work with the system. It is, therefore, one of the primary responsibilities of the systems analyst to design adequate controls for any data processing system which is implemented.

STUDENT ACTIVITIES—CHAPTER 8

1. Define the term "system controls."

2. What are the basic reasons for establishing system controls?

3. List the basic types of controls.

    1)
    2)
    3)

4. Explain the term "batching" and the use of a "batch control form."

5. What is the basic purpose of input controls?

6. Explain the types of input controls which should be utilized when designing on-line systems.

7. Explain the term "processing controls."

8. Explain the use of "group controls."

9. List the basic types of tests and checks that may be utilized for individual record checking.

   1)
   2)
   3)
   4)
   5)
   6)
   7)
   8)
   9)

10. Why are auditing controls important?

DISCUSSION QUESTIONS

1. An executive from one of the leading accounting firms in the country recently remarked ''It is virtually impossible for accountants in our firm to properly audit any large-scale data processing financial or inventory system. For one thing, most systems are not designed for any type of efficient auditing and, even if they are, there is such a multitude of data which can be processed on a large computer that it would take hundreds of man-years to audit the system.''

   What do you think of his opinion?

2. A vice-president of an east coast firm is of the opinion that ''all the checking and validation in the world cannot stop someone who wants to defraud any data processing system. The only way to stop fraud in data processing systems is to screen your personnel who will be working with these systems to be sure there are no crooks, felons, or others with a history of stealing. If we can keep the crooks out of our business, and I'm sure we can, then we can stop wasting so much time and money on security systems and checking programs and get on with the business of producing useful management information from our data processing systems.''

   Is this vice-president right?

3. A president of a fire alarm company stated ''The biggest danger to the data in a data processing department is the destruction by fire, since there are so many electronic components around. Magnetic tape is especially susceptible to heat of any kind. Therefore, data processing managers should worry and protect against fire and stop spending so much time and money worrying about programmers or clerical personnel gaining access to the data and using it for their own purposes!''

   Is this person just trying to sell fire alarms or is he correct?

CASE STUDY PROJECT—CHAPTER 8
The Ridgeway Company

INTRODUCTION

After designing the systems processing for the billing systems of the Ridgeway Country Club the analyst must formally define and document the controls which are to be incorporated with the system to ensure that all transactions are accounted for and that "only valid data is accepted and processed completely and accurately."

STUDENT ASSIGNMENT 1
CONTROL OF SOURCE DOCUMENTS

1. Prepare a narrative description of the procedures that are to be followed in establishing a system of controls for the source documents that serve as input to the billing system.

2. Design all required control forms, including any necessary batch control forms.

STUDENT ASSIGNMENT 2
INPUT CONTROLS

1. Prepare a narrative description of the input controls that are to be established.

STUDENT ASSIGNMENT 3
PROCESSING CONTROLS

1. Define the editing which is to take place for each of the fields of the transaction records by the "edit program."

STUDENT ASSIGNMENT 4
OUTPUT CONTROLS

1. Define any output controls which are necessary for the output to be produced from the billing system.

SYSTEMS DESIGN APPROVAL - PRESENTATION TO MANAGEMENT

CHAPTER 9

SYSTEMS DESIGN APPROVAL - PRESENTATION TO MANAGEMENT

9

INTRODUCTION

Throughout the entire systems design phase of the systems project the analyst should have been involved in ''selling'' the system, that is, working closely with all levels of management and supervisory and operations personnel in the development of an effective systems design. Proper ''selling'' of the system during the systems design phase has two significant advantages:

1. Proper selling and involvement of personnel associated with the new system will lead to a systems design that is greatly improved over the original system because of the analyst's interaction with operations and management personnel at all levels.

2. Systems implementation will normally be made much easier when the system is ''sold'' during the systems design phase.

At this point, however, two important steps remain in the systems design phase of the systems project. These steps are:

1. The final systems design must be formally agreed upon and approved by users of the system and by management.

2. A formal presentation must be made to top management in order to obtain approval to enter the Systems Development Phase of the systems project.

SYSTEMS APPROVAL

In obtaining approval for the overall systems design from users and management, the following steps should be taken to ensure complete knowledge and agreement as to the design of the system:

1. Obtain approval of all output reports.

2. Obtain approval of system input.

3. Obtain approval of the file design and processing.

4. Obtain approval of the overall system design from data processing management.

5. Make a presentation to top management to obtain approval to enter the Systems Development Phase of the systems project.

Systems Approval - Output Reports

Although various members of the user departments should have been consulted during the systems design phase concerning the formats of the output reports, it is likely that the analysts will have prepared the final design of the reports on printer spacing charts somewhat independently of the persons in the user departments. Therefore, it is desirable for the analyst to review the final design of all reports with the users so that they can be assured that all data required has been included. The format of the report is also important because it may be that the changing of the location of a field or other data on the report may make the use of the report an easier task.

There are problems which can occur in report design that may be obvious to report users but easily overlooked by the systems analyst. For example, there may be binders in which certain reports are to be placed of which the analyst was not aware; and it may be desirable to allow more space on the side or top of the report so that the report will be more readable when placed in the binder. If this factor was overlooked in the initial design, a change in the format of the report can easily be made by preparing a new printer spacing chart. Although not a change of any significance at this time, a change such as this after the program to produce the report has been written would require some program modification which, although not major, could cause additional implementation cost and delay.

When a particular document has been reviewed by the user and management it is many times a good idea to have them indicate their ''buy-off'' of the document by signing or initialing the form so that if controversies develop at a later time, the analyst has evidence that approval had been given. This step is important because, as noted previously, it is best if there are no changes after final approval of the system is obtained from management and the users. An example of this technique is illustrated in Figure 9-1.

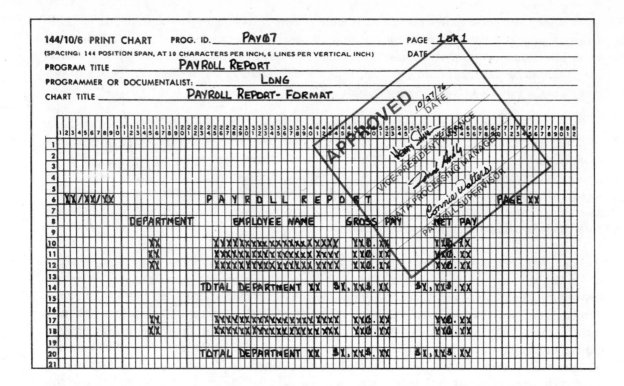

Figure 9-1 Example of Approval

Note from the example in Figure 9-1 that the format of the report has been approved by all persons who are directly involved in the use of the report, including user personnel, data processing management, and higher level management responsible for the control of the data contained on the report. Thus, if any questions arise when a report is produced out of the system, the analyst has evidence that the format of the report had been approved prior to the implementation of the system.

Systems Approval - Input

The same type of approval should be given to the input as is given to the output. The input can be basically broken down into two categories:

1. Source Documents

2. Input Records

All source documents should be formally approved by responsible persons within the originating department. It is important for the format of the source documents to be approved since these documents will be used in the various phases of the system and should be designed for ease of use. As noted previously, the source documents are designed to supply the information which will be necessary to create the output desired from the system. Therefore, the user departments must check carefully to ensure that they feel the information available on the source documents is sufficient for the output required and that the information to be recorded on the source document is readily available.

The newly designed source documents should also be reviewed for ease of understanding. It does little good to have a source document with all possible information on it if someone cannot fill in the document properly. Thus, when the user departments are examining the source documents to be used within the system, they should be made aware that this is one of the aspects of the input which must be approved.

The formats of the input records, such as punched cards, are primarily of interest to those who will be involved in transcribing the data from the source document to a machine processible form. As noted previously, one of the considerations when designing the input records is the ease in transcribing the data from the source document to the input record. The supervisor of the data entry section in the data processing department or in the department where the input data is to be prepared should be able to approve the input record formats based upon the information in the source documents.

In addition, there should be approval of the medium to be used to prepare the input records. If the analyst has specified key-to-tape, but the data processing management feels that key-to-tape would be too expensive in light of the savings it may afford, then there would have to be agreement to switch to another type of input medium. It must be noted that although many of these topics seem to be relatively minor and of little importance, it is disagreements during implementation in such areas as record formats and whether cards or tape or disk should be used which can critically affect the implementation schedule. Therefore, it is by far the best policy to have all disagreements and controversies settled well before the implementation of the data processing system. If everyone knows what is going to take place, then there is a strong possibility of a smooth implementation of the new system.

Systems Approval - File Design And Processing

The approval of the file design and the processing which is to take place within the system is perhaps more critical to the data processing department management than to the users. It is important to point out to the user, however, the variations in the processing of the manual or old system as compared to the new system. Most likely, there will always be some type of scheduling differences which the user must be aware of and perhaps changes in job duties. Since the analyst has performed an extensive investigation and analysis, it is to be hoped that there will be a minimum of misunderstandings of procedures between the user and the system which has been designed. There are occasions, however, where the analyst feels that he has designed a system which will fit into the plans of the user department and department management disagrees. At this point, it is necessary to get the differences resolved because, as noted previously, changes after implementeation can be costly and very error-prone. Cases have even been reported that dissatisfied users have "sabotaged" the system to prevent effective implementation.

In most cases, the user will not be particularly interested in the file design of any master or transaction files, but he will be interested in how they will be able to process the data to give him the results he needs. For example, if a department manager feels that some type of fast response is required because there are terminals in the system which will be used to make inquiries into a master file, he is not interested whether the files are direct organization or indexed sequential organization, but he is extremely interested in the time it will take the computer to respond to his inquiry. Thus, when presenting the system to the user, the analyst must consider the type of information which the audience will want to know and then tell them what the system will do, pertinent to their interests.

Systems Approval - Data Processing Management

Another important aspect of the final approval of the systems design is the approval which should be forthcoming from the data processing departmental management. No computer system will ever be successfully implemented if there is not complete backing and cooperation from those in charge of the data procesing department. There may be hardware problems which must be solved before the system can be implemented and the management of the data processing department should be aware of any hardware needs and should approve them.

The file design developed by the analyst is always of great interest to the data processing department management. This is because there may be programming or operational difficulties with certain types of file design with which the analyst is not familiar. In addition, there may be hardware acquisitions which must take place and data processing management should be aware of what and how the files are to be processed.

Personnel requirements may also pose significant problems to data processing management and these must be resolved. It does little good, for example, for the analyst to propose a complex on-line system utilizing terminals if programming personnel currently employed do not have the knowledge or background to program such a system or if top management is not willing to train or hire the required personnel.

Thus, prior to making a presentation to management the analyst should gain as wide approval as possible from those using and affected by the system.

Summary - Systems Approval

Again it should be pointed out that an important reason for obtaining approval of source documents, input, processing, and output of a newly designed system is to assist in providing a smooth implementation and overcoming the natural ''resistance to change'' that always occurs when the activities and methods of doing business are changed. A system that is ''forced'' upon the user from the systems department is much more likely to fail than a system in which all users have been involved from preliminary investigation to final implementation.

If the analyst has made it quite clear that when the final systems design is approved there will be no changes to the system, then there is a good possibility that this is what will occur. If, on the other hand, the presentation to the user has not been understood by all who are involved and there has not been agreement between all parties on the systems design, then the likelihood of requests for changes and the likelihood that these requests must be honored is much stronger.

Formal approval of the systems design is extremely important because changes which must be incorporated into a system after the system has been designed and either partially or completely implemented are extremely expensive to include. They are expensive in terms of time because changing something which is already in existence is more difficult than including it in the original plan. For example, adding a field to a master file after the master file has been created requires a complete recreation of the total file. Changing this file may also require changes to any program which uses the file because the file definitions in the program will not be correct. There will also have to be changes made to certain routines because with the addition of a field, there will probably be additional processing which will have to take place. For example, if a field is added to a record and that field must appear on a report, the heading routines in the program would have to be changed to include the new field as well as new routines written to place the field on the report. Thus, with the single addition of a new field, there may have to be changes and corrections made to many programs, in addition to reorganizing the file.

Another reason that changes after the system has been implemented are bad is that whenever a program is "patched" with a new routine, the chances for processing errors within the program greatly increase. If the programmer can plan for all of the processing which is to take place within a program when he designs the logic of the program, it is relatively easy to place the routines in their proper place within the program in a logical order and efficient manner. When routines are added to a program at a later time, this efficiency is largely lost and so is the main logic which the programmer has used within the program. It has been repeatedly found that programs which are patched with new routines take a great deal longer to test and debug and they fail at a much higher rate than those programs which include all of the processing in their original design.

PRESENTATION TO MANGEMENT

It will be recalled that the last formal presentation which the systems department made to management was when approval was given to enter the systems design phase of the systems project. Once the system has been designed, it is necessary for the analyst to again make a formal presentation to management.

The primary objective in making a presentation to management after completion of the systems design phase of a systems project is to obtain management approval to continue with the next phase of the systems project, that of Systems Development. During each phase of the systems project allocation of resources increases in terms of time and personnel; therefore, management must be informed at each phase of the project and their approval obtained.

The presentation to management after the systems design should include the following:

1. A management oriented abstract briefly reviewing the problem and summarizing progress. This review should include time, personnel, and costs expended thus far on the project, a brief overview of the systems design, and the analyst's recommendations.

2. A detailed review of the overall operation of the proposed system and an explanation of the benefits of the new system.

3. A projected time schedule for the implementation of the new system indicating the time required for each of the major steps in the systems development phase.

4. An estimate of the costs for each of the steps in the next phase of the systems project, the Systems Development phase.

5. Comparison of the costs of operating the ''old'' system as compared to the new system when the new system is implemented.

6. An objective recommendation from the analyst relative to the value of the new system.

The analyst, in making the presentation of the systems design to management, should follow the same basic presentation techniques as explained in Chapter 3; that is, the audience should be defined, the objectives of the presentation defined, the presentation should be well organized, and the analyst should utilize the necessary audio-visual aids to make the presentation effective and easily understood. The analyst should **not** enter this stage of the systems project under the assumption that management will readily accept and approve the system, as it must be realized that management may terminate any project at any time if they feel that other priorities are more important or if the system under review does not contribute significantly to overall profit structure or efficient operations of the business.

THE WRITTEN PRESENTATION

In addition to an oral presentation to management, the analyst should prepare a written report for management review. The first section of this written report should be a management abstract. With projects involving large expenditures of company resources in terms of money and personnel, the presentation of the systems design will normally be made to top management personnel at or near the vice presidential level. In some smaller organizations the company president may also be involved. It should be recalled that the higher the level of management, the more summarized the presentation or report should be; however, this presentation must be designed to provide enough information about the system for management to make an intelligent decision relative to the effect of the system on the profitability or operating success of the company.

The Management Abstract

The management abstract is designed to provide top management with an overview of the factual information needed relative to the origination and current status of the systems project, the current and projected costs of the study and the overall operation and projected benefits of the new system. This abstract should be a brief summary of the details of the presentation that are about to be made.

Review of the New System

The new system which has been designed is then described at the management level. Emphasis should be placed on the managerial advantages of the new system, such as more accurate information and reduced costs, rather than the detailed data processing operating procedures. Narrative descriptions of major inputs, outputs and main processing should be included. A systems flowchart may be included to show the major operations performed. A description of the detailed contents of each of the fields in a record or a detailed description of the records in a file should be avoided unless such information is requested. Personnel requirements and equipment requirements should also be discussed in relation to the new system, as management is always cost conscious. Changes in organizational structure should also be reviewed if the new system will result in new job duties or responsibilities.

Review of the Systems Development and Implementation Schedule

After management has gained an overview of the history of the project and has an understanding of the new system, the schedule for systems development and implementation must also be reviewed. Major tasks include the programming of the system, the system testing, the documentation of the system, the file conversions, and actual implementation and cut-over to the new system.

The bar or GANTT CHART is quite useful for management presentations relative to scheduling as the Gantt Chart is easily read and presents the pertinent information in which management is interested. The activities listed on the chart can be in any detail desired. For management presentations, only major activities are normally illustrated. Figure 9-2 illustrates a Gantt Chart.

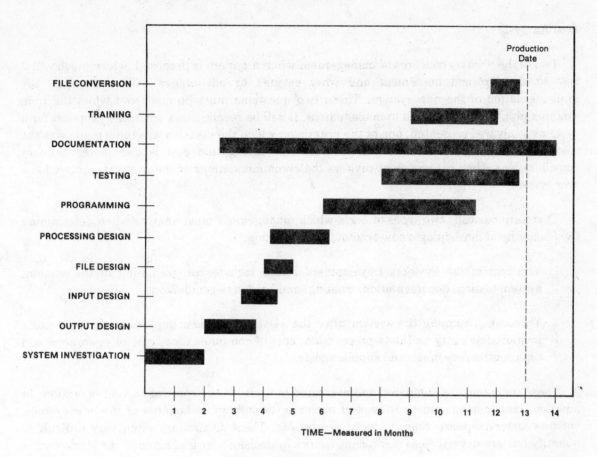

Figure 9-2 Gantt Chart

In the Gantt Chart illustrated in Figure 9-2 it can be seen that the activity to be accomplished is listed vertically on the chart and the elapsed time which the activity will take is listed across the bottom. The bars on the chart illustrate when each activity is to begin and when each activity is scheduled to be completed. Thus, from looking at the chart it is immediately apparent when an activity begins, when it ends, and the elapsed time required to complete the activity. It should be noted that the bars do not indicate the amount of work which is required in order to complete an activity. For example, the programming of the system is indicated to begin at the start of the seventh month of the systems project and continue through the eleventh month. This indicates that an elapsed time of five months will be required to complete the programming. It may be, however, that ten programmers are assigned to the project during these five months and, therefore, the manpower required is 50 man-months. In a like manner, the documentation is scheduled to begin in the second month and continue through the 14th month. However, the documentation of the system may only require 2 man-months worth of effort. Thus, the Gantt Chart is used to illustrate the elapsed times required to complete a project on schedule and does not necessarily illustrate the actual amount of work, in terms of time, required to complete the project.

In addition, it is simple to see from the Gantt Chart which activities are taking place concurrently. For example, it can be seen that the programming is to begin at the start of the seventh month and the program testing will begin at the start of the ninth month. Thus, there will be two months elapsed time of programming effort before any program testing will begin. Since, however, some programs can be tested while others are still being written, the programming efforts and the testing efforts will run concurrently for about 3-1/2 months. Then the programming effort should be completed but there is still significant testing to be done before the system is implemented.

Cost Analysis

Two of the primary concerns of management when a system is proposed is how much will it cost to develop and implement and what savings or advantages will result from the implementation of the new system. These two questions must be answered when the final systems proposal is presented to management. It will be recalled that when the proposal for a systems study was presented, one of the criteria on which the decision was to be made was the cost of performing the analysis. It goes without saying that cost is one of the primary considerations when management reviews the recommendations of the systems analyst for a new system.

There are basically two types of costs which management must analyze when determining the feasibility of developing a new system. These include:

1. The cost of the Systems Development Phase includes programming of the system, system testing, documentation, training, and hardware acquisition.

2. The cost of running the system after the system has been implemented; these costs include data entry and data preparation, cost of computer time, cost of operations and data control personnel, and supplies costs.

These costs must be compared and evaluated in terms of the costs of current operation. In addition, management must review and evaluate the intangible benefits of the new system, such as faster response time, reduced errors, etc. These factors are often very difficult to quantify but are nevertheless significant factors in decisions by management to implement a system.

It should again be pointed out that if, in the opinion of management, the cost of continued systems development and implementation is excessive as related to the benefits gained, it is possible that the system may be "scrapped" at this stage of the project.

The costs associated with systems development and implementation are relatively easy to identify and estimate if the analyst has performed a realistic estimate of the systems development and implementation schedule. The costs of systems development include the cost for personnel that will be involved in systems development including the programmers and analysts who will be working on the system.

The costs of implementation includes the costs of computer hardware, the number of hours or minutes of computer usage required for the processing of the data within the system, the cost of data entry, the cost of operations personnel, and the cost of materials such as cards, printer forms, magnetic tape, and related supplies. If new hardware is being acquired care must be taken to include such factors as housing of the hardware, remodeling of the facilities, and special electrical wiring requirements. The costs of such factors should be included in the management presentation.

The analyst in determining costs should estimate, as closely as possible, what all costs will be. This can be done by estimating the time which will be required to develop and implement the system plus the time which will be required to run the system. These times multiplied by the salaries of the personnel required plus the cost of using the various pieces of hardware for these times should provide an estimate of the system costs. Similarly, the analyst should be able to accurately forecast the amount of materials which will be required and to estimate these costs over a given period of time.

It should be pointed out, however, that the ability to estimate costs is directly associated with the ability to develop a realistic schedule relative to the time it takes to program, test, and implement a system. Historically, analysts have failed to accurately estimate systems development and implementation time by far underestimating time requirements, resulting in projects being late and costs far exceeding original estimates.

Recommendations of the Analyst

In submitting a final recommendation the analyst should state the advantages of the new systems design. Normally these advantages are based upon three basic factors.

1. The new system will reduce cost by reducing personnel, eliminating overtime, or preventing the necessity of hiring additional personnel.

2. The new system will provide management with more timely information.

3. The new system will eliminate errors and provide greater control.

The cost advantage of reducing personnel, eliminating overtime, etc., is often the easiest to quantify. The cost of the "old" system relative to personnel, time and material can easily be measured and compared to the newly developed system and the results objectively analyzed. However, factors such as improving accuracy, better control and more timely information are often times difficult to cost justify, but nevertheless are important factors in obtaining management's approval for the systems design. Such factors should be emphasized in the presentation to management.

Alternative Systems

In some cases, it is advisable for the systems analyst to prepare several alternative systems for the inspection of management and the users. It is not normally necessary to go into a great deal of detail for the alternative systems. For example, the analyst could prepare a detailed report concerning the system which he recommends as the system which should be implemented. He could then prepare just quick sketches of alternatives which may be available.

In general, the proposal of an alternative system would be necessary when there are cost considerations which must be taken into account. For example, it may be that an installation has four tape drives and only two direct-access devices. The system which the analyst feels would be most beneficial includes files which would be stored on direct-access devices, but it would require the rental or purchase of two more devices. He should present an alternative system which could be processed on the hardware which is currently available. Thus, if for reasons of cost or lack of availability of new hardware, the recommended system could not be implemented, there would be an alternate system which would be implemented. It should be emphasized, however, that the analyst should very strongly recommend the proposed system when he feels it would offer greatest benefit to the company. The alternative system should only be implemented when there is no way in which the recommended system could be implemented because of costs or other considerations.

SUMMARY

After the design of the system, it is imperative that the analyst present the system to the users of the system and to management to get their final approval of the remaining steps to be performed for implementation of the system. It is critical that all persons involved in the system give their approval at this point so that the programming, testing, and implementation of the system can proceed without constant changes and revisions.

CASE STUDY - JAMES TOOL COMPANY

In the payroll system for James Tool Company, the entire system in terms of input, output, files, and processing has been designed by the systems analysts on the project. It remains for the management of James Tool Company to give their approval to enter the systems development phase of the systems project.

The formal presentation to management is to be made to Howard James, President; Alice Hilbert, Vice President of Personnel; Arnold Henderson, Vice President of Financial Affairs; and David Green, Data Processing Manager.

Prior to making the formal presentation to management, the analysts Don Mard and Howard Coswell, made sure that all phases of the system had received approval by the users and management associated with the system.

Case Study - Systems Approval - Output Reports

All output reports to be prepared by the new system have been approved by users and related management personnel. Note in Figure 9-3 that the printer spacing chart for the Union Deduction Register has been approved by the appropriate individuals within the James Tool Company, including Delbert Donovan, Manager of Union Relations; David Green, Manager of Data Processing; Jim McKeen, Manager of the Payroll Department; and Arnold Henderson, Vice President of Financial Affairs. Approval is indicated by requiring their signature to a copy of the printer spacing chart.

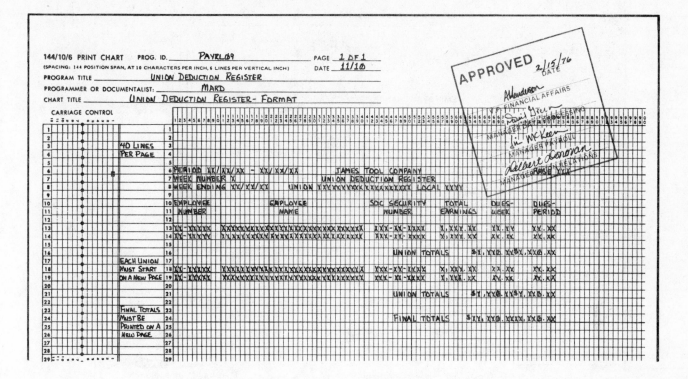

Figure 9-3 Approval - Systems Output

The printer spacing charts for all other reports to be prepared as a part of the new system would require similar approval.

Case Study - Systems Approval - Input

Because the original source documents have been redesigned it is essential that these documents be formally reviewed and approved by individuals in the personnel department and by the supervisor of the Data Entry Department.

In the James Tool Company Alice Hilbert, Vice-President of Personnel, and Harry Denton, the Employment Manager, reviewed the newly designed New Employee Form and Employee Change Form. It is important that these forms be reviewed by responsible individuals within the personnel department to ensure that the formal use of the form is easily understood, that the form can be filled in accurately, and that the form contains all required data. Approval of the forms was also obtained from Jane Miller, Manager of Computer Operations, and from Wilma Brown, data entry supervisor. Approval by the manager of computer operations and the data entry supervisor assures that the information on the source document can easily be transcribed by the data entry operators. Figure 9-4 illustrates the New Employee Form and Employee Change Form with the required approval.

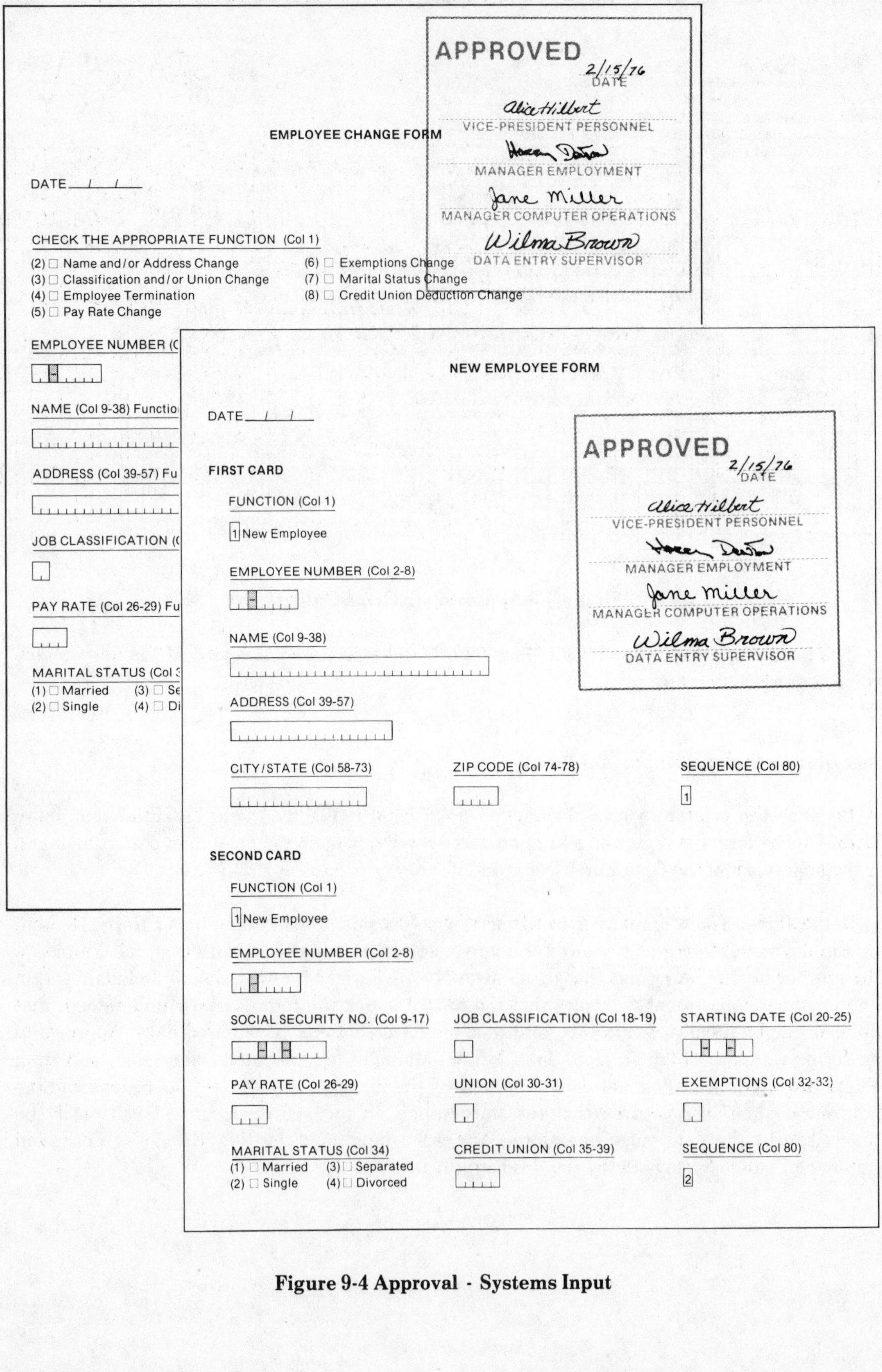

EMPLOYEE CHANGE FORM

APPROVED 2/15/76
DATE
alice Hilbert
VICE-PRESIDENT PERSONNEL
Harry Dutton
MANAGER EMPLOYMENT
Jane Miller
MANAGER COMPUTER OPERATIONS
Wilma Brown
DATA ENTRY SUPERVISOR

DATE___/___/___

CHECK THE APPROPRIATE FUNCTION (Col 1)

(2) ☐ Name and/or Address Change   (6) ☐ Exemptions Change
(3) ☐ Classification and/or Union Change (7) ☐ Marital Status Change
(4) ☐ Employee Termination   (8) ☐ Credit Union Deduction Change
(5) ☐ Pay Rate Change

EMPLOYEE NUMBER (C

NAME (Col 9-38) Functio

ADDRESS (Col 39-57) Fu

JOB CLASSIFICATION (

PAY RATE (Col 26-29) Fu

MARITAL STATUS (Col
(1) ☐ Married   (3) ☐ Se
(2) ☐ Single   (4) ☐ Di

NEW EMPLOYEE FORM

DATE___/___/___

FIRST CARD

FUNCTION (Col 1)

☐1 New Employee

EMPLOYEE NUMBER (Col 2-8)

NAME (Col 9-38)

ADDRESS (Col 39-57)

CITY/STATE (Col 58-73)   ZIP CODE (Col 74-78)   SEQUENCE (Col 80)
                                                              ☐1

APPROVED 2/15/76
DATE
alice Hilbert
VICE-PRESIDENT PERSONNEL
Harry Dutton
MANAGER EMPLOYMENT
Jane Miller
MANAGER COMPUTER OPERATIONS
Wilma Brown
DATA ENTRY SUPERVISOR

SECOND CARD

FUNCTION (Col 1)

☐1 New Employee

EMPLOYEE NUMBER (Col 2-8)

SOCIAL SECURITY NO. (Col 9-17)   JOB CLASSIFICATION (Col 18-19)   STARTING DATE (Col 20-25)

PAY RATE (Col 26-29)   UNION (Col 30-31)   EXEMPTIONS (Col 32-33)

MARITAL STATUS (Col 34)   CREDIT UNION (Col 35-39)   SEQUENCE (Col 80)
(1) ☐ Married   (3) ☐ Separated   ☐2
(2) ☐ Single   (4) ☐ Divorced

Figure 9-4 Approval - Systems Input

Likewise, the design of the systems input, such as the punched cards which are specified on the Multiple Card Layout form, should be formally approved. The approval of the Payroll Change Records are illustrated in Figure 9-5. These forms have been approved by David Green, Manager of Data Processing; Jane Miller, Manager of Computer Operations; and Wilma Brown, Data Entry Supervisor.

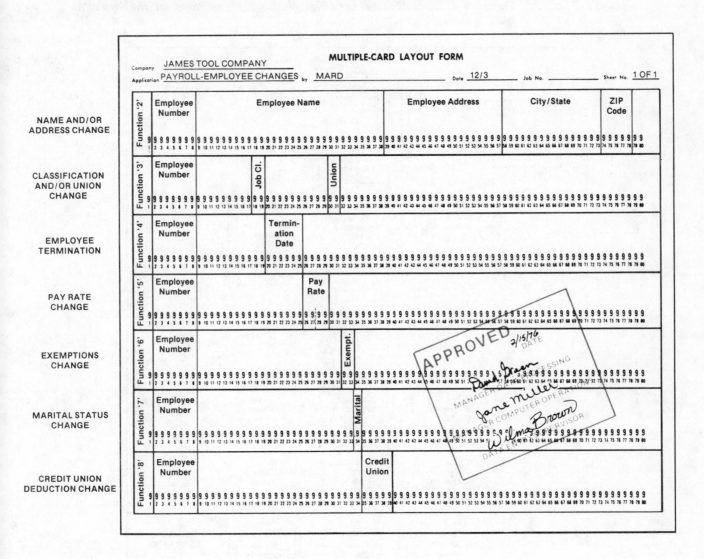

Figure 9-5 Approval - Input

Case Study - Systems Approval - File Design and Processing

Approval of the File Design and Processing and overall systems design was obtained from the manager of the data processing department. In the James Tool Company, David Green, Manager of Data Processing, approved the overall systems design after reviewing the detailed systems flowchart. Figure 9-6 illustrates the approval of the first segment of the systems flowchart.

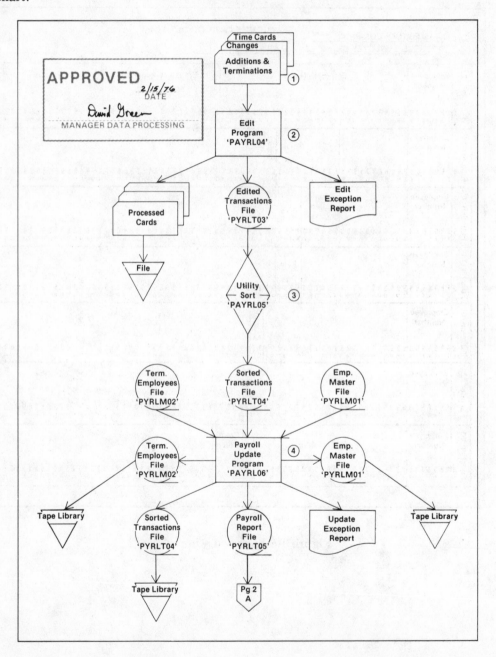

Figure 9-6 Approved Systems Flowchart

Thus, prior to the final presentation to management, the systems analysts had obtained widespread approval for all aspects of the system.

Case Study - Management Report

The following pages illustrate the report presented to management by the systems analysts as a result of the systems design phase of the system project for the payroll system.

JAMES TOOL COMPANY
SYSTEMS DESIGN
PAYROLL SYSTEM

February 25, 1976

ABSTRACT

On September 10, 1975, a request was received by the systems department to investigate problems occurring within the payroll system of the James Tool Company. Excessive overtime was being paid and errors were occurring in the deduction of union dues. Upon receipt of the complaint a preliminary investigation was undertaken. As a result of the preliminary investigation, it was determined that the complaints were valid and it was recommended by the systems department that a detailed investigation of the payroll system be undertaken to determine alternative solutions to the problem. The cost of the detailed investigation was estimated at $5,350.00. Permission to conduct the detailed investigation was given by management on October 4, 1975.

The detailed investigation was undertaken immediately. Work was completed on Noverber 2, 1975, at a cost of $4,950.00, four hundred dollars under the amount budgeted for the study. Several alternative solutions to the payroll problem were presented, including revising the current system and designing a new system using the computer system currently available to the James Tool Company. Management endorsed the recommendation to develop a computerized payroll system. It was estimated that the total cost of systems design and production of the payroll for two years would be $42,000.00. It was estimated that it would take three months for the systems design phase of the project and an additional five months to program and implement the system.

The systems design phase of the project was completed on February 15, 1976.

The new systems design provides for the efficient computerized production of all required reports including a Payroll Register, Union Deductions Register and Weekly Payroll Checks. The system includes numerous checks and controls not found in the manual system, and will eliminate the need for overtime in the payroll department.

It is recommended that approval be given to begin the systems development phase of the systems project leading to the planned implementation of the new payroll system. Estimated completion time is five months.

Figure 9-7 Management Report (Part 1 of 7)

-2-

OPERATION OF THE PROPOSED SYSTEM

The proposed system is to be implemented on the IBM System/370 computer currently being leased from IBM by the James Tool Company. No additional hardware will be required. The operation of the computerized payroll system includes the following steps:

1. Information obtained from personnel and payroll departments is read into the computer.

2. During the week the personnel and payroll departments will complete various forms from which data for the system will be prepared. These include information concerning new employees, changes for each employee such as pay rate changes, and information concerning terminated employees.

3. Also, each employee will use a computer-prepared time card for clocking in and out.

4. At the end of the week, all the information will be assembled and transmitted to the data entry area in the data processing department where, after some initial checking, it will be transcribed into a machine-processible format.

5. The data will be input to the computer system on Tuesday evening. Extensive checking will take place in the initial program in the payroll system to ensure that all data entering the system is valid and complete. Any errors which are found will be corrected before further processing takes place.

6. The master files which contain all of the payroll information for each employee will be updated to reflect the weekly earnings of each employee as it is calculated.

7. After the pay for each employee has been calculated, the various reports from the system will be prepared. In addition, new time cards will be prepared.

NOTE: A detailed systems flowchart is contained in the Systems Documentation Manual.

Figure 9-8 Management Report (Part 2 of 7)

-3-

SYSTEMS DEVELOPMENT SCHEDULE

During the systems development phase the following major activities must be undertaken:

1. Programming the system to produce the required results.
2. System testing.
3. System documentation.
4. Conversion of manual files to magnetic tape files.

Programming, documentation and operations can be handled by the current staff. It is estimated that 10 man months of effort will be required to program and fully test the system. By assigning two programmers to the project the payroll system should be ready to be implemented within five calendar months.

The following chart illustrates the projected systems development schedule:

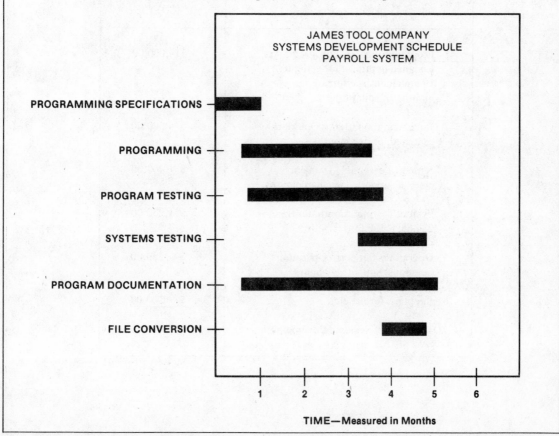

Figure 9-9 Management Report (Part 3 of 7)

-4-

SYSTEMS DEVELOPMENT COSTS

The following is a summary of the costs estimated to occur during the Systems
Development phase of the systems project:

SYSTEMS DEVELOPMENT COST ANALYSIS

PAYROLL SYSTEM

ACTIVITY	ESTIMATED COSTS
1. Programming - Two programmers assigned to project for estimated 10 man months effort. (Five months calendar time.)	$14,000.00
2. Documentation Clerk - One clerk assigned for five months.	$ 4,000.00
3. Support Personnel - Data entry, clerical support. Two man months.	$ 1,400.00
4. System Testing - Computer time, 25 hours, $50.00 per hour.	$ 1,250.00
5. Operations Support - Operations personnel support, 40 hours.	$ 250.00
6. Supplies, forms, etc.	$ 500.00
TOTAL Cost of Systems Development	$21,400.00

Figure 9-10 Management Report (Part 4 of 7)

-5-

IMPLEMENTATION COSTS

After the system has become fully operational and implemented the following monthly costs may be anticipated for the preparation of a weekly payroll for the James Tool Company.

PROJECTED MONTHLY COST
PREPARATION OF WEEKLY COMPUTERIZED PAYROLL

ACTIVITY	MONTHLY COSTS
1. Data Entry, Operations - (35 hours per month, $6.00 per hour)	$ 210.00
2. Computer Cost - (8 hours at $50.00 per hour)	$ 400.00
3. Supplies - Checks, Forms	$ 530.00
4. Manager, Payroll Department	$1,400.00
5. (2) Payroll Clerk, Data Control	$1,500.00
TOTAL MONTHLY COSTS	$4,040.00

Figure 9-11 Management Report (Part 5 of 7)

-6-

CURRENT COSTS

At the present time the following monthly costs are incurred in the preparation of the weekly payroll.

SALARIES	MONTHLY COSTS
(1) Manager Payroll	$1,400.00
(1) Chief Payroll Clerk	$ 800.00
(2) Payroll Clerks - ($700.00 per mo.)	$1,400.00
(1) Timekeeper	$ 675.00
TOTAL COSTS	$4,275.00

Figure 9-12 Management Report (Part 6 of 7)

-7-

SUMMARY - RECOMMENDATIONS

An analysis of the monthly cost for the preparation of the computerized payroll system as compared to the current manual system reveals a projected monthly savings of $235.00.

This savings is achieved by the reduction of three units of personnel within the existing payroll department. Based upon the projected savings the costs associated with the systems analysis and development phase of the systems project should be regained in approximately four years.

The computerized payroll system will eliminate all overtime now occurring in the payroll department and provide for the processing of numerous new employees as the James Tool Company expands with little increase in costs.

Internal controls within the system and the establishment of formalized procedures for processing Union Dues should eliminate errors which have been occurring within the manual system relative to Union Dues.

It is recommended, therefore, that the computerized payroll system be approved so that the systems development and implementation can proceed as soon as possible.

Donald A. Mard
Senior Analyst

Howard Coswell
Systems Analyst

Figure 9-13 Management Report (Part 7 of 7)

SUMMARY

The approval of the systems design is a critical step in the systems project because it is at this point that management and the users of the system will see the product which will solve the problems which instigated the systems project in the first place. Approval from all required levels is mandatory at this point because once the Systems Development Phase of the project begins, it becomes extremely costly to change the design of the system. Therefore, the analyst must be sure that all persons who have any responsibility over the system whatsoever have been contacted and have given their approval to the design. From this point on, changes must be avoided if at all possible so that increased costs are not incurred.

STUDENT ACTIVITIES—CHAPTER 9

REVIEW
QUESTIONS 1. List the steps which should be taken in obtaining approval of a new systems design prior to making a presentation to management.

1)

2)

3)

4)

2. Discuss the purpose in making a presentation to management upon completion of the systems design phase of the systems project.

3. How should final approval of an output report be signified by the user?

4. Who should approve the design of source documents within a company?

5. Who should approve the design of input records? Why?

6. Who should approve the file design and processing?

7. Discuss those factors related to systems approval which are of particular interest to data processing management.

8. Why is formal approval of the systems design essential prior to implementation?

9. List six items which should be a part of the presentation to management.

    1)

    2)

    3)

    4)

    5)

    6)

10. What is the purpose of a Gantt Chart?

11. Discuss the cost factors which management must consider when determining the feasibility of developing a new system.

12. A new systems design will normally offer certain advantages over the present system. List these advantages.

    1)

    2)

    3)

DISCUSSION QUESTIONS

1. After spending eighteen months on the systems analysis and design phase of a systems project the final presentation to management is to be given in two days to the company president and two vice-presidents. You, as the senior systems analyst, have just been informed of a major new hardware announcement. Upon investigation you **know** that this new product will result in substantial improvement both in cost and effectiveness of the system just designed.

   However, the new product will necessitate a substantial redesign of the "new" system and require at least eight months of additional effort to redesign the system using the latest hardware.

   What would you do?

2. "One of the traits of a good systems analyst is that he must be a 'salesman'. During the presentation to management he must 'sell' the system he has developed; otherwise it will never be implemented. I've seen a number of good systems that were never implemented because the systems analyst failed to sell his ideas to management."

   Do you agree with this statement?

3. One senior level manager says "When a presentation is made to management relative to a new systems design I prefer to have a written report only; no oral presentation is necessary. I want the facts. The costs, the time to implement, and the resulting benefits are easily recorded in a written form.On the basis of the facts I'll make a decision as to whether the system is to be implemented. Systems analysts don't get paid to make management level decisions and that's exactly what they try to do when they try to 'sell' a system in an oral presentation. I've seen too many systems implemented because 'smooth talking' systems analysts have convinced management that a system is an absolute necessity when the recorded facts just don't justify it."

   Do you agree with this point of view?

CASE STUDY PROJECT—CHAPTER 9
The Ridgeway Company

INTRODUCTION

After the billing system for the Ridgeway Country Club has been completely designed and after the output reports, systems input, file design and processing has been approved by all users and data processing management, it is necessary to make a presentation to management to obtain approval to enter the systems development phase of the systems project.

STUDENT ASSIGNMENT 1
PRESENTATION TO MANAGEMENT

1. Prepare a detailed report to the management of the Ridgeway Country Club relative to the new billing system which has been designed. This report should include:

    1) A management-oriented abstract briefly reviewing the problem and summarizing progress to date.

    2) A detailed review of the overall operation of the proposed system.

    3) A projected time schedule for implementation of the new system.

    4) An estimate of the costs of the systems development phase of the project.

    5) Comparison of costs of operating the old system as compared to the new system.

    6) A recommendation from the analyst.

PHASE IV

SYSTEMS DEVELOPMENT

PROJECT MANAGEMENT, SCHEDULING AND CONTROL

CHAPTER 10

PROJECT MANAGEMENT, SCHEDULING AND CONTROL

10

INTRODUCTION

Once the systems design phase of a systems project has been completed and the systems design has been approved by management, the systems project then enters the SYSTEMS DEVELOPMENT phase. During the systems development phase, the analyst must: 1) establish a project plan for the development and implementation of the system; 2) develop the detailed specifications necessary for the programming of the system; 3) program and test the system; 4) develop the final documentation for the system; 5) conduct necessary training and file conversion to properly implement the system.

PROJECT MANAGEMENT

The management of this phase of the systems project is extremely important, for this phase of the project will require much more company resources in terms of time, personnel, and money than any of the previous phases of the systems project.

The management of the systems development effort will not normally be the sole responsibility of the analyst, as the senior programmer on the project will also share in these responsibilities. Since, however, the ultimate responsibility for the implementation of the system rests with the systems analyst, the analyst must be closely involved with all phases of the systems development effort. Many companies will select a PROJECT MANAGER, who has the responsibility to coordinate all steps in the systems development to ensure that the system is implemented on schedule. The project manager will normally be the senior systems analyst who has been working closely with all phases of the systems project.

PROJECT PLANNING

After the systems design phase has been approved by management one of the first tasks of the project manager is to develop a project plan for the systems development and implementation phases of the project.

Project planning refers to the management technique of carefully defining the tasks to be accomplished, the schedule for accomplishing these tasks, and defining the manpower and resources required to accomplish the activities within the specified time allowed. The basic steps in project planning include the following:

1. Define the major tasks required to complete the project.
   A. Break the major tasks down into subtasks.
   B. Estimate the time required to complete each of the subtasks.

2. Establish the schedule and relationship of all subtasks.

3. Develop a manpower schedule with an indication of the technical and clerical skills required, when the skills are required, and for how long.

4. Establish a method of monitoring the progress of the project.

The history of the development of major data processing systems has often been characterized by failure. Much of this failure can be attributed to a lack of adequate project planning. The following statements contained in a data processing management periodical summarize what too often has occurred in systems projects.

"The process of conceiving, designing, installing, and operating any major data processing system is inherently a difficult, time-consuming and frustrating task."

"A carelessly planned project will take three times longer to complete than expected. A carefully planned project will take only twice as long."

"If anything can go wrong it will . . . if nothing can possibly go wrong it will anyway."

"When things are going well, something will go wrong."

"When things appear to be going better, you have overlooked something."

"Projects progress quickly until they are 90 percent complete, and then they remain 90 percent complete forever."

"The system finally installed will be installed late and won't do what it is supposed to."

"The benefits will be smaller than initially estimated, if estimates were made at all."

"No major computer project is ever installed on time, within budget, with the same staff that started it, nor does the project do what it is supposed to . . . it is highly unlikely that yours is going to be the first." *

Although the comments are undoubtedly intended to be humorous, they do reflect the problems which all too frequently have occurred in the past in major systems projects and the statements tend to emphasize the need for careful project planning.

---

*"Maximizing Return on EDP Investments," **Journal of Data Management**, September, 1972.

DEFINING THE MAJOR TASKS TO BE ACCOMPLISHED

The first step in the systems development phase of the project is to define the major tasks to be accomplished during the systems development. It will be recalled that in the presentation to management after the systems design phase, the systems analyst prepared an estimate of the time required for implementation of the system. At this time the analyst must prepare an even more detailed analysis of the time to complete the project, for it is essential that control be exercised over the project and that the project be carefully monitored to assure its progress toward implementation. If in this detailed analysis of the project the time or cost of system development far exceeds the original estimate presented to management during the systems design phase, the analyst should return as soon as possible to management with revised estimates.

The major tasks common to all systems projects in the systems development phase include:
1. Programming
2. Program Testing
3. Documentation
4. Training
5. File Conversion

If new hardware is required for the project, hardware and software acquisitions must also be considered during the systems development phase.

Each of these major tasks can readily be scheduled and monitored through the use of a Gantt Chart as illustrated in the previous chapter. Careful analysis reveals, however, numerous subtasks which also should be defined, coordinated and controlled.

Defining the Subtasks

To establish a means of control over the systems project, each of the major tasks must be further segmented into subtasks. These subtasks should be defined to the level at which control is desired.

Some of the common subtasks associated with systems development include:
1. Develop programming specifications
2. Prepare program flowcharts
3. Code and desk check program
4. Compile programs
5. Prepare test data for each program
6. Test and debug each program
7. Document the programs
8. Test all programs within the system which are inter-independent
9. Order supplies
10. Order new hardware and software, if required
11. Prepare computer site if new hardware is required
12. Install new hardware
13. Create system test data
14. Perform system tests
15. Conduct user training
16. Convert Files
17. Develop final system documentation

From the previous subtasks it can be seen that some tasks cannot begin until a previous task is completed while, in other cases, several of the subtasks may be taking place concurrently. For example, subtask 2, the preparation of the program flowcharts, cannot begin until subtask 1, the development programming specifications, has been completed. It is possible, however, to order supplies sooner or later than specified in the list above.

Thus, it is essential that the project manager define the major tasks, and then further define the sequence, time frame, and interrelationship of all subtasks to provide for the development of a realistic schedule, as it is easy to underestimate the complexity of a job if subtasks are not defined.

PROJECT SCHEDULING - PROGRAMMING

Prior to giving the programming staff the documentation and the go-ahead to begin programming the system, it is normally necessary for the systems analyst to establish a schedule of activities which will ultimately lead to implementation. One of the primary, and probably most critical, function to be scheduled is the programming of the application programs which are to comprise the system. In most systems, the programming effort is extremely time consuming and all other activities which must be performed in the systems development are scheduled within the framework established by the time required to program the system.

Unfortunately, there is no general rule which can be used to determine the amount of time which will be required to ready the programs for production. There are, however, some factors which will enter into the formula for determining the time required to program a particular application. In order to accurately estimate the production of the programming project the analyst must have knowledge of the following factors:

1. The overall system to be developed.

2. The number of programs within the system.

3. A detailed knowledge of the input, output, type of routines, and the complexity of each of the programs.

4. The number of programmers assigned to the project.

5. The experience level of the programmers assigned to the project.

6. Special or unique requirements such as the use of new programming languages or new hardware, or the use of on-line terminals.

In further analyzing program development time studies have shown that a reasonable allocation of time for programming activities is as follows:

ACTIVITY	TIME
Development of Program Logic, Flowcharting	35%
Coding of the Program	25%
Testing and Debugging of the Program	35%
Final Documentation	5%

These percentages should be useful in developing a work schedule on a particular application. It should be noted, however, that the method of estimating and scheduling programming projects often rests with the subjective evaluation of the project manager, based upon experience with similar projects.

In general, the time required for the logic preparation and the coding can be fairly accurately estimated. These times will normally be dependent, of course, not only on the size and complexity of the program but also upon the skill and experience of the programmer. In most business applications, there are basic processing techniques, such as editing or updating, which are utilized over and over and once a programmer has become adept at handling these techniques he can usually take the processing required in a new program and fit it to one of the general techniques he has used. If the programmer is inexperienced, however, it may take considerably longer for him to develop the logic which will be required to solve the problem. The same holds true for the coding of the program—the more experienced programmer should be able to code his program more efficiently, both in terms of the time required and the number of errors which will be made in the coding, than the inexperienced programmer. These factors must be taken into consideration when scheduling program development time.

To program a system efficiently and accurately, it is critical that the programmer understand exactly what is called for in the programming specifications so that the program can be designed to encompass all of the processing required. The analyst must make himself totally available to the programmers prior to, and during, the programming of the system so that any questions which arise can be handled quickly. If the programmer is unable to fully understand the programming specifications or has a question concerning processing within the system, and the analyst is not available for a day or two, it is likely that there will be some delay in the schedule. This type of occurrence must be considered when determining the time required for the programming of the system.

In addition to the analyst being available, there are times when there may be a need for the programmer to consult directly with the user of the system. Problems may arise relative to the values which can be found in certain fields within a record or the frequency of occurrence of certain data in the input records. This type of information may be required by the programmer in order to write an efficient program yet may not be known by the analyst or the analyst may not be readily available. In these cases, the user may be consulted by the programmer in order to get answers. Therefore, the analyst should ensure that the user knows the programmer and that the programmer has access to any knowledge the user may have. As has been mentioned before, the analyst is normally the liaison between the user and the data processing department so he must ensure that there is complete cooperation between all parties involved in order that the project proceed on schedule.

PROJECT SCHEDULING - PROGRAM TESTING AND DEBUGGING

The amount of time to schedule for program testing and debugging is commonly difficult to estimate. Typical problems include:

1. Testing is dependent upon the availability of the computer hardware to do the testing.

2. Results which are obtained from the tests are often correct but the program must be changed because there was a lack of understanding in the problem analysis stage of the programming.

3. Some programmers are not skilled in the ability to adequately and efficiently test a program.

Whenever a program is to be compiled or tested, a computer must be available to the programmer. In many installations, there are standard procedures that can be followed for testing which ensure that the programmer will have adequate time to test programs. In other installations, however, the testing of programs receives very low priority and is often subject to various delays. Unscheduled production runs, production reruns, hardware failures, etc. can result in a delay in program testing which can cause a delay in the completion of a program or perhaps in the completion of the entire project. Thus, the analyst in conjunction with the data processing manager must schedule adequate test time for the programming staff working on a new project.

The analyst must also plan for any new hardware required for the implementation of the system and must ensure that the hardware which has been ordered will be available when required for the implementation of the system. In many instances, however, the new hardware will not be available when program testing begins. Therefore, alternate action must be taken. Perhaps the most common method of circumventing this problem is to test the programs at a remote site which has the required hardware. In general, the manufacturer of the hardware will have machines available for testing by a customer who is ordering new equipment. If so, the analyst or senior programmer must schedule computer time with the manufacturer. This, of course, puts more pressure on the programmer to be ready to test his program because if he is not ready, then the test time will normally be lost and must be rescheduled.

Thus, if the hardware is not available at the installation for testing, an important consideration for the analyst is to make arrangements for testing either at the manufacturer's site or some other remote location.

The second problem in testing and one of the major reasons that testing and debugging many times takes longer than anticipated is that the results from the program, although technically correct, do not conform to the programming specifications or at least what the analyst anticipated from the program. In most cases, this is a result of either the programmer not understanding the programming specifications or the programming specifications not adequately representing what the analyst wanted from the program. Unfortunately, in most circumstances, these problems do not arise until the testing portion of the program and this means that while the program is changed and retested, time is passing, resulting in a possible delay in the implementation date.

There are several ways in which this problem can be overcome but the most obvious and easiest way is for the analyst to take the time initially to define the program in the programming specifications with much detail and accuracy and then work with the programmer to ensure that the details are clearly understood. This would involve such things as reviewing the programming specifications with the programmer in detail prior to the programmer preparing the program logic, checking the program flowchart which the programmer has prepared prior to beginning the coding, checking the coding if the analyst knows the programming language, and keeping in close contact with the programmer during the testing of the programs so that any errors or misunderstandings which occur can be detected early in the testing procedure.

The skill of the programmer in debugging and testing a program is something over which the analyst has relatively little control. Again, when determining the schedule, the skill of the programmer must be taken into account. Although the analyst will normally not be directly involved in debugging the program, he should, as much as time will allow, be reviewing the results of the testing. The senior programmer working on the project will also have responsibilities in the area of program testing and debugging and the analyst should also work closely with the senior programmer assigned to the project.

In summary, the analyst, when scheduling the time required for testing and debugging, must plan for adequate test time on available hardware or arrange for test time at some remote site; the analyst must make sure that the programming specifications are thoroughly understood by the programmer assigned to the project; and the analyst must evaluate the capability of the programmers assigned to the project.

PROJECT SCHEDULING - DOCUMENTATION

The documentation of a system normally begins in the systems design phase with the design of the system output. This documentation includes the planning of the output reports through the use of printer spacing charts and continues through the design of input documents, records and files, and the systems flowchart.

During the systems development phase of the systems project additional documentation must be developed. This documentation includes:

1. Programming specifications
2. Program flowcharts
3. Source listings of all programs with related test data and job control
4. Operations documentation manuals for computer operators
5. User's manual for data control and related clerical staff

The preparation of this documentation is a continuing process during the systems development phase. It is important, however, that the analyst provide for sufficient time for the programmer or related personnel to prepare supporting documentation. Far too often programmers code a problem, test and debug the program and then move to the next programming assignment. It is the job of the systems analyst to assure that the programmer not only codes the program but supplies supporting documentation as well. The time required to prepare this documentation must be considered when setting up the schedule for systems development.

PROJECT SCHEDULING - TRAINING

Whenever a new system is implemented on a computer, there is frequently a requirement for personnel with some type of specialized training. In some instances, new personnel may be hired. In other situations, it may be better to train someone who is already on the job in another capacity. Regardless of which method is used, the analyst must be aware of the personnel requirements to implement a new system and take the steps to ensure that the required personnel will be available and properly trained.

Hiring New Personnel

The analyst, when determining the personnel requirements of the new system, will have two primary areas of concern. They are:

1. Defining programming personnel required to implement the system

2. Defining the clerical and support personnel required to implement the system

The analyst must be able to define the personnel who will be needed to successfully implement the system. In most cases, this will involve the programming personnel who will be needed to write the programs. Unless the system is a major, large system, the programming personnel in the department will often suffice. If, however, considerable man months or years will be required for the programming effort, or if there is an area of specialty required for the programming, such as the need for a programmer with experience with on-line systems or data communications, then the analyst will define his needs and in many cases participate in the selection process of the individuals to be hired or promoted to required positions from within the company.

The analyst must also define any additional personnel that will be needed for the operation of the system once it is implemented. Normally this will involve the hiring of additional data preparation or clerical personnel. The data preparation personnel include the keypunch operators, key-to-tape operators, and others who will take source documents and prepare input to the system. Again, depending upon the size of the system, there may be no persons required or a large number of persons required. The clerical personnel will normally include those who will check the source documents for errors, will take batch totals to be used for checking after the computer run, and similar activities.

Another area of concern for the analyst is in defining the personnel who will be required for the departments which will be using the new system. Again, these people will probably fall into the data control and clerical positions. It is quite important for the analyst to define these jobs and give an accurate estimate of the number of people who will be needed because in many instances, the management of the user department is not data processing oriented and will have no real feel for the people they will need. Guidance must come from the systems analyst who designed the system.

Training Personnel

The answer to personnel requirements is not necessarily hiring new people. Many companies have a policy of "promoting from within," that is, providing individuals currently employed within the company the opportunity to advance to new jobs of increasingly greater responsibility. As related to computer programming personnel, some managers feel that it is easier to train an employee with a thorough knowledge of the company in the area of computer programming than it is to train an experienced computer programmer in the complexity of a particular business.

If current employees are to be trained, it is the responsibility of the systems analyst to define the training needs of the company relative to the development of the new system. There are five basic sources of training commonly used. These include:

1. Manufacturer's training schools

2. Self-study courses using programmed instruction textbooks

3. In-house training programs using a variety of audio-visual methods including video tape courses

4. Public education programs at adult education schools, community colleges, and universities

5. Short courses and seminars offered by a variety of private companies and professional associations

Regardless of the method of training used, it is the responsibility of the analyst to define and schedule the training required to assure the successful implementation of the system.

The analyst must also be concerned with the training of people who will actually operate the system in its day-to-day operation. This training will normally consist of one or more training sessions for those who operate the system. This training will normally be conducted by the systems analyst or by someone thoroughly familiar with the new system. In order for the training to take place properly it will normally be necessary to have some type of documentation which can be distributed to the trainees from which they will be able to learn their duties in the new system. Thus, this training must normally be scheduled after the operations documentation and testing is completed.

Management Training

As noted, it is very important for the people who will be processing the data within a system to be thoroughly trained so that the system will work smoothly. It is also necessary to train those who will be actually using the output from the system. In some cases, these will be clerical personnel who use the daily reports as working papers. In other cases, however, the reports will contain information which is used by supervisory and management personnel and it does little good to have a system producing good information if management is not aware of the information being produced and how to use it.

It is, therefore, one of the functions of the analyst and those working on the project to educate management in the use of the information which will be produced from the system. It is surprising to many that although management will give its approval to a system, they are somewhat ignorant of the results which may be of benefit to them. Most persons in management positions will be grateful to the analyst for pointing out ways in which the information can be used to their benefit. It is also true, of course, that many times the formats of the reports must be explained so that management personnel will feel comfortable with the information supplied. Therefore, the systems analyst should schedule some time for meetings with management and user personnel to explain the output of the system and to answer any questions regarding the operation or benefits of the system.

PROJECT SCHEDULING - FILE CONVERSION

When converting from a manual system to a data processing system, it is necessary for the systems analyst to provide for the necessary time to convert the existing records and files to a machine processable form such as punched cards, magnetic tape or magnetic disk. For files with many thousands and even hundreds of thousands of records this can consume a substantial period of time. The primary effort in file conversion will come from the data entry section, that is, that section of the company which is charged with the responsibility of converting information in source documents to machine processable form. Additionally, however, considerable computer time may be required and the analyst must determine the required time.

The analyst must also schedule time for file conversion and should closely monitor the progress to assure all files are prepared for processing under the new systems plan.

PROJECT SCHEDULING - PROCUREMENT OF HARDWARE, SOFTWARE, AND SUPPLIES

The programming and testing of the system, the documentation of the system, and the planning and scheduling of personnel training will in many applications consume a good portion of the analyst's time as he supervises the progress being made on the system. There are other areas, however, which require attention to ensure that the system will be implemented on schedule. These include acquisition of the necessary hardware, software, and other supplies which may be necessary for the operation of the system.

Hardware

With many new systems there may be new hardware required to implement a system. This means, therefore, that the analyst must schedule the ordering of the required hardware and also follow up with the manufacturer or supplier to ensure that the hardware will be delivered on time and will be functioning when the system is scheduled to go into production. Often there will be a six to twelve month "lead time" required when ordering computer hardware from a major manufacturer, that is, new hardware must be ordered six to twelve months prior to delivery.

The sources for hardware are varied and there are also various types of agreements which can be made for the acquisition of the hardware. Hardware is commonly acquired by four different methods:

1. Leasing from the manufacturer
2. Purchasing from the manufacturer
3. Leasing from a "third party"
4. Purchasing from a "third party"

Historically, most companies have leased hardware directly from the manufacturer. Because of the rapidly changing technology, hardware purchased from the manufacturer often becomes obsolete within a few years; therefore, many companies have avoided purchase. In addition, in the past, the cost of purchase of a computer system was extremely high, often in excess of several hundred thousands of dollars for a medium scale system. Because of the large amount of capital needed for the purchase of a computer and because of the danger of obsolescence, leasing a system became the dominant method for acquiring a computer system for most companies.

During the past decade "third party" leasing has been used by many companies to acquire computer hardware. With third party leasing, the computer systems needed are leased from a company who has previously purchased the computer from a major manufacturer. Leasing from a third party normally is less expensive than leasing directly from the manufacturer. Primary disadvantages result from a loss of "support" from the manufacturer in terms of systems programming assistance, training, educational materials, new product announcements and software such as new programming languages.

The determination of what route will be followed in procuring is a function of the money required, the use to which the hardware will be put, and a myriad of other factors. What is important to note at this time is that by whatever means the equipment is to be obtained, the analyst must carefully schedule the ordering, delivery, and installation of hardware.

There are other factors which are associated with hardware which must be considered. In some circumstances, there will be a requirement that additional facilities be added. For example, new air conditioning, electrical power or new flooring may be required. The analyst must be aware of these requirements and ensure that they are scheduled well enough in advance of hardware delivery that all will be in readiness when the hardware is delivered. These are the types of tasks which may be overlooked when the more apparent problems of programming are being attended to but if they are not properly handled, they will contribute to a late project just as any other area of the system.

Software

If additional hardware is required for the new system, it is quite likely that new or altered software will also be required. If for no other reason, the operating system will normally have to be updated to be able to control new devices. When these software changes are required, it will be the responsibility of the analyst to contact the systems programming staff to inform them of the needed changes and to supply other information which may be required. These software changes must not be left to the last moment. In many instances, a great deal of planning and coordination is required in order to implement changes in an operating system and if there is any possibility of these changes being required, the analyst must notify the systems programmers early and must work with them in scheduling and implementing the changes.

Besides the operating system, there may also be software which must be acquired which is to perform a special function within the system. For example, it may be required that a special "sort" or other utility program be available and there may be some length of time between when the program is ordered and the program is delivered. In addition, even after the program is delivered there may be some modifications which must be made to it and, of course, it must be incorporated into the operating system. Thus, the analyst must make plans to have all of the required software tested and running in the installation well before the system is scheduled for implementation.

Supplies

There are numerous "supplies" which normally must be ordered and obtained well before system implementation and, for the most part, it is the responsibility of the systems analyst to ensure that these supplies are ordered. Thus, he must define what must be purchased and also any lead times which may be required so that there is a coordination of all of the supplies.

Another area which is important in terms of supplies are those which are commonly in use within the department even without the new system. Common among these items are punched cards and standard-sized printer forms. When a new application is put on the computer, it may be that more cards or more forms will be required than have been previously used in the installation. This is especially true if the system to be placed on the computer was previously a manual system which required no supplies from the computer installation whatsoever. The analyst must be aware of the requirements of the new system and must be sure that the normal quantities of these standard supplies are increased, if necessary, to satisfy the new requirements of the new system.

The analyst must also forecast the need for extra disk packs and/or magnetic tape reels which may be required for the new system. The number of tapes and disk packs will, of course, depend upon the number of files which are to be included within the system and will also depend upon the retention periods for the files. As noted in Chapter 7, files may be retained for some period of time to be used as input to another cycle of the system and backup files may be created in case a file is accidentally destroyed. When these files are retained on either disk or tape, the disk pack or reel of tape cannot be used for any other purpose. Since these reels or packs are out of circulation, there must be packs and reels to replace them in order for other systems on the computer to be processed. Therefore, the analyst must determine the number of disk packs and tape reels which will be necessary for the system, which can be dependent upon how long files are to be retained and how many backup files are to be created.

It should be noted that disk packs and reels of tapes cannot, in many instances, be delivered as readily as cards and paper. In addition, disk packs and most tapes must be "initialized" with certain constant information which is stored on all disks and tapes. For the most part, this information consists of the labels which are stored directly on the tapes or disk pack. Therefore, the analyst should allow some lead time between delivery and use so that this initialization procedure can be followed.

SUMMARY - PROJECT SCHEDULING

For the previous discussion it is apparent that the analyst is faced with a large and important task in defining and scheduling the many major tasks and subtasks related to the system development phase of a systems project. A lack of attention to any one of many factors can lead to a late schedule and even failure of the system to be implemented on time.

SCHEDULING TECHNIQUES

There are a variety of techniques which may be utilized to schedule and control systems development tasks. Two common tasks include:

1. GANTT Chart
2. PERT or Critical Path Method

A Gantt chart of the major tasks in systems development is illustrated in Figure 10-1. This is the same type of chart as illustrated in Chapter 9 and used as a basis for presenting the implementation schedule to management.

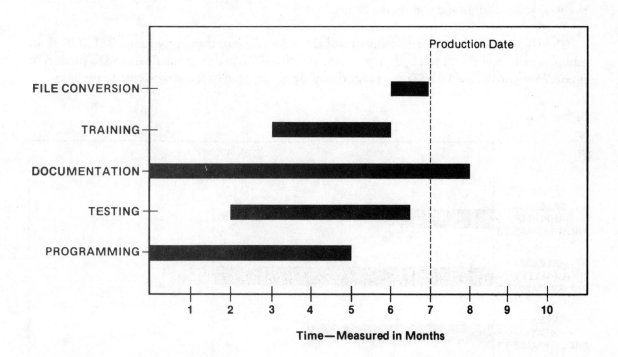

Figure 10-1 Example of Gantt Chart

It should again be noted that the Gantt chart illustrated does not indicate the resources required to accomplish the tasks, but only illustrates the beginning and ending time periods. For example, in the chart illustrated in Figure 10-1 it can be seen that the programming is to be completed within a five-month period of time. The chart does not indicate the number of programmers assigned to the programming project nor the number of man hours to be expended.

As can be seen from Figure 10-1, the estimated time between the beginning of the programming effort and the planned date for production of the system is seven months. The testing of the programs will begin two months after the programming has begun and will continue until several weeks before implementation. The training of the people who will be operating and using the system will take several months. The file building and conversion will take approximately one month and the documentation of the system and the programs will be a continuing effort from the start of the programming until a month after the system is in production.

It will be noted that the chart illustrated in Figure 10-1 is quite general, that is, none of the detailed programming or testing which must take place is depicted.

The Gantt chart can be used at whatever level or contain whatever details are deemed necessary to schedule and control the project.

The Gantt chart in Figure 10-2 illustrates the scheduling of three programs. SALES01 is an edit program. SALES02 is an update program and SALES03 is a program designed to produce a report from the updated file. Three programmers are assigned to the programming projects.

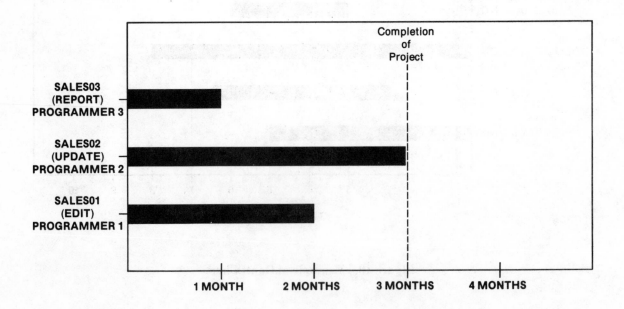

Figure 10-2 Gantt Chart

In the Gantt chart illustrated in Figure 10-2 it can be seen that the edit program SALES01 will take 2 months to complete; the update program, SALES02 will take 3 months to complete and the report program, SALES03 will take 1 month to complete. If three programmers were assigned to the project the schedule could appear as illustrated in Figure 10-2. It should be noted, however, that from a scheduling point of view it does little good to complete the report program, SALES03, at the end of the first month as this report program requires the output from the update program, SALES02, for final systems testing.

A simple revision in the schedule could easily free one of the programmers for another project and have no effect on the overall completion time for the project. This revised schedule is illustrated in Figure 10-3.

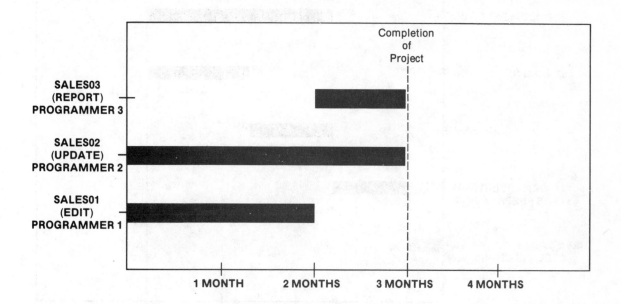

Figure 10-3 Gantt Chart

In the example in Figure 10-3 it can be seen that the programming for the report program SALES03 does not begin until the second month and is completed by the third month. By rescheduling this program effort at a later time, one of the programmers is freed to work on another project with no delay in final system testing.

It can be seen from the example above that scheduling a large number of projects can develop into a complex task if the most efficient use of programming personnel is to be achieved.

Thus, in addition to scheduling the overall time frame in which a project is to be complete, the analyst must have a detailed knowledge of the manpower available to complete a task, and schedule this manpower in the most efficient manner possible.

For greater control, the scheduling which is illustrated in Figure 10-3 can be further broken down to reflect the schedule associated with each program within the system, because it is important to be able to determine what the schedule for a program is so the programmer is aware of the time frame within which he must work. The Gantt Chart for the edit program, SALES01, scheduled for completion date in 2 months, is illustrated in Figure 10-4.

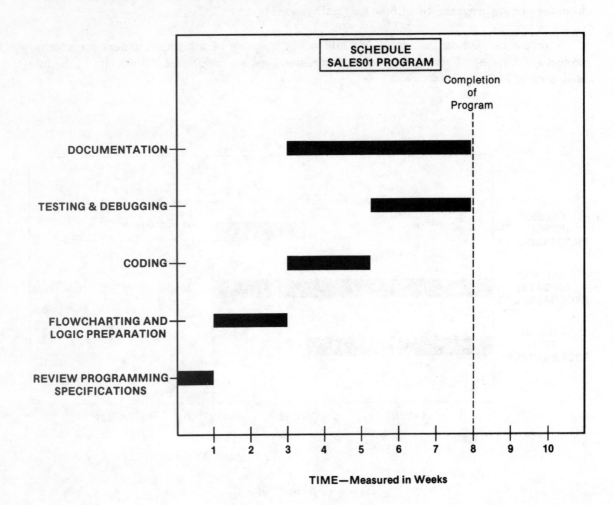

Figure 10-4 Example of Detailed Gantt Chart

Note from the chart that the various elements of the programming procedure are broken down and the estimated time to complete each task is scheduled. These elements—Review Programming Specifications, Flowcharting and Logic Preparation, Coding, Testing and Debugging, and Program Documentation—include most of the functions which must be performed by the programmer from the time he receives the programming assignment from the analyst to the time the program is ready to be put into production.

One aspect in the scheduling of the programming portion of the implementation is consultation with the programmer. A schedule which is devised without the advice and consent of the programmer has a much smaller chance of being met than one which is approved by the programmer as well as the analyst or senior programmer in charge of the project. By consulting with the programmer, it can be determined if any aspects of the project have been overlooked by the analyst and if the programmer feels that the time allocations are realistic. In addition, the programmer is made aware of the schedule which must be met and it is more likely that with his approval, the schedule will be met. Thus, prior to setting final dates for the schedule, the analyst and the programmer must agree on the tasks to be performed and the estimated times required.

The systems analyst must be extremely careful when scheduling so as not to underestimate the time required for the programming and other steps which must be performed in systems development phases of the systems project. The tendency of most inexperienced analysts and many experienced analysts as well is to underestimate the time required to perform any given task. Historically, this has seemed to be especially true in data processing systems projects as many systems which have been designed and implemented on a computer have been implemented well past the deadline established in the original schedule. Missed deadlines can be extremely costly to most business organizations.

In systems requiring additional hardware, the manufacturer of the hardware will normally ensure that the equipment is installed as scheduled. Thus, if the system is not implemented as scheduled the company must pay for the additional hardware not being used. A delay of several months can conceivably cost the business thousands of dollars in costs for equipment not being utilized.

Project Control

One of the basic purposes in the utilization of a Gantt chart is not only to schedule tasks but also to provide a means of control over the progress of the entire project.

It is, of course, one thing to establish a schedule and another to follow the schedule. As noted previously, a common characteristic of newly designed data processing systems is that they are frequently ''late.'' The underestimating of the time required is certainly one of the reasons but another important reason is the inadequate management of the project once the activities on the schedule are under way. It is extremely important that the analyst supervise each of the activities in the schedule to the point where there is control over what is taking place.

As has been noted, communication with those who are working on a project is imperative if the project is to be completed properly and on time. The analyst, therefore, should schedule regularly held meetings at which time a review of the work accomplished to date can take place and any discrepancies which are beginning to show between the schedule and the point at which the project is can be resolved. These meetings should take place at least weekly, especially on a project whose duration is not long term, so that problems which have been encountered can be resolved before they become so major as to cause delays in the planned schedule. At these meetings, an ''activity'' chart should be kept so that the progress can be seen. An example of an activity chart which can be made from a Gantt Chart is illustrated in Figure 10-5.

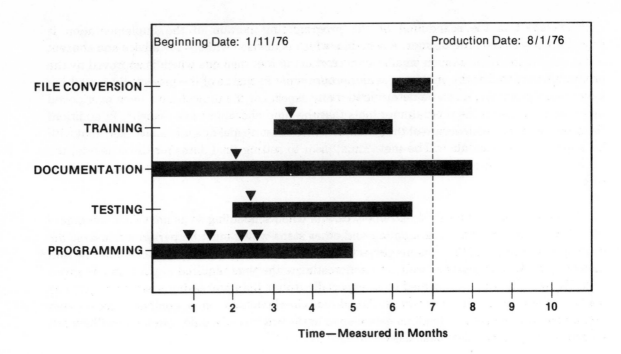

Figure 10-5 Example of Activity Chart

Note from the example above that the Gantt Chart is used because it represents the schedule which is to be accomplished. The arrowheads (▼) are used to indicate what percentage of the given task is accomplished to date. They are not used to indicate where in time the project is. For example, the last arrowhead on the programming bar indicates that the programming is approximately 50 percent completed, and this much programming should be completed in about 2-1/2 months, as the programming effort is scheduled for 5 months. If this chart was prepared after only 2 months since the beginning of the project, the project would be ahead of schedule as far as the programming was concerned. On the other hand, if 3-1/2 months had passed since the project was begun, the project would be seriously behind schedule. It is important to note that the arrows represent that percentage of the project which has been completed, not the amount of time which has elapsed.

The use of the Gantt Chart in the manner illustrated above is an estimate because no one can say exactly what percentage of the task is completed. It has been known to happen that the programming is 90 percent completed in only 50 percent of the time alloted to programming but the remaining 10 percent of the programming went far beyond the scheduled completion date. Therefore, it can be seen that it does no good to have activity charts like the one illustrated in Figure 10-5 if it is not accurately maintained and the estimates are not honest ones made by those involved in working on the project.

There are other methods which may be employed in determining·whether the project is on schedule, such as oral reports or estimates by the personnel involved of the time required to complete the project. Whatever method is used, however, it is vital to maintaining a project on schedule that any time the project is behind schedule, corrective action be taken. This action may involve placing additional people on a particular task or altering the schedule so that the schedule is more realistic relative to the problems at hand. In any case, a project should not continue to be behind schedule for long or the deadline for implementation will most likely not be met.

OTHER SCHEDULING TECHNIQUES

Although the bar or Gantt Chart is useful in scheduling, these types of charts do not readily show the interdependencies among tasks or people. To show interdependencies an "activity network chart" is often used.

To provide management with control over complex projects the concept of network analysis was applied to scheduling problems. In 1958 PERT (Program Evaluation and Review Techniques) was developed by the Navy Special Projects Office to control the development of a missile program involving over 2,000 contractors. Concurrently, the management control technique known as Critical Path Method or CPM was developed in private industry to meet similar management needs. Both PERT and CPM use the concept of a network of activities and events as a model for an actual project for the purpose of assisting in the scheduling of a complete project.

A PERT network shows the various "activities" which must be accomplished before another activity can be undertaken. It also illustrates various activities which can be carried on simultaneously. Figure 10-6 illustrates a simplified PERT network.

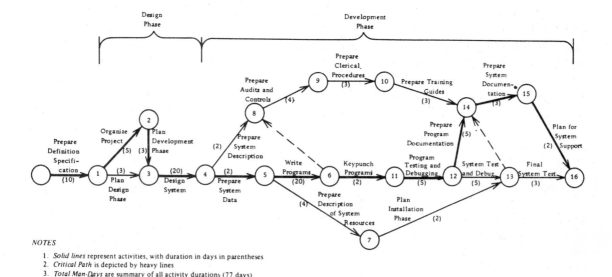

NOTES

1. *Solid lines* represent activities, with duration in days in parentheses
2. *Critical Path* is depicted by heavy lines
3. *Total Man-Days* are summary of all activity durations (77 days)
4. *Dashed lines* reflect time dependency between nodes

Figure 10-6 PERT Chart

The PERT CHART is used with projects for controlling several related procedures making up what can be termed a project or network activity. A chart is prepared to assist in visualizing the total tasks making up a project. From the beginning to the end of the project there are typically several paths of work sequences that can be followed. Using the PERT chart as a guide the time required for the longest sequence of operations is completed. This is known as the critical path because this path controls the completion of the entire network. A delay in any of the tasks along this path would necessarily delay completion of the entire network. In contrast, other activities of the project not included in the critical path could be delayed without delaying the completion of the entire project. Usually 80 to 90 percent of the activities fall into this category; thus, only 15 percent of the jobs are critical in content and sequence to assure the completion of the entire project within a stated period. In other words, PERT highlights the key or critical job or work to be accomplished.

The first step in establishing a PERT or CPM control system for a project is the definition of the project in the form of a network model. Any project consists of a collection of individual ACTIVITIES. An activity is defined as any operation which requires time and resources and which has a definite beginning and end. An activity is indicated by an arrow. An activity should represent the smallest unit of work over which control is desired.

PREPARE SYSTEMS
DEVELOPMENT SCHEDULE

Figure 10-7 Activity Lines

Activities are combined with EVENTS to form the basis of the PERT chart. An EVENT is a point in time within the project which has significance to management. An event represents the beginning or ending of an activity and is portrayed by a circle or "node." No expenditure of manpower or resources may be associated with an event. The following example illustrates the charting of events and activities in a PERT chart.

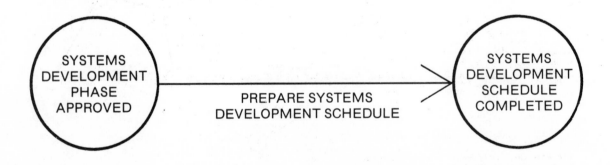

Figure 10-8 Events and "Activities"

Note in the example that the "events" are the "Systems Development Phase Approved" and the "Systems Development Schedule Completed." The activity, which requires time and resources, is "Prepare Systems Development Schedule." Using activities and events as building blocks a "network" of activities may be constructed which illustrate the dependencies between the activities of a project. Each activity has certain constraints with respect to its starting time. Either it may start as soon as the project begins or else its start is dependent upon the completion of another activity or group of activities.

In many PERT charts events are represented by a number contained within a circle or "node" indicating either the beginning or the ending of an activity. Thus, if event 3 represented the beginning of the coding, then the arrow between event 3 and event 4 would represent the coding of the program and the circle at event 4 would indicate that the coding was complete. Since an event is merely a starting or terminating marker, it does not take any time nor work; it merely marks where an activity, represented by an arrow, begins or ends. This is illustrated in Figure 10-9.

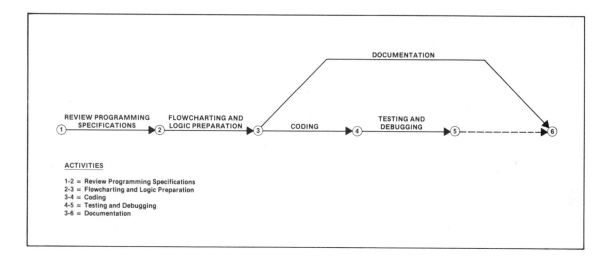

Figure 10-9 PERT Chart

As can be seen from Figure 10-9, the PERT chart represents the sequential steps that must be taken to get to event 6, which is the completion of the program. Note that the documentation, which is the activity represented by the arrow from event 3 to event 6, will be taking place at the same time the other portions of the programming activity are taking place. Thus, the PERT chart can illustrate several different activities which can be occurring at the same time.

Whenever an activity arrow goes from one event to another, such as going from event 2 to event 3, it means that event 2 must be completed before the next activity can begin. Thus, from the example it can be seen that the reviewing of the programming specifications must be completed before the flowcharting can begin because event 2 must be completed before any activity following event 2 can begin. The dotted line from event 5 to event 6 is termed a "dummy" or "zero-time" activity because there is no activity which must be performed to reach the end of the project which is dependent upon event 5 being completed. Therefore, the dotted line indicates that the completion of event 6, that is, the completion of the documentation and therefore the project, is not totally dependent upon the completion of the testing and debugging.

Specification of Time

It will be noted from the example in Figure 10-9 that there are no indications on the PERT chart of any times which are required to complete a given activity and it will be recalled that this is the primary reason for determining schedules. Therefore, when using a PERT chart, it is necessary to determine some times which each of the activities will require in order to project a completion date. The formula which is widely used to determine the time which a project requires is illustrated below.

$$\text{Time} = \frac{\text{Least Time} + 4(\text{Most Likely Time}) + \text{Longest Time}}{6}$$

In using the formula, the least time is that time which is estimated to be the fastest the particular activity can be accomplished. As a general rule it is estimated that the likelihood that the activity can be completed in this time is one in a hundred. The most likely time is the best estimate of the time which the activity will consume. The longest time is the greatest amount of time which can be foreseen for the activity and its likelihood should also be about one in a hundred.

Thus, for the activity of "flowcharting and logic preparation" it may be estimated that the fastest or least time required would be three days. The most likely time required for logic preparation would be 15 days and the longest time which would be required is 21 days. The time to be allocated to the activity is calculated below.

$$\text{Time} = \frac{3 + 4(15) + 21}{6} = \frac{84}{6} = 14 \text{ days}$$

Thus, from the calculation, the time allocated to the flowcharting and logic preparation activity on the PERT chart would be 14 days. The times for the various activities which are to be performed are then entered on the PERT chart, as illustrated below.

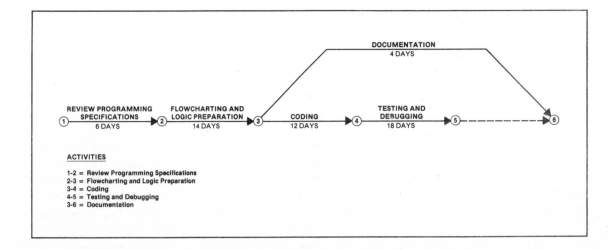

Figure 10-10 PERT Chart with Times

In the example in Figure 10-10, each of the activities is described on the activity line and the estimated days for the activity to be completed, as calculated from the previous formula, is given below the line. This method of documenting a PERT chart is somewhat standard although variations, such as listing both the estimated time and the three times which were used in the calculation, will sometimes be used.

Once the activities have been defined and the times for each activity estimated, a CRITICAL PATH can then be established. A critical path is that path through the PERT chart which allows no slack time, that is, it must be done either before or on schedule in order for the entire project to be completed on time. In the example in Figure 10-11 it can be seen that the critical path is events 1, 2, 3, 4, 5, 6 because if any of those activities is late, or takes longer than the estimated time, then the project will not be completed on schedule. Note that the sum of the days required for the completion of the project is 50 days. The activities between event 3 and event 6 will take 30 days. The time for documentation is estimated to take 4 days. Therefore, there are 26 days of slack time in the documentation, that is, the documentation of the system can take 26 more days then estimated and the project will still be completed on schedule.

The PERT chart shown in Figure 10-10 is extremely simple and in fact is not using PERT to its full advantage. PERT becomes useful when there are many activities which must be completed and many of the activities can be carried on simultaneously. The example in Figure 10-11 shows a more complex PERT chart with 12 events specified.

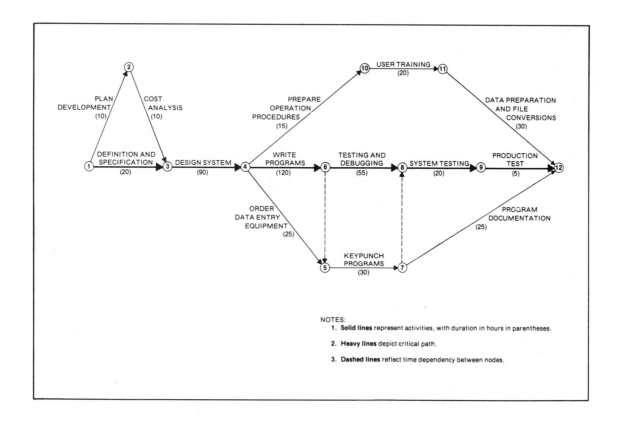

Figure 10-11 Example of Complex PERT Chart

MANAGEMENT REPORTING

At the same time that the analyst is controlling the scheduling, programming, and training activities which must be accomplished prior to system implementation, he must also keep management informed of the status of the system. There are likely to be several levels of management to whom the analyst will report.

First, he will report to his immediate supervision, normally the project manager for the systems project. This reporting will normally be some type of oral report and will take place continually throughout the entire project. It is done so that the project manager will be aware of any problems which may occur.

In addition, the data processing manager should usually be informed of progress within the systems project because in some instances, when problems appear, it will be of assistance to the analyst to have the data processing manager aid him in applying whatever pressure may be required to rectify the problems. In many cases, the request of the analyst or project manager who is daily working on a project will not carry nearly as much weight as the request of the data processing manager. Therefore, the data processing manager should be kept abreast of the progress of the system both because he should know the status from a management standpoint and because he may be of considerable aid in helping the project reach a successful conclusion.

The management of the user departments should also be kept closely abreast of what is transpiring in the course of the project. They must know because during the time of preparing for the implementation of the system, some or many of their people will be performing tasks directly related to the new system and the management of the departments must know how the personnel will be scheduled and what work they will be doing. In addition, of course, the user management is anxious to know what is taking place because they are anxious to implement the system and have the advantages and money savings of the new system.

Top level management within a company should also be informed of the progress being made. Although their need to know is primarily because of their management responsibilities concerning people and money, there is also the desire on the part of top management to see a money-saving system implemented and, of course, they are also aware that any delays in the implementation of the system will be costly in terms of time and money. Additionally, if some problems do occur, top level management will have the ability to dictate what is going to happen in case a lower-level employee is balking at some task which must be accomplished.

In general, therefore, it is necessary for the analyst to keep management on all levels informed of the progress of the system. Not only will he keep them happy because they know what is happening, he may also be building some strong allies in case of problems with the implementation of the system.

SUMMARY

The analyst must prepare for the implementation of the new system by meticulously scheduling the tasks to be performed and by arduously following the progress of these jobs so that the entire effort will dovetail together when the deadline date for implementation of the system arrives.

When all of the various jobs have been performed or are being performed, the analyst still has one very large task to be completed. Quite simply, all of the effort is for nought if the system does not work in the manner desired. Thus, the analyst must very carefully design the testing schedule, the test data, and the various techniques which are required to ensure that when the system goes "on the air," it functions smoothly. In addition, he must be concerned that all of the facts of the system are documented both for the needs of the operational staff and for any programmers or analysts in the future who will be called upon to make changes in the system.

CASE STUDY - JAMES TOOL COMPANY

When approval was given to the systems design for the payroll system for the James Tool Company, the analysts Mard and Coswell had to determine the detailed schedule which would be followed in order to implement the system in the five months which was allocated. The first step which the analysts performed was to define the major tasks facing them. These tasks included the following:

1. Writing the programs which make up the payroll system.

2. Testing the programs which are written.

3. Documenting the programs and also the related systems activities.

4. Training the data processing and user personnel in the use of the system.

5. Converting the manual payroll files to the master files which are to be stored on magnetic tape.

Within these tasks are the many subtasks which must be done before the payroll system can be implemented. The analysts Mard and Coswell defined these substasks. Some of the subtasks which they defined are listed on the following page.

1. Develop the program specifications for each program in the system. Although a good portion of this is done since the file worksheets, printer spacing charts, and so on are completed, there is still some detailed narratives which must be developed prior to submitting these specifications to the programmers.

2. Determine the two programmers who are to be assigned to the project and the time it will take for the programming effort.

3. Develop the standards and programming languages which are to be used in programming the payroll system.

4. Develop test data which is to be used in the systems testing to be performed after all of the programs have been individually tested.

5. Determine the number of additional magnetic tape reels, if any, which must be ordered since the payroll system will be using magnetic tape and order these tapes.

6. Contract with printers to have the New Employee Forms, the Employee Change Forms, and the payroll checks printed. Determine the time when this activity must be completed.

7. Document the procedures to be followed in the data processing and user departments to perform the daily payroll function.

8. Perform the systems testing to test not only the programs but also the procedures which have been developed.

9. Train personnel in both the data processing department and the user departments in the use of the new forms and the schedule which is to be followed in preparing the payroll.

10. Convert the data used in the manual payroll system to data which can be used in the new payroll system. This includes the development of Employee Numbers with the new check digit and the conversion of the data from the manual files to the new Employee Master File.

11. Conduct management and user level meetings to ensure that they are aware of the progress being made in the system, any problems which have arisen, and also to train them in the use of the new system.

Case Study - Developing Program Specifications

In order for programmers to write programs, they must be supplied with detailed specifications indicating the processing which is to take place in each program. Mard and Coswell knew that much of the work had been done in this area since the file layout worksheets, printer spacing charts, and multiple card layout forms had been completed during the systems design phase. However, a detailed narrative must still be prepared for each of the programs for distribution to the programmers. The analysts estimated that the entire process would consume a total of three weeks. They further determined, however, that as each program specification was completed, it could be given to the programmer so that three weeks will not elapse before programming could begin. Instead, approximately one week will elapse before programming of the payroll system could begin.

Case Study - Determine the Programmers

The analysts decided that they needed experienced business applications programmers on this project if it was going to meet its schedule. In talking with Clyde Harland, the manager of systems and programming for the James Tool Company, it was decided that two of the more experienced programmers on the staff, Harvey Waller and Nancy Kohler, would be assigned to the project. By knowing the capabilities of the programmers who would be working on the system, the analysts were better able to accurately schedule the events which must transpire before implementation of the system. They estimated that the programming, testing, and documentation should take a total of 4-3/4 months.

Case Study - Develop Standards and Programming Languages

The analysts must also develop the programming standards and decide upon the languages which are to be used within the system. Since the data processing installation at James Tool Company had previously established many standards and had a standards manual, the analysts estimated that only several days would be required to determine in detail the methods to be used for the payroll system.

Case Study - Develop Test Data

One of the critical aspects of any system is the systems testing which takes place in order to ensure that the entire system works accurately before implementation. Knowing that this is a critical operation and that the development of test data and the testing techniques can be a difficult task, Mard and Coswell determined that three weeks would be required to adequately design the test data and testing procedures.

Case Study - Ordering Magnetic Tape

Since the master files of the payroll system of the James Tool Company are to be stored on magnetic tape, the analysts knew that there was the possibility that additional reels of magnetic tape might be required when the system was implemented. In checking with Jane Miller, the manager of computer operations, however, the analysts found that she had ordered a number of extra reels of tape several months ago because the manufacturer had offered a special discount on quantity orders. Therefore, after assessing the needs of the payroll system, the analysts, together with Ms. Miller, determined that no additional reels of tape needed to be ordered to satisfy the needs of the payroll system.

Case Study - New Forms

The analysts contacted a number of outside vendors concerning the acquisition of the new forms required for the payroll system. After considerable negotiating, they were assured that all of the forms could be supplied in the quantities required four weeks after the order was received by the forms company if the forms were completely designed and ready for production. Since the New Employee Forms, the Employee Change Forms, and the payroll checks were all designed in the systems design phase of the project, the analysts allocated four weeks for receiving the forms from the time they submitted the orders.

Case Study - Documenting Procedures

The analysts were aware from experience that documentation is a time-consuming effort. The procedures must be determined and then documentation material must be developed that is error-free and extremely clear. When documenting for persons who are not data processing oriented, great care must be taken to ensure that all contingencies are covered and that no questions can arise concerning the procedures which are described in the documentation. Therefore, the analysts decided that a total of 4 man-weeks would be required in order to develop the required documentation describing the procedures to be followed in the payroll system.

Case Study - Systems Testing

As has been noted previously, systems testing is quite critical in the successful implementation of a data processing system. It is here that any misunderstandings between the analysts, programmers, and user are normally uncovered. In addition, successful systems testing must be undertaken before any system can be implemented. Therefore, since systems testing is such an important function, and since there must be time allowed for corrections to be made to the system during the testing, the analysts allocated one month for systems testing.

Case Study - Training

The analysts knew from past experience that adequate training of the users of a data processing system can determine whether the system is successfully utilized after implementation. They were also aware, however, that the personnel using the new payroll system would be the same personnel who had previously been processing the payroll under the manual system. Therefore, portions of the system would be familiar to those in the payroll department; thus, the analysts allocated one week for the actual training of the personnel to use the new system.

Case Study - File Conversion

As has been noted, it is always necessary when converting from a manual system to a data processing system to plan for the conversion of the data found in the manual system to a form which is able to be processed in the new computer system. In the James Tool Company payroll system, two main "conversion" efforts must be scheduled. First, employee numbers with check digits must be developed for all employees in the company. It will be recalled that the employee number is to consist of a two digit department number, a four digit seniority number, and a one digit check number. The analysts estimated the time required to determine the department number and seniority number for each employee and to punch this information on punched cards to be 1 week. Additionally, they estimated that the time required to write a program to determine the check digit and the time required to run the program would be one week. Therefore, it is estimated two weeks would be needed to generate the employee numbers.

Once the employee numbers were generated, the data in manual files could be transcribed onto the New Employee Form so that the data could be prepared to load the new Employee Master File on magnetic tape. In addition, the data concerning the year-to-date earnings of each employee would have to be prepared. The preparation of this data and the computer time to run the programs to create the new master files was estimated at three weeks.

Case Study - Management and User Meetings

As has been noted, keeping management and the users informed of the status of a system being implemented is quite an important task. In addition, it is critical that those in management who are going to be using the system be knowledgeable in the information which is produced by the system. The analysts estimated that a total of one week would be required to conduct the meetings necessary to both inform management and users of the status of the system and also to train them in the use of the information being produced from the new payroll system.

Case Study - Measurement of Progress

As noted previously, once the tasks have been defined by the systems analyst and the time estimates have been developed, it is necessary to develop documentation of the project. One of the most useful methods is the Gantt chart or a milestone chart. The analyst prepared the milestone chart in Figure 10-12.

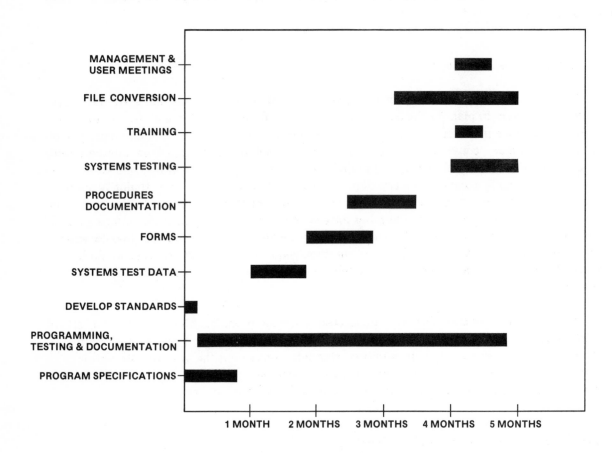

Figure 10-12 Milestone Chart for Payroll System

Note from the milestone chart illustrated in Figure 10-12 that each of the steps to be accomplished in the payroll system are specified and their estimated times for completion are also listed. This chart would be used to document the progress which is being made toward completion as the systems project is worked on. As noted previously, each of the activities on the chart can be further broken down so that a more detailed evaluation of each step can take place. The chart illustrated in Figure 10-13 illustrates a breakdown of the steps to take place in the preparation for the file conversion.

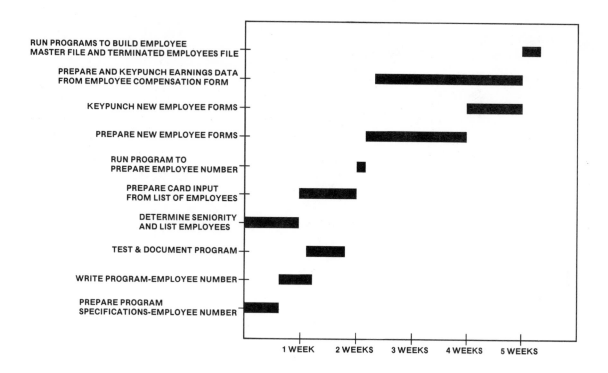

Figure 10-13 Milestone Chart for File Conversion

Note from the chart in Figure 10-13 that each step which must be done in preparing and converting the manual files to the computer files is specified together with the estimated time required to perform the tasks. This chart would then be used to indicate the amount of work which had been accomplished for this job and also what work is left to do.

SUMMARY

The scheduling of the tasks to be performed in the Systems Development phase of the systems project can be quite critical as all elements of the phase must be closely coordinated in order for the system to be implemented at the expected date. In the James Tool Company payroll system, the analysts would break down into units all of the tasks which must be accomplished and keep close control over the time which was being expended and the progress which was being made. If at any time the schedule appeared to "slip," corrective action would have to be taken in the form of more manpower, overtime, or other actions which would lead to following the schedule so that the payroll system would be implemented at the expected time.

STUDENT ACTIVITIES—CHAPTER 10

1. List the activities that occur during the systems development phase of a systems project.

    1)

    2)

    3)

    4)

2. What is meant by the term "Project Planning"?

3. List the basic steps in project planning.

    1)

    2)

    3)

    4)

4. List the major tasks common to all systems projects in the systems development phase.

    1)
    2)
    3)
    4)
    5)

5. List the subtasks associated with the systems development project.

    1)

    2)

    3)

    4)

    5)

    6)

    7)

    8)

    9)

    10)

    11)

    12)

    13)

    14)

    15)

    16)

6. What factors should be considered in order to accurately estimate the time required to program the applications within a project?

7. As a general rule program development studies have indicated that a reasonable allocation of time for programming activities is as follows:
(Indicate the correct percentage.)

    Development of Program Logic   _____%

    Coding the Program   _____%

    Testing and Debugging   _____%

    Final Documentation   _____%

8. Discuss the factors which influence the scheduling of program testing and debugging.

9. What documentation should be developed during the systems development phase of a systems project?

10. Discuss the factors with which the analyst should be concerned relative to the training of personnel.

11. List five sources of training for data processing personnel.

    1)
    2)
    3)
    4)
    5)

12. What portion of the data processing department is faced with the task of file conversion?

13. List four common methods of acquiring hardware.

    1)
    2)
    3)
    4)

14. Discuss the advantages and disadvantages of "third party leasing" as a method of hardware acquisition.

15. List two common techniques which may be utilized to schedule and control systems development tasks.

    1)
    2)

16. Explain the purpose of a PERT Chart.

17. What is meant by the "critical path" in a PERT Chart?

DISCUSSION QUESTIONS

1. "A good data processing manager should be able to accurately schedule programming tasks by keeping a detailed history of the time previous programs have begun and the time the programs are completed. Dividing this time into the number of lines of code written, he should be able to accurately estimate the number of lines of code per day which his staff can generate."

   "It's nearly impossible to accurately schedule programming activities. Programming is a creative task and 'You can't schedule creativity'."

   Which position do you support?

2. As data processing manager you currently have one programmer/analyst on your staff. You have been given an additional yearly budget of $17,000.00 for the hiring of one or more programmers.

   Would you hire two programmer trainees at $700.00 per month or one additional senior programmer at $1,400.00 per month?

CASE STUDY PROJECT—CHAPTER 10
The Ridgeway Company

INTRODUCTION

Once the systems design phase of a systems project has been approved by management the project then enters the systems development phase. In the Ridgeway Country Club, Mr. Ridgeway and other members of top management reviewed the design of the billing system and gave their approval to begin the systems development phase of the project, including the actual programming of the system.

STUDENT ASSIGNMENT 1
PROJECT PLANNING

1. Develop a project plan for the systems development phase of the billing systems project of the Ridgeway Country Club.

   1) Prepare a list of the major tasks which must be accomplished.
   2) Prepare a detailed list of the subtasks which must be accomplished before the billing system can be implemented.

STUDENT ASSIGNMENT 2
PROJECT SCHEDULING

1. Design a Gantt Chart to assist in scheduling the major tasks in the systems development of the billing system. Include estimated times for the completion of each major task.

STUDENT ASSIGNMENT 3
TASK SCHEDULING

1. Develop a schedule for the subtasks defined in Student Assignment 1. Design a Gantt Chart for those activities which require one.

STUDENT ASSIGNMENT 4
PROGRAM SCHEDULING

1. Develop schedules and Gantt Charts for the programs which are to be written for the billing system.

PROGRAMMING ASSIGNMENTS AND SPECIFICATIONS

CHAPTER 11

PROGRAMMING ASSIGNMENTS AND SPECIFICATIONS

11

INTRODUCTION

Once management and the users have given their approval to the system which is to be implemented and the project manager has developed a schedule of the activities to be undertaken during systems development, it remains to transform the systems design from flowcharts and ideas to an actual working system which can process data and produce results.

One of the most important aspects of systems development which can lead to fast, accurate implementation and to efficient, reliable changes and maintenance to the system is standardization of the programming effort. This standardization should include uniformity in programming languages used, uniformity in programming techniques, and uniformity in program documentation. In many systems, the analyst is the person who will dictate the methods to be used in the programming of the system. Therefore, it is important to examine some of the various techniques which are used to ensure a systematic approach to programming the system.

The major task to be accomplished at this step of the systems development phase is to produce computer programs that perform in a reliable and error-free manner and that are easily maintained.

The basic steps which must be followed in producing the computer programs required to implement the systems design include:

1. Determine the programming language to be used.

2. Develop detailed programming specifications.

3. Analyze the program specifications and develop the programming logic and flowcharts required to solve the problem.

4. Code and desk check the programs.

5. Prepare test data.

6. Test and debug the programs.

7. Prepare programming documentation.

PROGRAMMING LANGUAGES

One of the first considerations which must be made concerning the method of programming is the selection of the programming languages to be used. Factors to be considered in selecting a programming language include the following:

1. Capabilities of the programming language
2. Knowledge of the current programming staff
3. Cost of the compiler
4. Ease of programming
5. Efficiency of the language
6. Compatibility of the programming language

Capabilities of the Programming Language

When considering a programming language, the analyst must first have defined the functions to be performed within the system, that is, he must have determined the type of file processing which is to be performed, the types and complexity of the calculations which are to take place, the kinds of reporting which is to be output from the system, whether any special data communications is to take place, and so on. This is necessary because certain programming functions can be performed in one language yet cannot be performed in another language. For example, FORTRAN does not have the ability to perform data communications techniques such as polling terminals, so these functions would normally be done in a language such as Assembler Language which has these capabilities.

Thus, before any other consideration, the language capabilities must be considered. It is a distinct possibility that the programs within a system can be written in a number of different programming languages. It is necessary, however, to eliminate those which cannot be used because of the requirements of the programs within the system.

Knowledge of Programming Staff

The next consideration in determining the language to be used is the personnel who are available for the programming of the system. Obviously, if the programmers available for a project cannot program in a given language, it does not matter what language is the best for the particular application unless extensive training is undertaken. Therefore, one of the first tasks of the analyst is to identify those members of the programming staff who will be available for the project and what languages they are capable of using. It should be noted that the decision concerning who will be working on the system and what programming language they will be using is not normally the exclusive decision of the systems analyst. In most cases, the programming manager or a senior programmer will assist in determining who will be working on the system and will inform the analyst of their recommendations concerning the languages to be used in the system.

Cost of the Compiler

In 1971, computer manufacturers began the separate pricing of computer hardware and software. A monthly licensing fee of a few dollars per month to hundreds of dollars per month is normally charged for newly developed software from major manufacturers. Software developed before separate pricing went into effect is normally available free of charge. Thus, the analyst is faced with the task of evaluating the capabilities of newer software against the fee charged for the use of this software.

Ease of Programming

The ease and speed with which programs may be written is a factor which must be considered when selecting a programming language. The ease of programming also directly influences the problem of program maintenance. Changes are often necessary at some future date and a program that is easy to change and maintain may significantly influence future costs.

Efficiency of the Language

High-level programming languages, such as COBOL, are normally less efficient than low-level programming languages such as Assembler Language; that is, more machine instructions are generated when using high-level programming languages. Therefore, more internal storage is required for programs written in high-level programming languages and the programs do not normally execute as fast as those written in low-level languages because of the additional machine instructions. Although storage and execution times are less important today than a decade ago, the analyst must be certain that the compiler for the language selected can itself be executed in the available storage and that it will produce programs which can be stored and executed within the available storage.

Compatibility of the Programming Language

It is sometimes desirable to utilize a programming language that is compatible with other computers available from the same or different manufacturers. As the data processing needs of the business expand, the programs developed for one system should be able to run on a larger and faster system. Perhaps equally as important, is the compatibility of the language among different manufacturers. If hundreds of man hours are spent in programming in a language unique to one manufacturer, it is difficult, if not impossible, to change computer systems at some future date.

Types of Programming Languages

There are a variety of programming languages available today which are utilized in programming for business applications. Some of these programming languages include:

1. COBOL
2. Assembler Language
3. RPG
4. FORTRAN
5. PL/I

Although there are many other programming languages available, the five languages above dominate the programming that takes place for business oriented applications, and normally the analyst will select the programming language used for the system from the list above.

Evaluating Programming Languages

In addition to the factors mentioned previously, the language should be evaluated on the basis of the technical factors explained below.

1. INPUT-OUTPUT CAPABILITIES—The language ideally should support a variety of storage devices such as the console typewriter, disk, tape, graphic display devices, and so on.

2. DATA-MANIPULATION CAPABILITY—For business applications the language should provide the facility to process numeric and alphabetic data. It should also contain the ability to edit the output with special characters such as the dollar sign, the comma, and so on. Accuracy and precision in rounding decimal numbers may also be a consideration.

3. FILE-ORGANIZATION TECHNIQUES—The type of file-organization structure which the language supports is an important factor. In the business environment, sequential files as well as direct and indexed files should be supported. The types of record formats, that is, fixed-length and variable-length records, handled by the programming language may be significant.

4. DEBUGGING AIDS—Diagnostics generated by the compiler on error conditions should be meaningful and should assist the programmer in correcting these error conditions.

Numerous other related factors should be considered when selecting the specific programming language or languages that are best suited to meet the needs of the business. Some of the additional questions to consider include:

1. How long does it take to learn the language?
2. Are adequate instructional manuals available?
3. Can the program be written and keypunched accurately and quickly?
4. Does the compiler generate efficient object programs?
5. How long does it take to compile?

Although the selection of a programming language may seem to be a complex task recent surveys have shown that COBOL is by far the most dominant programming language for business-type applications. Nevertheless, the analyst should be aware of the advantages and disadvantages of the major programming languages and give consideration to the use of that language that will lead to the most efficient system implementation.

COBOL

The most widely used programming language for business applications is COBOL (COmmon Business Oriented Language). This language was developed by a group of computer users and manufacturers in 1960 to provide a "high-level, business-oriented language." An example of a segment of a COBOL program is illustrated in Figure 11-1.

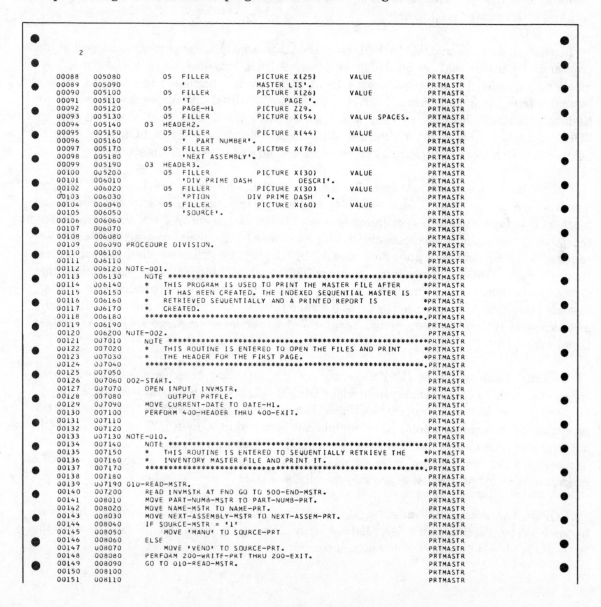

```
     2

00088   005080        05   FILLER        PICTURE X(25)      VALUE          PRTMASTR
00089   005090        *                  MASTER LIS'.                      PRTMASTR
00090   005100        05   FILLER        PICTURE X(26)      VALUE          PRTMASTR
00091   005110        *'T                PAGE '.                           PRTMASTR
00092   005120        05   PAGE-H1       PICTURE ZZ9.                      PRTMASTR
00093   005130        05   FILLER        PICTURE X(54)      VALUE SPACES.  PRTMASTR
00094   005140   03   HEADER2.                                            PRTMASTR
00095   005150        05   FILLER        PICTURE X(44)      VALUE          PRTMASTR
00096   005160        *    PART NUMBER'.                                   PRTMASTR
00097   005170        05   FILLER        PICTURE X(76)      VALUE          PRTMASTR
00098   005180        'NEXT ASSEMBLY'.                                     PRTMASTR
00099   005190   03   HEADER3.                                            PRTMASTR
00100   005200        05   FILLER        PICTURE X(30)      VALUE          PRTMASTR
00101   006010        'DIV PRIME DASH          DESCRI'.                    PRTMASTR
00102   006020        05   FILLER        PICTURE X(30)      VALUE          PRTMASTR
00103   006030        'PTION     DIV PRIME DASH   '.                       PRTMASTR
00104   006040        05   FILLER        PICTURE X(60)      VALUE          PRTMASTR
00105   006050        'SOURCE'.                                            PRTMASTR
00106   006060                                                            PRTMASTR
00107   006070                                                            PRTMASTR
00108   006080                                                            PRTMASTR
00109   006090   PROCEDURE DIVISION.                                      PRTMASTR
00110   006100                                                            PRTMASTR
00111   006110                                                            PRTMASTR
00112   006120   NOTE-001.                                                PRTMASTR
00113   006130        NOTE *************************************************PRTMASTR
00114   006140        *    THIS PROGRAM IS USED TO PRINT THE MASTER FILE AFTER   *PRTMASTR
00115   006150        *    IT HAS BEEN CREATED. THE INDEXED SEQUENTIAL MASTER IS *PRTMASTR
00116   006160        *    RETRIEVED SEQUENTIALLY AND A PRINTED REPORT IS        *PRTMASTR
00117   006170        *    CREATED.                                             *PRTMASTR
00118   006180        ***********************************************************.PRTMASTR
00119   006190                                                            PRTMASTR
00120   006200   NOTE-002.                                                PRTMASTR
00121   007010        NOTE *************************************************PRTMASTR
00122   007020        *    THIS ROUTINE IS ENTERED TO OPEN THE FILES AND PRINT  *PRTMASTR
00123   007030        *    THE HEADER FOR THE FIRST PAGE.                       *PRTMASTR
00124   007040        ***********************************************************.PRTMASTR
00125   007050                                                            PRTMASTR
00126   007060   002-START.                                               PRTMASTR
00127   007070        OPEN INPUT  INVMSTR,                                PRTMASTR
00128   007080             OUTPUT PRTFLE.                                 PRTMASTR
00129   007090        MOVE CURRENT-DATE TO DATE-H1.                       PRTMASTR
00130   007100        PERFORM 400-HEADER THRU 400-EXIT.                   PRTMASTR
00131   007110                                                            PRTMASTR
00132   007120                                                            PRTMASTR
00133   007130   NOTE-010.                                                PRTMASTR
00134   007140        NOTE *************************************************PRTMASTR
00135   007150        *    THIS ROUTINE IS ENTERED TO SEQUENTIALLY RETRIEVE THE *PRTMASTR
00136   007160        *    INVENTORY MASTER FILE AND PRINT IT.                  *PRTMASTR
00137   007170        ***********************************************************.PRTMASTR
00138   007180                                                            PRTMASTR
00139   007190   010-READ-MSTR.                                           PRTMASTR
00140   007200        READ INVMSTR AT END GO TO 500-END-MSTR.             PRTMASTR
00141   008010        MOVE PART-NUMB-MSTR TO PART-NUMB-PRT.               PRTMASTR
00142   008020        MOVE NAME-MSTR TO NAME-PRT.                         PRTMASTR
00143   008030        MOVE NEXT-ASSEMBLY-MSTR TO NEXT-ASSEM-PRT.          PRTMASTR
00144   008040        IF SOURCE-MSTR = '1'                                PRTMASTR
00145   008050             MOVE 'MANU' TO SOURCE-PRT                      PRTMASTR
00146   008060        ELSE                                                PRTMASTR
00147   008070             MOVE 'VEND' TO SOURCE-PRT.                     PRTMASTR
00148   008080        PERFORM 200-WRITE-PRT THRU 200-EXIT.                PRTMASTR
00149   008090        GO TO 010-READ-MSTR.                                PRTMASTR
00150   008100                                                            PRTMASTR
00151   008110                                                            PRTMASTR
```

Figure 11-1 Example of COBOL Program

COBOL was intended to be "English-like" in design and to be self-documenting. As can be seen from Figure 11-1, COBOL is written in "English-like" statements and it is relatively easy to read the source listing.

COBOL is considered to be an easy language to learn and program because it is a high-level language and is not concerned with the internal structure and machine language instruction formats of the computer.

COBOL can be used for all types of business problems. The language has the capabilities of processing any type of data file which is normally used in a business environment and can process data in any format in which data can be stored on the files. It has the capabilities of using most programming techniques which are applicable to business programming, including subroutines written in either COBOL or another language and it offers most of the features required for structured programming.

In general, it can be said that COBOL gives the business applications programmer all of the programming power that he needs. In addition, COBOL was developed to be "machine-independent," that is, programs written in COBOL can be compiled and executed on any computer from any manufacturer. In 1968, a "standard" COBOL, developed by the American National Standards Institute (ANSI), was released with the intention of standardizing all COBOL compilers from all manufacturers. With the advent of the ANSI COBOL, most COBOL programs, with minor modifications, can be compiled and executed on any machine.

There are some disadvantages in using COBOL, however. Perhaps the biggest disadvantage is that COBOL is a relatively inefficient language because of the number of machine language instructions generated for each COBOL statement. As was noted previously, the programmer need not be concerned with the machine language instructions which are required to manipulate data when he writes a COBOL program. The compiler, however, must translate each COBOL sentence into a series of individual machine language instructions, that is, the instructions which the machine can "understand" and execute. This translation process, in some cases, will generate a sequence of machine language instructions which are not nearly as efficient as if they had been directly written in a lower level language such as Assembler Language.

Another disadvantage of COBOL is due to the fact that there are some processing capabilities which a computer may have which COBOL will not be able to initiate. For example, on the System/360 and System/370, certain machine language instructions provide the programmer with the capability of manipulating bits within a byte. When using COBOL, however, the programmer does not have this capability. Although this ability may not be critical in most business applications, if the bit manipulation was required for a program, then the programmer would have to use a language which allowed this.

COBOL has become the most widely used business programming language, however, because it is relatively versatile, is a relatively easy language to learn and use, and because the language is compatible with a wide range of computer systems.

Assembler Language

On most computers, some type of Assembler Language is available for use. An assembler language is a symbolic programming language in which symbolic operation codes are used to represent the actual machine language instruction which is desired. The compiler merely translates the symbolic instruction into a machine-language instruction, normally on a one-for-one basis. Thus, one of the ways in which an assembler language differs from a high-level programming language such as COBOL is that a single statement in COBOL will generate a series of instructions whereas a symbolic instruction in an assembler language will generate one machine-language instruction. Figure 11-2 illustrates a segment of an assembler language program for the System/360.

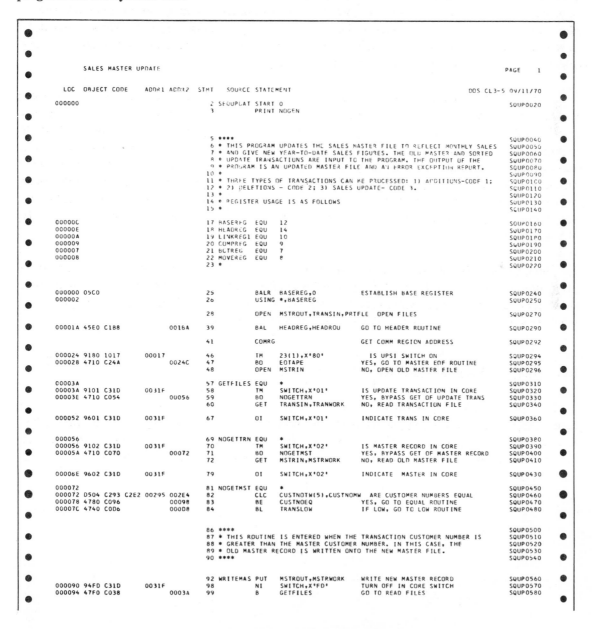

Figure 11-2 Assembler Language Program

Most symbolic instructions in an assembler language directly reflect the machine language instructions which are available on a computer system and when writing programs in assembler language, the programmer can take advantage of the entire repertoire of instructions to perform whatever processing he wishes. Thus, there are no restrictions upon the operations which can be performed as there might be with a high-level programming language such as COBOL. In addition, assembler programs are normally more efficient than any other language both in terms of storage usage and in execution time because the programmer need code only those instructions which are absolutely necessary.

It is somewhat common to find in a complete system that the majority of the programming is done in COBOL but that there are certain subroutines which are coded in Assembler Language in order to take advantage of certain efficiencies of Assembler Language or in order to perform certain tasks which cannot be done in the high-level programming language. Thus, regardless of the type of hardware which is being used, there is almost always an assembler language which can be used with the computer to take advantage of its features.

There are some disadvantages which are associated with Assembler Language, however, which in many cases prohibit it from being used in a large majority of business systems. It is a relatively difficult language in which to code a program, certainly more difficult than COBOL. It is more difficult because instructions more closely resemble the machine language of a system and normally more statements must be written to accomplish a particular operation than with a high-level language. In addition, since Assembler Language programs are normally more difficult to code, they will usually take longer to code than programs written in a high-level programming language and they may be more prone to error because of the inherent difficulty of the language.

Thus, the analyst and the senior programmers who will be involved with the system to be programmed must consider the advantages and disadvantages of Assembler Language and then decide whether the system should be written in this machine-oriented language or whether a high-level language such as COBOL is more desirable.

RPG - (Report Program Generator)

RPG is a high-level programming language that was originally developed for small scale computer systems to be used primarily for the preparation of printed reports from input files. Extensions to the language now make RPG particularly useful in more sophisticated applications such as file maintenance where files are to be updated and processed.

```
    DOS/360*RPG*CL 3-9                      NO NAME                              PAGE 0001

          01 010 H                                                                AVGSLS
    001   01 020 FSLSCDS  IP  F   80   80         READ40 SYSIPT                   AVGSLS
    002   01 030 FREPORT  O   F  132  132         PRINTERSYSLST                   AVGSLS
    003   02 010 ISLSCDS  AA  01                                                  AVGSLS
    004   02 020 I                               1    40SLSMAN                    AVGSLS
    005   02 030 I                               5    24 NAME                     AVGSLS
    006   02 040 I                              32   392CURSLS                    AVGSLS
    007   02 050 I                              40   472YTDSLS                    AVGSLS
    008   02 060 I                              48   490MONTH                     AVGSLS
    009   03 010 C     01     YTDSLS    ADD  CURSLS   NEWYTD  82                   AVGSLS
    010   03 020 C     01     NEWYTD    DIV  MONTH    AVGSLS  82H                  AVGSLS
    011   03 030 C     01     TOTCUR    ADD  CURSLS   TOTCUR  92                   AVGSLS
    012   03 040 C     01     TOTYTD    ADD  NEWYTD   TOTYTD  92                   AVGSLS
    013   03 050 C     01     TOTOYT    ADD  YTDSLS   TOTOYT  92                   AVGSLS
    014   04 010 OREPORT  D  1       01                                           AVGSLS
    015   04 020 O                               SLSMANZ  7                        AVGSLS
    016   04 030 O                               NAME    31                        AVGSLS
    017   04 040 O                               CURSLS  46 '    ,    0.  '         AVGSLS
    018   04 050 O                               YTDSLS  61 '    ,    0.  '         AVGSLS
    019   04 060 O                               NEWYTD  75 '    ,    0.  '         AVGSLS
    020   04 070 O                               MONTH Z 81                         AVGSLS
    021   04 080 O                               AVGSLS  97 '    ,    0.  '         AVGSLS
    022   04 090 O     T  2      LR                                                 AVGSLS
    023   04 100 O                               TOTCUR  46 ' ,     ,   0.  '        AVGSLS
    024   04 110 O                               TOTOYT  61 ' ,     ,   0.  '        AVGSLS
    025   04 120 O                               TOTYTD  75 ' ,     ,   0.  '        AVGSLS
```

Figure 11-3 RPG Source Listing

RPG is based upon a fixed-logic built into the compiler which dictates what processing will occur. Through the use of special "indicators" the programmer can specify the logic which will be used in conjunction with the fixed RPG logic to solve a problem. This fixed logic, however, somewhat limits the functions which can be performed with RPG because the programmer often cannot cause an operation to occur whenever he wishes. Therefore, there are some applications which are not easily implemented using the RPG language.

On the other hand, RPG is probably the simplest of the business-oriented languages to code for the preparation of routine business reports. The programmer merely specifies the functions which he wishes to be performed by making entries on specialized coding forms and the RPG compiler constructs the logic of the program and the machine-language instructions required to execute the program. Since the compiler generates the machine-language instructions, the inefficiencies which are found in a high-level programming language, such as was pointed out when discussing COBOL, are also found in RPG. In fact, RPG is probably more inefficient than COBOL because there are certain fixed routines which are contained within the RPG program which is generated by the compiler which are not necessary in the COBOL program. Also, since the RPG compiler generates more instructions than COBOL or Assembler Language, it follows that the program will take longer to execute than the others. Therefore, if efficiency is the primary consideration in determining the programming language to be used, RPG is probably not the best choice; but if ease of programming and the speed with which the programmer can write the program is quite important, then strong consideration should be given to RPG as the programming language.

Although RPG was originally developed by International Business Machines, many other manufacturers have implemented this programming language and provide RPG as a part of their systems software.

FORTRAN

The first widely used high-level programming language was FORTRAN. Although primarily a language which is used for scientific, mathematical and engineering applications, there are some limited business applications which require the mathematical capabilities of the language and which can be more easily and efficiently programmed in FORTRAN than in another language such as COBOL or RPG. In particular, complex calculations or calculations which may involve large matrixes or very large or very small numbers may be effectively solved using FORTRAN.

FORTRAN, however, was not designed for business applications and therefore, unless needed for special processing, FORTRAN is not normally used in business programming. Its file handling capabilities are limited and the definitions of input and output records, especially when dealing with alphabetic data, are burdensome and subject to error.

Unless the entire program requires the attributes of number handling and the calculations for which FORTRAN is designed, the best method is to use a FORTRAN subroutine to process those applications which are especially adaptable to the techniques which are available in FORTRAN. Figure 11-4 illustrates a FORTRAN source listing.

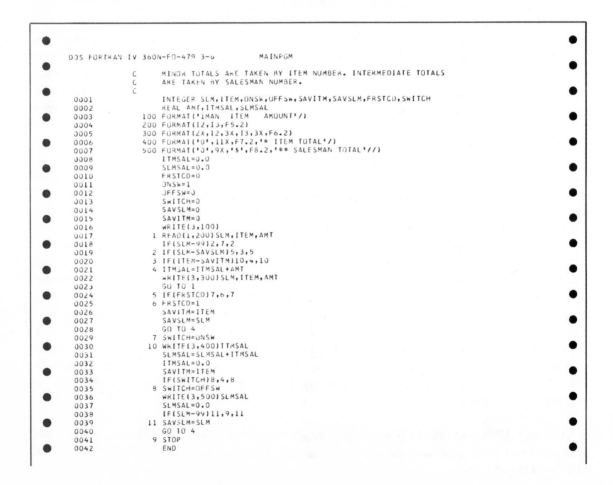

Figure 11-4 FORTRAN Source Listing

A particular advantage of FORTRAN is that it is the most widely implemented programming language, that is, nearly all manufacturers of computer equipment provide a FORTRAN compiler as a part of the systems software.

PL/I

Recognizing that there were advantages to using COBOL and FORTRAN, depending upon the application, IBM developed PL/I to attempt to take advantage of the best part of both FORTRAN and COBOL. PL/I allows many of the mathematical computations which are processed when using FORTRAN but it also allows data to be defined in a manner similar to COBOL.

While PL/I can be used for both scientific and business applications, it has some of the same disadvantages of these two languages, primarily that it is quite inefficient in terms of the coding generated. In addition, since PL/I was developed by IBM, it is currently available only on IBM computers and several other manufacturer's systems. Although PL/I is a very powerful language it has not been widely implemented on a national level. A particular advantage of PL/I is that it contains all of the instructions required when using a structured programming approach to solving problems. Figure 11-5 illustrates a PL/I program.

```
        DOS PL/I COMPILER   360N-PL-464  CL3-8                              PAGE  002

/* CARD TO TAPE */

                /* CARD TO TAPE */                                              TAPE0010
                                                                               TAPE0020
    1           PGM10:  PROCEDURE OPTIONS (MAIN);                               TAPE0030
                                                                               TAPE0040
                        /* CARD INPUT FILE DEFINITION */.                       TAPE0050
                                                                               TAPE0060
    2                   DECLARE   CARDIN FILE RECORD                            TAPE0070
                                  INPUT SEQUENTIAL                              TAPE0080
                                  ENVIRONMENT (MEDIUM (SYS004,2540)             TAPE0090
                                  F(80) CONSECUTIVE);                           TAPE0100
                                                                               TAPE0110
                                                                               TAPE0120
                        /* CARD INPUT AREA */                                   TAPE0130
                                                                               TAPE0140
    3                   DECLARE   1  CARD_INPUT   BASED (PTR),                   TAPE0150
                                  3   NAME_IN        CHARACTER (25),            TAPE0160
                                  3   ADDRESS_IN     CHARACTER (25),            TAPE0170
                                  3   CITY_IN        CHARACTER (25),            TAPE0180
                                  3   FILL1_IN       CHARACTER (5);             TAPE0190
                                                                               TAPE0200
                                                                               TAPE0210
                        /* TAPE OUTPUT FILE DEFINITION */                       TAPE0220
                                                                               TAPE0230
    4                   DECLARE   TAPOUT FILE RECORD                            TAPE0240
                                  OUTPUT SEQUENTIAL                             TAPE0250
                                  ENVIRONMENT (MEDIUM (SYS006,2400)             TAPE0260
                                  F(750,75) CONSECUTIVE);                       TAPE0270
                                                                               TAPE0280
                                                                               TAPE0290
                        /* TAPE OUTPUT AREA */                                  TAPE0300
                                                                               TAPE0310
    5                   DECLARE   1  TAPE_OUT   BASED (PTR_T),                   TAPE0320
                                  3   NAME_OUT       CHARACTER (25),            TAPE0330
                                  3   ADDRESS_OUT    CHARACTER (25),            TAPE0340
                                  3   CITY_OUT       CHARACTER (25);            TAPE0350
                                                                               TAPE0360
                                                                               TAPE0370
                        /* POINTER DEFINITION */                                TAPE0380
                                                                               TAPE0390
    6                   DECLARE   PTR     POINTER,                              TAPE0400
                                  PTR_T   POINTER;                             TAPE0410
                                                                               TAPE0420
                                                                               TAPE0430
                        /* ESTABLISH END OF FILE PROCESSING */                  TAPE0440
                                                                               TAPE0450
    7                   ON ENDFILE (CARDIN) GO TO END_CARDS;                    TAPE0460
                                                                               TAPE0470
                                                                               TAPE0480
                        /* START OF PROCESSING */                               TAPE0490
                                                                               TAPE0500
                        /* OPEN FILE */                                         TAPE0510
                                                                               TAPE0520
    8                   OPEN FILE (CARDIN), FILE (TAPOUT);                      TAPE0530
                                                                               TAPE0540
                                                                               TAPE0550
                        /* READ CARDS AND WRITE TO TAPE */                      TAPE0560
                                                                               TAPE0570
    9           READ_CARDS:                                                     TAPE0580
                        READ FILE (CARDIN) SET (PTR);         /* READ A CARD   */TAPE0590
   10                   LOCATE TAPE_OUT FILE (TAPOUT) SET (PTR_T); /*FIND OUT BUFFER */TAPE0600
   11                   NAME_OUT = NAME_IN;                   /* MOVE INPUT FIELDS TO*/TAPE0610
   12                   ADDRESS_OUT = ADDRESS_IN;             /* OUTPUT FIELDS  */TAPE0620
   13                   CITY_OUT = CITY_IN;                                     TAPE0630
   14                   GO TO READ_CARDS;                     /* GO TO READ NEXT CD */TAPE0640
                                                                               TAPE0650
                                                                               TAPE0660
                        /* END OF CARDS PROCESSING */                           TAPE0670
                                                                               TAPE0680
   15           END_CARDS:                                                      TAPE0690
                        CLOSE FILE (CARDIN), FILE (TAPOUT);   /* CLOSE FILES  */TAPE0700
   16                   END PGM10;                                             TAPE0710
```

Figure 11-5 PL/I Program

Other Programming Languages

There are numerous other programming languages which may be used on special occasions. A great majority of the other languages are used for mathematical and scientific work, but there are several which may prove useful in a business data processing environment. The determination of whether a particularly specialized programming language should be used for a given application must be made by the analyst or senior programmer based upon the work which must be accomplished. Some of the specialized languages must be purchased from an outside vendor and this fact may also be a consideration in determining whether a special language should be used.

DEVELOPING THE PROGRAMMING SPECIFICATIONS

After determining the programming language or languages which are to be used within the system, the analyst must also communicate to the programmer specific details relative to what each program is to do in order to process the data. The communication of this information is one of the more vital steps in the entire process of systems analysis and design because even if the system is the finest system ever designed, it will never be properly implemented if the programmers do not write the proper programs. In addition, if the program specifications are communicated unclearly and the programs are written without all of the information known, then program modifications will have to be made and, as discussed previously, whenever modifications must be made to a program, there is a high probability of errors, unreliability, and delay in implementation. Therefore, it is incumbent upon the analyst to take as much time and care as required to be sure that all information concerning the programs is accurately described to the programmer in the programming specifications.

These programming specifications should include the following:

1. A brief description of the system
2. A system flowchart
3. A listing and brief explanation of the programs within the system
4. The format of the input, output and files to be created and processed
5. A detailed definition of the processing to take place in each program

Description of the System

The programming specifications presented to the programmer should consist of a bound ''packet'' of instructions providing detailed information relative to the programming of the system.

The first portion of the programming specifications should contain a brief description of the overall project in a narrative form so that the programmer will have a general understanding of the purpose and operation of the entire system.

It should be noted that a single programmer may be involved only in the programming of one segment of the system, one program within the system, or even a portion of one program in a large system; nevertheless, an understanding of the overall system is often of value to the programmer so that he may have a feeling for the scope of the project.

Systems Flowchart

Regardless of whether there is one or many programmers who will be involved in programming the system, they should all be aware of the flow of data within the system and with the processing in general which will be accomplished by the system. Therefore, the programmer documentation will normally begin with the systems flowchart, a brief explanation of the processing which is to take place, and the function of each program and file in the system. This is done so that the programmer can see where in the flow of the system the program or programs he will write fit in the system and the relationships of the files in the system to the files with which he will be working. Figure 11-6 illustrates a systems flowchart.

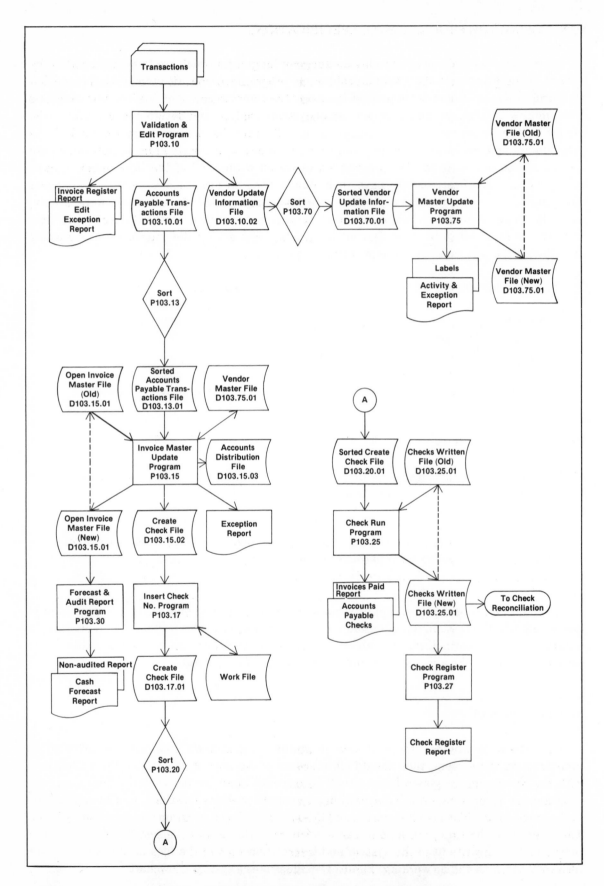

Figure 11-6 Accounts Payable Systems Flowchart

Program Descriptions

The documentation which the analyst attaches with the systems flowchart need not be extensive, but merely enough to explain the basic functions of the programs and the files. The functions of the programs which are in the system which may be prepared by the analyst is illustrated in Figure 11-7.

PROGRAM DESCRIPTION

1. P103.10 - VALIDATION AND EDIT PROGRAM - This program edits input data to the system. Creates two files: Accounts Payable Transaction File (D103.10.01) and Vendor Update Information File (D103.10.02). Creates two reports: Edit Exception Report and Invoice Register Report.

2. P103.13 - SORT - This utility program sorts the Accounts Payable Transaction File into Vendor Number, Document Number, Line Item Number, and Transaction Code Sequence. Creates the Sorted Accounts Payable Transaction File (D103.13.01).

3. P103.15 - INVOICE MASTER UPDATE PROGRAM - The purpose of this program is to update the Open Invoice Master File (D103.15.01) with transactions input from the accounts payable edit program. Check records are created and distribution is made for input to the general ledger.

4. P103.17 - INSERT CHECK NO. PROGRAM - The purpose of this program is to separate the Accounts Payable checks into three groups: those with net amounts less than $3,000, those with a net amount of $3,000 and over, and those with one or more negative remittance advises.

Figure 11-7 Example of Program Descriptions

Note from the example in Figure 11-7 that the programs which are contained within the systems flowchart are named and the function of each is briefly described. It must be noted again that the purpose of this description is not to detail the processing which is to take place within each program but merely to give an overview of the processing so that the programmer can see where within the system his program fits.

Definition of Input and Output

As a part of the programming specifications the multiple card layout forms and printer spacing charts previously defined in the Systems Design Phase of the systems project should be included.

These forms should provide the programmer with the required information relative to field size of input records, and the spacing and editing that is to occur in the output reports.

All of the data which is to be contained on a report, including the headlines of the report, must be illustrated. In addition, any special instructions which are appropriate must be included, such as the number of lines on a page and the skipping that is to occur. The programmer must be able to look at the printer spacing chart and know exactly what is required on the report and in what format the data is to be printed. It is not normally necessary to have a narrative with the printer spacing chart because the data which is contained on the report will come from either transaction records or master records and these will be described with the appropriate file description. The processing instructions to derive the data to be printed will normally be contained in the narrative or flowchart used to describe the processing which is to take place in a given program.

File Definitions

One of the major aspects of the file which must be given to the programmer is the format of the records which will be stored in the file. As was noted in Chapter 6, the Record Layout Worksheet is one of the aids which can be used for this purpose. These worksheets, which should have been created during the systems design phase, must be included with the programming specifications for use by the programmer.

In addition to the Record Layout Worksheet, some type of narrative explanation will normally accompany the file definition to be used by the programmer. This narrative is used to explain the contents of each field within the file and also to more precisely identify the processing which will be done with the file. A portion of the narrative which could accompany the file definitions is illustrated in Figure 11-8.

SORTED CREATE CHECK FILE

File Name: D103.20.01
Storage Medium: 2314 Direct Access Device
Record Length: 200 Characters per record
Blocking Factor: 17 Records/Block
Block Length: 3400 Characters
Approximate Size: Minimum 250 Records; Maximum 1450 Records
Programs Where Used: P103.25

The Sorted Create Check File contains a record for each invoice which is to be paid in the run. One vendor may have more than one invoice which is to be paid, so more than one record can exist with the same vendor number. The file is stored in check number and item number sequence. The following is a description of the fields within the record.

1. VENDOR NUMBER - The identifying number assigned to the vendor who is going to be paid with this transaction.

2. VENDOR NAME - The name of the vendor to be paid by the transaction.

3. INVOICE DATE - This is the date on the vendor invoice.

4. DISCOUNT DATE - This is the date on which the invoice must be paid in order to receive any discount.

5. DUE DATE - This is the date on which the invoice must be paid if no discount is taken.

Figure 11-8 Example of File Description

The remaining fields in the master record would be similarly described. The format of the field must always be included, for example, a field is a dollars and cents field. Also, whenever there is a numeric or alphabetic code employed in a field, then the codes and their meanings must be spelled out so that the programmer is aware of what processing must be performed based upon these codes.

In short, all of the factors which may affect the processing of the field within a record must be specified in the narative and Record Layout Worksheet describing the master record so that there is no mistake on the part of the programmer concerning the processing which must be programmed.

Program Definitions - Narrative Form

Probably the most critical of the documentation which is to be given to the programmer by the analyst deals with the processing which is to take place within a program. As noted previously, if the programmer cannot determine from the documentation what processing is to take place, then there is no chance that the program will be written correctly. An example of the programming specifications which could be established for the edit program for the accounts payable system is illustrated in in Figure 11-9.

PROGRAM: P103.10

I.   Program Name: P103.10

II   Program Function: Edit transaction cards; create edited Accounts Payable Transaction File (D103.10.01); Vendor Update Information File (D103.10.02); Edit Exception Report; Invoice Register Report.

III. Input and Output:

    A. The input to the program are the Accounts Payable Transaction cards (see Exhibit 1).

    B. The output from the program is the Accounts Payable Transaction File (see Exhibit 17), the Vendor Update Information File (see Exhibit 18), the Edit Exception Report (see Exhibit 4), and the Invoice Register Report (see Exhibit 5).

IV. Processing

    A. <u>General</u>

        1. Transactions are input to the validation program in batches, with a batch control card in front of each batch. The batch control card will be balanced to the transactions within the batch, and results of this balancing will be recorded as shown on the printer layout.

        2. If a record has an error, do not output the record on the transaction file. Report it on the edit exception report, and report it on the invoice register with a "*" in print position 132.

    B. <u>Invoice Processing</u>

        1. If a TC 10 (invoice) is input with no TC 13 (distribution) items, check for errors, and output invoice.

        2. If a TC 13 is input with a zero amount field, it has been input only for the "remarks" field. If the only TC 13 input with a TC 10 has a zero amount field, move the "remarks" field to the output record and process as if the TC 13 had not been input.

        3. If a TC 13 is input with "03" in the tax code field, distribute the tax over each distribution record (by dollar amount) for the transaction.

    C. <u>Control Record Processing (TC 70 - TC 78)</u>

        1. Each control record stands alone. For all transactions except TC 71, TC 73, and TC 76, if there is a space in code column 22, place a "1" in the plus/minus field of the output record. Otherwise, place a "0" in the field.

    D. Change transactions (TC, C1, C2, C3, C4) may be input with change information in a field, blank field, or a "N" in the first position of the field. If the field is blank or has change information, output the field as it is input. If the field has "N" in the first character position, and the rest of the field is blank, spread "unpacked" nines in the field when outputing the record.

Figure 11-9 Example of Programming Specifications

In the example in Figure 11-9 it can be seen that the instructions to the programmer are specified in a "narrative" form. The input and output used with the program are specified and the processing which is to be performed by the program is then indicated. As noted previously, it is important that this processing be absolutely clear to the programmer so that the program can be written correctly the first time.

Program Definition - Flowchart

The narrative form of presenting the program is widely used and in many problems will suffice to explain all of the processing required in the program. In some applications, however, it may be well to either replace or augment the narrative with one of several other forms of presenting the same information. One method which could be used is the flowchart. In this case, it would not be the systems flowchart but rather a flowchart which expresses the decisions which are to be made as specified in the narrative. The example in Figure 11-10 illustrates a flowchart which depicts the decisions to be made in part C, Control Record Processing, of the narrative in Figure 11-9.

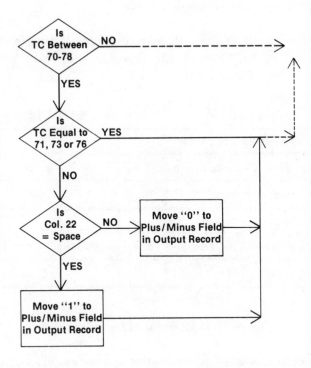

Figure 11-10 Example of Logic Flowchart

From Figure 11-10 it can be seen that the decisions which must be made as specified in the narrative of the program are indicated in flowchart form. It should be noted that this flowchart is not intended to be a program flowchart from which the programmer can directly code. It is used merely to depict the logical decisions which must be made concerning the editing portion of the program which must be written. The programmer will develop the logic of the program depending upon how he wishes to process each record. This flowchart is merely a pictorial representation of the narrative and is intended to augment the narrative for clarity only.

DECISION TABLES

Another documentation method which is sometimes used to indicate to the programmer the logical decisions which must be made within the program is the use of DECISION TABLES. A decision table is a graphical representation of the logical decisions which must be made within a program. For example, the following rules may apply to the deduction of payments for insurance from an employee paycheck.

RULE 1—If the employee is eligible for insurance, and he has not requested insurance, do not deduct the insurance payment.

RULE 2—If the employee is eligible for insurance and he has requested insurance, deduct the insurance payment.

RULE 3—If the employee is not eligible for insurance and he has requested insurance, write a letter telling him he is not eligible for the insurance.

RULE 4—If the employee is not eligible for insurance and has not requested the insurance, the insurance payment is not deducted.

Although not a complex set of rules in narrative form the rules do become somewhat difficult to remember as a group. To graphically represent these rules a decision table may be used. A decision table is illustrated in Figure 11-11.

─────STUBS───── ‖ ─────ENTRIES─────					
Insurance Deduction Table	1	2	3	4	} RULES
Eligible For Insurance	Y	Y	N	N	
Insurance Requested	N	Y	Y	N	
Deduct Insurance Pmt.		X			
Do Not Deduct Ins. Pmt.	X			X	
Write Letter			X		

CONDITIONS { Eligible For Insurance, Insurance Requested

ACTIONS { Deduct Insurance Pmt., Do Not Deduct Ins. Pmt., Write Letter

Figure 11-11 Example of Decision Table

The decision table is used to present alternative actions based upon conditions which may arise. Therefore, both the CONDITIONS which must be tested and the ACTIONS which will be taken must be specified. The conditions which may occur are placed in the upper left portion of the decision table in an area which is designated the CONDITION STUB. In the example in Figure 11-11 it can be seen that there are two entries in the condition stub—"Eligible for Insurance" and "Insurance Requested." Thus, there are two questions which are going to be answered by the decision table and, depending upon the answers to these questions, different action will be taken. The questions are, "Is the employee eligible for insurance?" and "Has the employee requested insurance?" Dependent upon the answers to these questions, different action is going to be performed.

The action which is to take place, dependent upon the conditions which arise, is specified in the ACTION STUB, which is contained in the lower left of the decision table. In the example in Figure 11-11, there are three different actions which will be taken—either the insurance payment will be deducted, or the insurance payment will not be deducted, or a letter will be written to the employee which tells him that even though he requested insurance, he is not eligible for insurance. These are the actions which will be taken, dependent upon the answers to the questions which are asked in the Condition Stub portion of the decision table.

The results of both the condition and the action are called RULES, that is, for each condition which can occur and its associated action, there is a rule. In the example, there are four rules. These four rules specify what processing is to take place dependent upon the answers to the questions posed in the conditions. The vertical columns in the right-hand portion of the decision table are the ''rule'' columns. Thus, as can be seen from Figure 11-11, there are four rule columns to correspond to the four rules specified.

The answers to the questions posed are contained in the CONDITION ENTRY portion of the table, that is, the upper-right portion of the decision table. These boxes contain either yes (Y) or no (N) answers to the questions posed in the condition stub portion of the table. The actions to be taken dependent upon the answers are in the ACTION ENTRY portion of the decision table, which is the lower right area. In order to utilize the decision table, the rule column is followed from top to bottom and the action specified in the action entry portion of the table is performed based upon the answers in the condition entry area of the table. This process is illustrated in Figure 11-12.

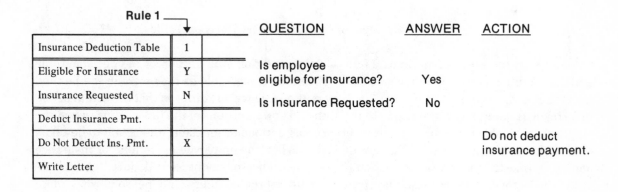

Figure 11-12 Example of the Use of the Decision Table

Note from Figure 11-12 that Rule 1 is illustrated and translated. The first question asked in Rule 1 is ''Is the employee eligible for insurance?'' and the answer, specified in the condition entry, is ''Yes.'' The second question, as specified in the Condition Stub, is ''Is insurance requested?'' and the answer in the condition entry is ''No.'' Therefore, the ''X'' is placed in the action entry opposite the action ''Do not deduct insurance payment,'' which is specified in the Action Stub. It will be noted that this action corresponds to the rule which is specified previously for Rule 1. In a similar manner, all of the rules which may pertain to a particular application can be specified in the decision table.

When the programmer receives a decision table such as illustrated in Figure 11-11, he can determine both the questions or decisions which must be made by his program and also the actions to be taken, based upon the answers to those decisions. Thus, it can be seen that a decision table can be used to conveniently present, in tabular form, the decisions which must be made within the program.

The table which is illustrated in Figure 11-11 is called a "Limited Table" because the entire condition and the entire action are contained in the "stub" of the table and the condition entry contains only a yes or no answer. This is the simplest decision table which can be designed and will many times suffice in illustrating to the programmer the decisions which must be made within his program.

Another type of decision table which will sometimes prove useful is the form of a table called an Extended Entry, where part of the condition or action is expressed in the entry side of the table. This is illustrated in Figure 11-13.

Insurance Deduction Table	1	2	3	4
Eligible For Insurance	Y	Y	N	N
Insurance Requested	N	Y	Y	N
Insurance Payment	Do Not Deduct	Deduct		Do Not Deduct
Letter			Write	

Figure 11-13 Example of Extended Entry Decision Table

Note from the example in Figure 11-13 that the condition stub portion of the decision table and the condition entries are the same in the Extended Entry table as they were in the Limited Entry table. The Action Stub and Action Entry portions, however, differ in that the actual action to be taken is determined by using both the stub and entry portions of the table. Thus, when the employee is eligible for insurance but insurance has not been requested, the action to be taken on the "insurance payment" is "do not deduct." When the employee is eligible for insurance and has requested insurance, the action of the "insurance payment" is "deduct." As can be seen, the extended entry table may provide a bit more flexibility and be somewhat more compact because, for example, if there were four or five different actions which should be taken concerning the insurance payment, dependent upon the conditions found, they could all be stated with the one action stub entry "insurance payment" whereas if the limited entry table was used, the five actions to be taken would have to be stated in the action stub with five different entries.

There is no limitation that a limited entry table or an extended entry table must be used exclusively, that is, a decision table can combine the two types. This is illustrated in Figure 11-14.

Insurance Deduction Table	1	2	3	4
Eligible For Insurance	Y	Y	N	N
Insurance Requested	N	Y	Y	N
Insurance Premium	Do Not Deduct	Deduct		Do Not Deduct
Write Letter			X	

Figure 11-14 Example of Mixed Table

In the example in Figure 11-14 it can be seen that there is a combination of a limited entry table, which is used by the "write letter" action, and an extended entry table, which is used by the "insurance premium" action. The decision on whether to use a limited entry table, an extended entry table, or a mixed table is normally based on the processing to be illustrated by the decision table and the relative merits of each method to clearly explain the processing which must be accomplished. It must be recalled that the reason for using a decision table in the documentation for the programmer is to explain, as clearly as possible, the logical decisions which must be made within a program. If the decision table tends to confuse, or if it is difficult to structure the processing which is to be accomplished within the decision table, then the narrative or flowchart or some other means should be used.

SUMMARY

The documentation which is passed to the programmer from the analyst must describe, in detail, all of the processing which is to take place within the system. In addition, each of the programming assignments must be detailed so that the programmer is aware of his responsibilities concerning the programming tasks for the system.

CASE STUDY - JAMES TOOL COMPANY

Once the schedule for the development of the payroll system was completed, the analysts Mard and Coswell turned their attention to developing the programming specifications for each of the programs in the payroll system.

Case Study - Programming Language

The analysts had little difficulty in determining the programming language to be used for the payroll system since it had been decided when the System/370 was installed that COBOL would be used in all systems unless there were certain programming techniques required which could not be done in COBOL. Since it appeared to the analysts that the processing which was to be accomplished in the payroll system could be done using the COBOL language and since the programmers who were assigned to the system knew COBOL, it was decided that COBOL would be used for all programs in the payroll system.

Case Study - Programming Specifications

The analysts then began documenting the programming specifications for each program which is to be written in the system. The final document which will be distributed to each programmer will include the following:

1. A brief explanation of the payroll system.
2. A systems flowchart.
3. A brief explanation of the purpose of each program in the system.
4. The format and layout of all files and reports to be used in the system.
5. A detailed explanation of each program in the system.

The following is a portion of the document distributed to each programmer. The example includes a brief explanation of the system and the detailed processing which is to take place in the edit program.

EDIT PROGRAM

PAYRL03

INTRODUCTION

The Payroll System has been developed to process the weekly payroll for the James Tool Company. Output from the system will include weekly paychecks, a payroll register, and a union deduction register. Input includes employee time cards, additions to an Employee Master File, deletions from the Employee Master File, and changes to the Employee Master File. The system includes a number of controls to ensure that the payroll is processed completely and accurately.

SYSTEMS FLOWCHART

The following pages contain the complete systems flowchart for the payroll system. (NOTE: See page 370 for an example of the systems flowchart.)

PROGRAM EXPLANATIONS

The following is a brief explanation of each program which is a part of the payroll system. (NOTE: See page 372 for an example of a program explanation.)

INPUT FILES - EDIT PROGRAM (PAYRL03)

The following are the formats of the input files to the edit program of the payroll system. (NOTE: See page 264 for an example of the format of the input files.)

OUTPUT FILES - EDIT PROGRAM (PAYRL03)

The following are the formats of the output files and reports from the edit program of the payroll system. (NOTE: See pages 321 and 323 for an example of the format of the output files.)

PROCESSING - EDIT PROGRAM (PAYRL03)

The following is an explanation of the processing which is to take place in the edit program.

Figure 11-14 Programming Specifications (Part 1 of 7)

-2-

Function Code 1

A function code of 1 indicates an Addition Record to the Employee Master File. A code 1 input record is made up of two cards - each card is identified by the sequence code (1 or 2) in column 80. It should be noted that these cards will not necessarily enter the edit program in consecutive order. Therefore, each card must be checked separately. If a record with a function code of 1 is read and the sequence code is not 1 or 2, the record should be written on the exception report and not written on the output transaction file.

Function Code 1 - Sequence 1

1. Employee Number - The Employee Number must contain all numeric data. A modulus 11 check should be performed on the field. If the field is not numeric or if the modulus 11 check fails, the record should be written on the exception report and not written on the output transaction file.

2. Name and Address - No checking will be performed on these fields.

3. ZIP Code - The ZIP Code must contain numeric information. If it does not, the record should be written on the exception report and not written on the output transaction file.

Function Code 1 - Sequence 2

1. Social Security Number - This field must contain numeric data. If not, the record should be written on the exception report and not written on the output transaction file.

2. Job Classification - The job classification must be one of the valid classifications. The table below contains the valid job classification codes together with the accompanying salary ranges.

2C - 2.80 - 3.90	3Y - 3.76 - 5.24
2G - 2.92 - 3.87	4E - 3.74 - 5.41
2N - 3.01 - 3.68	4H - 3.87 - 5.38
2Y - 3.23 - 4.02	4L - 3.92 - 5.56
3B - 3.21 - 4.09	4R - 4.07 - 5.68
3D - 3.35 - 4.51	4T - 4.28 - 6.24
3F - 3.42 - 4.62	5D - 4.58 - 6.98
3J - 3.38 - 4.71	6J - 5.27 - 7.83
3M - 3.53 - 5.01	6U - 5.78 - 8.45
3T - 3.64 - 4.99	6Y - 6.29 - 10.79

If the job classification does not fall into one of the valid codes, the record should be written on the exception report and not written on the output transaction file.

Figure 11-15 Programming Specifications (Part 2 of 7)

-3-

3. Starting Date - The month portion (first two characters) must contain the value 01 through 12. The day portion (second two characters) must contain the value 01 through the maximum number of days for the given month. For example, the maximum value for a month of 01 is 31 while the maximum value for the month of 09 is 30. The year (last two characters) must be numeric. If the data in the starting date field does not satisfy this criteria, the record should be written on the exception report and not written on the output transaction file.

4. Pay Rate - The data in this field must be numeric and must fall within the range dictated by the job classification code (see number 2). For example, a rate of 4.52 would be valid for code 3F but not for 3D.

5. Union Code - This code must contain the value 01 through 04 to reflect the unions in the company. If any other value is in this field, the record should be written on the exception report and not written on the output transaction file.

6. Exemptions - This field must contain a numeric value and not be larger than 12. If larger than 12, the record can be written on the output transaction file but the record should also be included on the exception report to note this unusual entry. If the field contains non-numeric data, the record should be written on the exception report and not written on the output transaction file.

7. Marital Status - This field must contain the value 1, 2, 3 or 4. If any other value is in this field, the record should be written on the exception report and not written on the output transaction file.

8. Credit Union Deduction - The value in this field must be numeric. In addition, if this field is greater than 30% of the hourly pay rate times 40, the record should be written on the exception report. In any case, if the value is numeric , the record can be written on the output transaction file. This processing is illustrated in the following decision table.

CREDIT UNION DEDUCTION	1	2	3
NUMERIC	N	Y	Y
LESS THAN .30 (PAY RATE x 40)	-	N	Y
WRITE ON EXCEPTION REPORT	X	X	
WRITE ON OUTPUT FILE		X	X

Figure 11-16 Programming Specifications (Part 3 of 7)

-4-

Function Code 2

A function code of 2 indicates a Name and/or Address change.

1. Employee Number - The Employee Number must contain all numeric data. A modulus 11 check should be performed on the field. If the field is not numeric or if the modulus 11 check fails, the record should be written on the exception report and not written on the output transaction file.

2. Name and Address - No checking will be performed on these fields. It will be noted that the Name field could be blank or the address field could be blank because this transaction can be used to change either the name field or the address field or both.

3. ZIP Code - If the ZIP Code field is not blank, it must contain numeric data. If it does not contain numeric data, the record should be written on the exception report and not written on the output transaction file.

Function Code 3

A function code of 3 indicates a change in the classification code or a change in the union code.

1. Employee Number - The Employee Number field must contain all numeric data. A modulus 11 check should be performed on the field. If the field is not numeric or if the modulus 11 check fails, the record should be written on the exception report and not written on the output transaction file.

2. Job Classification Code - If this field is not blank, it must contain a valid classification code. The valid classification codes are listed under Function Code 1 (2). If a valid classification code is not found in this field, the record should be written on the exception report and not written on the output transaction file.

3. Union Code - If this field is not blank, it must contain the value 01, 02, 03, or 04. If any other value is found in this field, the record should be written on the exception report and not written on the output transaction file.

4. Both fields blank - If both the job classification code field and the union code field contain blanks, the record is invalid. Therefore, it should be written on the exception report and the record should not be written on the output transaction file.

Figure 11-17 Programming Specifications (Part 4 of 7)

-5-

Function Code 4

A function code of 4 is used to indicate that an employee has terminated employment.

1. Employee Number - The Employee Number must contain all numeric data. A modulus 11 check should be performed on the field. If the field is not numeric or if the modulus 11 check fails, the record should be written on the exception report and should not be written on the output transaction file.

2. Termination Date - The month portion (first two characters) must contain the value 01 through 12. The day portion (second two characters) must contain the value 01 through the maximum number of days for the given month. For example, the maximum value for a month of 01 is 31 while the maximum value for the month 09 is 30. The year portion (last two characters) must contain numeric data. If the data in the termination date field does not satisfy this criteria, the record should be written on the exception report and not written on the output transaction file.

3. No Termination Date - If the termination date field is blank, the record should be written on the exception report and not written on the output transaction file.

Function Code 5

A function code of 5 is used to indicate a pay rate change.

1. Employee Number - The employee number field must contain all numeric data. A modulus 11 check should be performed on the field. If the field is not numeric or if the modulus 11 check fails, the record should be written on the exception report and should not be written on the output transaction file.

2. Pay Rate - The pay rate field should contain numeric data. If non-numeric data is found in the pay rate field, or if the field is blank, the record should be written on the exception report and should not be written on the output transaction file.

Function Code 6

A function code of 6 is used to indicate a change to the exemptions field in the master record.

1. Employee Number - The employee number field must contain all numeric data. A modulus 11 check should be performed on the field. If the field is not numeric or if the modulus 11 check fails, the record should be written on the exception report and should not be written on the output transaction file.

Figure 11-18 Programming Specifications (Part 5 of 7)

-6-

2. Exemptions - The exemptions field must contain numeric data and not contain a value greater than 12. If larger than 12, the record can be written on the output transaction file but the record should also be included on the exception report to note this unusual entry. If the field is blank or contains non-numeric data, the record should be written on the exception report and not written on the output transaction file.

Function Code 7

A function code of 7 is used to indicate a marital status change.

1. Employee Number - The employee number field must contain all numeric data. A modulus 11 check should be performed on the field. If the field is not numeric or if the modulus 11 check fails, the record should be written on the exception report and should not be written on the output transaction file.

2. Marital Status - This field must contain the value 1, 2, 3, or 4. If it contains any other value, the record should be written on the exception report and should not be written on the output transaction file.

Function Code 8

A function code of 8 is used to indicate a credit union deduction change.

1. Employee Number - The employee number field must contain all numeric data. A modulus 11 check should be performed on the field. If the field is not numeric or if the modulus 11 check fails, the record should be written on the exception report and should not be written on the output transaction file.

2. Credit Union Deduction - The value in this field must be numeric. If it is not numeric, the record should be written on the exception report and should not be written on the output transaction file.

Function Code 9

A function code of 9 identifies an employee time card.

1. Employee Number - The employee number field must contain all numeric data. A modulus 11 check should be performed on the field. If the field is not numeric or if the modulus 11 check fails, the record should be written on the exception report and should not be written on the output transaction file.

Figure 11-19 Programming Specifications (Part 6 of 7)

-7-

2. Week Ending Date - The first two digits (Month) in this field must contain the value 01 through 12. The second two digits (Day) must contain the value 01 through the maximum number of days for the given month. If the data in this field does not satisfy this criteria, the record should be written on the exception report and should not be written on the output transaction file.

3. Regular Hours Fields - The regular hours fields in the time card must contain numeric data and cannot contain a value greater than 8 hours or must be blank. If any of the seven separate regular hours fields violates this criteria, the record should be written on the exception report and not written on the transaction output file.

   After the regular hours fields have been checked and pass the edit test, the values should be totaled and the total should be placed in the time card output record.

4. Overtime Hours Fields - The overtime hours fields in the time card must contain numeric data and cannot contain a value greater than 16 or must be blank. If any of the seven separate overtime hours fields violates this criteria, the record should be written on the exception report and not written on the transaction output file.

   After the overtime hours fields have been checked and pass the edit test, the values should be totaled and the total should be placed in the time card output record.

Any Other Function Code

Any other value which is found in the function code field of the input records is invalid. If a code other than 1-9 is found, the record should be written on the exception report and not written on the transaction output file.

Figure 11-20 Programming Specifications (Part 7 of 7)

As can be seen from the previous example, the entire processing which is to take place in the Edit program of the payroll system is set out in detail. Each field and each step in the program is specified so that the programmer can write his program based upon the exact specifications given him by the analysts Mard and Coswell.

The remainder of the programs in the payroll system would have similar specifications prepared. When the programming specifications are complete, they are given to the programmers assigned to the project, who then analyze them to be sure that there are no questions concerning the processing which is to take place within the system. Once the programmers have the specifications, the programming on the project can begin.

SUMMARY

The programming specifications are quite important because all of the programs in the system will be based upon these specifications. If they are not accurate and complete, the programmer will have little chance of writing the programs which will process the data properly.

STUDENT ACTIVITIES—CHAPTER 11

1. List the steps to be followed in producing the computer programs required to implement a system.

   1)
   2)
   3)
   4)
   5)
   6)
   7)

2. List the factors which should be considered when selecting a programming language.

   1)
   2)
   3)
   4)
   5)
   6)

3. Which are the most widely used programming languages for business applications?

4. What technical factors should be considered when evaluating a programming language?

5. Which is the most widely used programming language?

6. Briefly discuss the advantages and disadvantages of COBOL as a programming language.

7. Briefly discuss the advantages and disadvantages of Assembler as a programming language.

8. Briefly discuss the advantages of RPG as a programming language.

9. FORTRAN is primarily used for which type of applications?

10. Briefly discuss the advantages of PL/I as a programming language.

11. List the items which should be included in the programming specifications.

    1)
    2)
    3)
    4)
    5)

12. What is the purpose of a Decision Table?

13. What is the difference between a ''Limited Decision Table'' and an ''Extended Entry Decision Table''?

DISCUSSION QUESTIONS

1. ''The most important factor to consider in selecting a programming language is 'ease of programming'. It makes little sense to spend 8 weeks programming an application in a low level language such as Assembler Language, when the same job could be programmed in a high level language such as COBOL in half the time. The few seconds saved when a program is executing because of fewer instructions generated in a low level language can never be cost justified!''

   Do you agree with this statement?

2. ''We program in Assembler Language in our installation, as we update nearly a million records monthly. Some of our programs contain over 25,000 statements in Assembler Language and require over 200,000 bytes of storage. To program in COBOL would require us to obtain a bigger, faster computer system with larger storage. We can't justify COBOL in our installation!''

   Could COBOL or any other high level language ever be justified in an installation of this type?

CASE STUDY PROJECT—CHAPTER 11
The Ridgeway Company

INTRODUCTION

After the schedule for the systems development phase of the billing systems project for the Ridgeway Country Club has been completed, then the systems analyst must develop the programming specifications for each of the programs within the system.

STUDENT ASSIGNMENT 1
SELECTING THE PROGRAMMING LANGUAGE

The data manager has requested a written recommendation relative to the programming language to be used in the programming of the billing system of the Ridgeway Country Club. The current programming staff is proficient in COBOL, RPG, and Assembler Language.

Prepare the written recommendation to the data processing manager. Be sure to justify your choice.

STUDENT ASSIGNMENT 2
PREPARING THE PROGRAMMING SPECIFICATIONS

1. Prepare the programming specifications for the program which is to be used to edit the transaction records of the billing system.

2. Prepare the programming specifications for the program which is to be used to prepare the statements which are to be sent to the members of the Ridgeway Country Club.

These programming specifications should include:

1. A brief description of the system.

2. A systems flowchart.

3. A listing and brief explanation of the programs within the system.

4. The format of the input, output and files to be created and processed.

5. A detailed definition of the processing to take place in each program.

PROGRAMMING, TESTING AND DOCUMENTATION

CHAPTER 12

PROGRAMMING, TESTING AND DOCUMENTATION 12

INTRODUCTION

After the systems analyst has prepared the programming specifications for the programmer, three basic tasks remain in the systems development phase of the systems project. These tasks include:

1. Programming
2. Program Testing
3. Final Documentation of the system

After these tasks have been completed the system should be ready to enter the Implementation Phase.

PROGRAMMING

The process of producing programs from a set of programming specifications may be divided into two basic subtasks. These include:

1. Development of the program logic
2. Coding of the program

Development of Programming Logic

After the programmer receives the programming specifications, a considerable portion of the programmer's time is normally required in reviewing the specifications and developing the logic required to perform the processing required in the program. This logic is normally defined by means of a PROGRAM FLOWCHART. This program flowchart will contain the detailed steps that are required to solve the problem. Figure 12-1 illustrates a program flowchart.

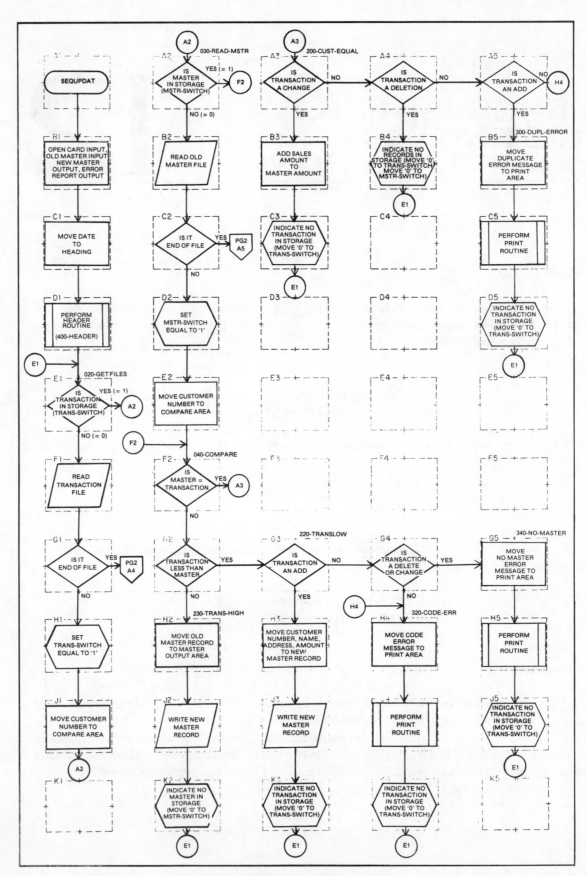

Figure 12-1 Program Flowchart

As can be seen from the flowchart in Figure 12-1, the detailed logic required in the solution of the problem is recorded using standard symbols similar to those used in the systems flowchart. The accuracy of the logic developed in the flowchart is extremely important for it is from this flowchart that the programmer will develop the actual code or computer instructions. As previously mentioned, this step in the programming process is commonly allocated 35 percent of the time scheduled for programming the application.

Coding the Program

Once the detailed program flowcharts have been developed the computer program is then written in COBOL or whatever language has been selected. Figure 12-2 illustrates a segment of the coding for a COBOL program.

```
    3

00175  009150                                                                      SEQUPDAT
00176  009160                                                                      SEQUPDAT
00177  009170  NOTE-020.                                                           SEQUPDAT
00178  009180     NOTE ********************************************************    *SEQUPDAT
00179  009190     *    THIS ROUTINE IS ENTERED TO READ EITHER THE                 *SEQUPDAT
00180  009200     *    TRANSACTION FILE OR MASTER FILE OR BOTH AND THEN GO         *SEQUPDAT
00181  010010     *    TO THE PROPER PROCESSING ROUTINE.                          *SEQUPDAT
00182  010020     ********************************************************         *.SEQUPDAT
00183  010030                                                                      SEQUPDAT
00184  010040  020-GETFILES.                                                       SEQUPDAT
00185  010050     IF TRANS-SWITCH NOT = 0                                          SEQUPDAT
00186  010060        GO TO 030-READ-MSTR.                                          SEQUPDAT
00187  010070     READ TRANSIN AT END GO TO 500-END-TRANS.                         SEQUPDAT
00188  010080     MOVE '1' TO TRANS-SWITCH.                                        SEQUPDAT
00189  010090     MOVE CUST-NUMB-TRANS TO CUST-NUMB-TRANS-C.                       SEQUPDAT
00190  010100  030-READ-MSTR.                                                      SEQUPDAT
00191  010110     IF MSTR-SWITCH NOT = 0                                           SEQUPDAT
00192  010120        GO TO 040-COMPARE.                                            SEQUPDAT
00193  010130     READ MSTRIN AT END GO TO 510-END-MASTER.                         SEQUPDAT
00194  010140     MOVE '1' TO MSTR-SWITCH.                                         SEQUPDAT
00195  010150     MOVE CUST-NUMB-MSTRIN TO CUST-NUMB-MSTRIN-C.                     SEQUPDAT
00196  010160  040-COMPARE.                                                        SEQUPDAT
00197  010170     IF CUST-NUMB-TRANS-C IS = CUST-NUMB-MSTRIN-C                     SEQUPDAT
00198  010180        GO TO 200-CUST-EQUAL.                                         SEQUPDAT
00199  010190     IF CUST-NUMB-TRANS-C IS < CUST-NUMB-MSTRIN-C                     SEQUPDAT
00200  010200        GO TO 220-TRANSLOW.                                           SEQUPDAT
00201  011010     GO TO 230-TRANS-HIGH.                                            SEQUPDAT
00202  011020                                                                      SEQUPDAT
00203  011030                                                                      SEQUPDAT
00204  011040  NOTE-200.                                                           SEQUPDAT
00205  011050     NOTE ********************************************************    *SEQUPDAT
00206  011060     *    THIS ROUTINE IS ENTERED WHEN THE CUSTOMER NUMBERS ON        *SEQUPDAT
00207  011070     *    THE TRANSACTION FILE AND THE MASTER FILE ARE EQUAL.         *SEQUPDAT
00208  011080     *    EITHER AN UPDATE OR A DELETION CAN TAKE PLACE. ANY          *SEQUPDAT
00209  011090     *    OTHER TYPE OF TRANSACTION IS INVALID.                       *SEQUPDAT
00210  011100     ********************************************************         *.SEQUPDAT
00211  011110                                                                      SEQUPDAT
00212  011120  200-CUST-EQUAL.                                                     SEQUPDAT
00213  011130     IF CODE-TRANS IS = '3'                                           SEQUPDAT
00214  011140        ADD SALES-AMT-TRANS TO SALES-AMT-MSTRIN                       SEQUPDAT
00215  011150        MOVE '0' TO TRANS-SWITCH                                      SEQUPDAT
00216  011160        GO TO 020-GETFILES.                                          SEQUPDAT
00217  011170     IF CODE-TRANS IS = '2'                                           SEQUPDAT
00218  011180        MOVE '0' TO MSTR-SWITCH                                       SEQUPDAT
00219  011190        MOVE '0' TO TRANS-SWITCH                                      SEQUPDAT
00220  011200        GO TO 020-GETFILES.                                          SEQUPDAT
00221  012010     IF CODE-TRANS IS = '1'                                           SEQUPDAT
00222  012020        GO TO 300-DUPL-ERROR                                          SEQUPDAT
00223  012030     ELSE                                                             SEQUPDAT
00224  012040        GO TO 320-CODE-ERR.                                          SEQUPDAT
00225  012050                                                                      SEQUPDAT
00226  012060                                                                      SEQUPDAT
00227  012070  NOTE-220.                                                           SEQUPDAT
00228  012080     NOTE ********************************************************    *SEQUPDAT
00229  012090     *    THIS ROUTINE IS ENTERED WHEN THE CUSTOMER NUMBER            *SEQUPDAT
00230  012100     *    IN THE TRANSACTION RECORD IS LESS THAN THE CUSTOMER         *SEQUPDAT
00231  012110     *    NUMBER IN THE MASTER RECORD. THE ONLY VALID                 *SEQUPDAT
00232  012120     *    TRANSACTION IS AN ADDITION (OR, IF THE FIRST-TIME,          *SEQUPDAT
00233  012130     *    A LOAD).                                                    *SEQUPDAT
00234  012140     ********************************************************         *.SEQUPDAT
00235  012150                                                                      SEQUPDAT
```

Figure 12-2 Example of COBOL Program

In small programs the coding for the entire program will be undertaken by a single programmer; in very large programs, however, the various routines may be broken down into individual modules and two or more programmers could be assigned the task of developing the code for a single program.

After the program has been coded it is then normally keypunched. The program should then be returned to the programmer and the program should be DESK CHECKED, that is, each card should be inspected individually for errors.

After desk checking the program and correcting any errors, the program is then ready for testing.

PROGRAM TESTING

Program testing and debugging is one of the most critical aspects of implementing a computer system because without programs which work, the system will never produce that output for which it was designed. Thus, before any system can be completely implemented on the computer, the analyst and programmers working on the system must be able to state unequivocally that the programs work exactly as they were designed to work and that any errors or mistakes which are found in any of the data which would be processed by the system will be handled properly.

As has been noted previously, one of the first steps is to establish a realistic schedule for testing. Since testing is quite unpredictable in terms of the results and, in some cases, the availability of the hardware required for testing, it is difficult to establish a day-to-day detailed testing schedule in advance. It should be possible, however, to estimate with some accuracy the time which will be required to test any given program. The most likely area for the "slipping" of deadlines in the implementation falls within the area of program and systems testing, so when testing schedules are established, there should be liberal time allowed. In fact, some experienced analysts state that the time required for testing and debugging should be closely estimated and then that time should be doubled in order to establish the scheduled time for testing. Although this time may seem excessive, experience shows that even the time computed through this method is many times not enough to complete the testing. Again, the lesson is that most persons tend to underestimate time requirements, and testing and debugging is probably that area which is most underestimated.

To effectively test a program the systems analyst should establish definite procedures which are to be followed. Basic rules for program testing and debugging are summarized below.

1. Individual programs should be compiled and all diagnostics removed prior to using test data.

2. Test data should be created by the programmer that first tests all main routines.

3. Additional test data should be created to assure that every routine and every instruction has been executed correctly at least once.

4. Program testing should include testing with as many types of invalid data as is likely to occur when the system is in production.

5. After each program has been individually tested and debugged, related programs in the system should be tested as a unified group. This is called "link" or "string" testing.

6. A final "systems test" should be performed using data prepared by the systems analyst and, in some cases, data which has been previously processed through the "old" system.

Testing Individual Programs

The testing of each individual program, sometimes called "unit testing," is normally handled by the programmer who has written the program. Although new program management techniques are emerging wherein certain programmers have responsibility for coding only and other programmers have the responsibility for testing the programs, at this point in time most installations have the programmer who is to write the program also debug it.

The amount of testing which is required to certify that a program is ready for production and the amount of realiability which can be given to a program are matters for controversy, since each installation may have its own standards. It can be said, however, that from a programmer's standpoint, a program should never enter systems testing and be put into production if the programmer has any doubt that the program will work.

Normally programs should first be compiled without using test data in order to eliminate all diagnostics in the program due to programming or keypunching errors.

After a "clean" compilation is obtained, that is, one without any compilation diagnostics, the programmer should then desk check his program. Desk-checking refers to the process of "playing computer" with the source listing, that is, following every step in the program and analyzing what will take place as the routine is processed. Desk-checking is probably the most useful tool in debugging a program and it is the most neglected and abused. Many programmers, immediately upon obtaining a compilation with no diagnostics, resubmit the program for a test run with test data. This is not good testing technique as time should be taken to review the source listing.

Desk-checking has an added benefit of refamiliarizing the programmer with his program. In a complex program, there may be a period of ten or more weeks between the time the program is started and the time it is compiled. During this time the programmer will forget some of the routines or other portions of his program which were written. When desk-checking, however, he must go completely through these routines again. Thus, the routines will be refreshed in his mind. This can be of great aid if the program fails on the computer because he is more likely to be able to isolate the problem faster and with more accuracy than if he had not reviewed the entire program in detail.

Creating Test Data

After the program has been desk-checked, it must be tested on the computer using test data. To properly test the program there must be good test data available. In almost all applications, the programmer should use data to test his program which has been designed specifically to test the routines within his program. Normally, test data will be designed to test main routines first. When it is found that the main routines produce the desired output, additional test data should be created that tests all other routines. This data should contain both valid and invalid data to test both the normal processing routines and the error routines of the program.

In addition, the test data should be designed so that limits within the program are tested. For example, data should contain both minimum and maximum values which can be contained within a field. The data should be designed to allow maximum values and minimum values to occur in any intermediate fields which are to contain totals. There should be variations in the various formats which the program can process so that all possibilities can be covered. All of the codes which can be used in records should be contained in the test data so that the various routines which are called based upon different codes can be tested.

Another important area which must be tested is the files which are to be used in the program. If an indexed sequential file is used, the file must be loaded and, in addition, must be tested using data to add records and delete and change the records within the file. When a direct file is used, the algorithm used to determine record addresses must be tested and also the routine which handles synonyms must be heavily tested. Any time data is stored on a file, such as magnetic tape or magnetic disks, whether it is sequential, indexed sequential, or direct, the data should be "dumped" by using some utility program so that the file contents can be examined in detail. A programmer cannot assume that the data is correct merely because the file was built or that the data on the file was used successfully as input to another program. The data must be closely examined.

It will be noted that most of this responsibility for program testing rests with the programmer who wrote the program. The programmer will normally design the test data, conduct the tests, and check the output from the tests. The analyst, however, can play an important role in the program testing by first attempting to ensure that good testing techniques are followed and then reviewing and making suggestions to the programmer concerning data to be tested. The analyst can look at reports and file dumps to ensure, early in the program testing, that the results correspond to what he expects. If there are variances, they can be corrected. The analyst can also look at the test data which is being used to determine if he sees any areas which should be tested and which have been overlooked by the programmer. It must be noted that the analyst should not dictate to the programmer what data should be used to test the program. This analyst merely serves in an advisory capacity. The only time this may not be true is if the programmer is having great difficulty with the program. Even in this case, however, it is likely that another programmer on the staff will be assigned to aid the original programmer and the analyst will merely continue in his advisory role.

Link or String Testing

After the individual programs have been tested using data designed by the programmer, the programs within the system which depend upon one another must be tested together. This is sometimes called ''link'' or ''string'' testing. For example, in a system with an edit program and an update program, the output of the edit program must be properly formatted and contain the correct data to be input to the update program. The programmer who wrote the edit program may not be the same programmer who wrote the update program. Thus, there may be some discrepancy between the format in which the data is created in the edit program and the format in which the data is expected in the update program. Although this may sound unlikely in view of the fact that the systems documentation clearly showed the correct formats, in actual practice it occurs somewhat frequently. Thus, for reasons such as these, the series of programs must be tested together.

Another reason for testing the series of programs is to ensure that the ''job stream'' is correct. In most computer systems there is an operating system and the features and functions of the operating system must be invoked through the use of the job control statements found in the job stream. The job stream normally consists of those job control statements which are necessary to invoke the proper programs to be processed, to define the files to be created and processed, and to designate the devices which are to be used for the files. Obviously, if files are defined incorrectly or programs are called improperly, the system will not function in the specified manner; therefore, the testing of the job stream is as important as the testing of the programs within the system.

The data which is used in string or link testing is normally developed by the systems analyst. This data should include data which will test certain conditions which may not have occurred in the program testing. For example, three programs may process a certain field and the data in this field is accumulated in the three programs. Although each program may be able to process a maximum size for the field, the accumulation in the three programs may cause the field to be too small in the last program. Each programmer when the programs were tested individually allowed the proper size field but when the actual processing takes place, it may be that the field size was improperly specified by the analyst. It must always be noted that not only is each program being tested by the string testing, but the system itself, and all of the specifications developed by the analyst, are also being tested. Therefore, data must be developed which not only tests the programs but also tests the specifications for the system which were designed by the analyst.

It is very important for the analyst to define what is to be tested during any given test run. In most systems, it is highly inefficient and very error-prone to attempt to test everything on the first test run. Therefore, the analyst should specifically designate portions of the system to be tested on each "pass" through the related programs within the system. In many batch processing systems, it may be necessary to design 10-20 "passes" through the system in order to adequately test each operation, each file definition and size, and all the other variances which can be found in data and in the processing.

In most cases, the testing procedure will run from the normal to the abnormal, that is, the first tests of the related programs within the system will process test data which represents those cases which occur most often within the system and present no abnormal situations. For example, in a payroll system, the first data tested may be for employees which work a standard 40 hour week with no overtime and with standard deductions. When this data is tested properly, the analyst may then mix this data with data which includes overtime and test the overtime routines. He may then switch to test data which has some invalid times in the time cards to be sure that this data is processed correctly.

It should be noted that it is not necessary nor even desirable that a large amount of data be tested on these runs. The analyst may be working with only 15 or 20 specially designed records to test the particular phase of the system. It must be recalled also that this is not primarily a program test but rather a test of the system interfaces. It must be assumed at this point that the programmer has thoroughly checked out his particular program and the analyst has approved the results. If this was not so, the program should never be in the string test stage.

System Testing

After the link or string testing is completed the entire system must be run through a new series of tests. In these "systems tests," which are normally **not** prepared or controlled by the programmer, the entire system is tested beginning with the preparation of source documents, converting the information on the source documents to computer input records, running all of the processing programs within the system, and disbursing the output generated from the tests. The main objectives of the systems test are as follows:

1. To perform a final test of the programs against the original programming specifications.

2. To ensure that the computer operations staff has adequate documentation and instructions to operate the system properly on the computer and properly handle the incoming and outgoing data from the system.

3. To ensure that the user departments are able to properly prepare data for use in the system and to properly disburse and use the output information.

4. To ensure that the "flow" of the system works properly, that is, the channels for the delivery of information from the user or other departments are established, the input data moves smoothly from the point where it is received in the data processing department through its preparation for use on the computer, and the output data is properly handled to allow for its distribution to the user departments.

As can be seen from the objectives of the system testing, more than just the testing of computer programs is involved, and, additionally, more personnel and departments are involved than just the data processing department. Because of this, there are a number of elements which must be considered by the systems analyst, including the following:

1. A determination must be made concerning what constitutes a satisfactory performance of the system, that is, at what point can the system be declared ready to go into production.

2. The methods of evaluation must be determined, that is, how is the performance of the system to be measured.

3. The different "cases" to be tested must be determined by the analyst and the data to test these cases must be designed and prepared.

4. The tests must be scheduled, both in the data processing department and in conjunction with the user departments which may take part in the tests.

5. All efforts in terms of data preparation, scheduling, and evaluation must be coordinated so that the testing takes place on schedule and both the data processing department and the user departments can make accurate determinations concerning the readiness.

In most of the systems testing, the test data is designed by the analyst exclusively to satisfy the objectives of the systems testing. After all of the testing has been satisfactorily completed, however, it may prove useful to run several complete systems tests using "live data," that is, data which has been previously processed through the currently existing system. In many systems, especially when manual systems are being replaced by computer systems, this will entail preparing data from the manual system which can be used in the computer system. Naturally the analyst will have to schedule this data preparation and also any file building with the data which must take place before the system can be processed on the computer. This system would then be run and the results are compared to the results which have been found with the current system. It should be noted that in this type of testing, large volumes of data are not normally used. Thus, in a payroll application, perhaps one department could be used in this phase of testing, that is, the "live" data from one department could be converted to the computer format and run using the new system.

Systems testing is one of the most important steps in ensuring that a reliable system is being placed in production yet is often one of the most neglected areas in testing a system. The primary reason it is neglected is that it is the last step before system implementation. Therefore, if any schedule slippage has occurred, it is likely that the system is behind schedule when systems testing is to take place and, to try to meet the schedule, systems testing is quickly performed, if at all. This can be a deadly mistake because it means the complete system has not been adequately tested and, therefore, any errors which are in the system will not be found until it is actually in production. Thus, it is imperative that complete systems testing be performed, even if it means the system will not be implemented exactly on schedule.

Backup and Restart Testing

As was noted in Chapter 6, one of the important aspects in the design of the system is preparing for the backup of files which may be inadvertently destroyed or otherwise rendered unusable and for the restarting of the system at some point other than the beginning if some type of disaster occurs during the running of the system. Neither the backup nor the restart systems can be assumed to be free of error when they are designed and there must be testing of both of these systems prior to implementation of the system.

The primary testing which is done on the backup system is to be sure that the files which are kept as backup are adequate and that the programs which are used to both copy the master files and other files which are saved and the programs used to reload the files work properly. Even though these programs are not a part of the system in terms of daily or weekly use, it is quite important that they work properly because without them, recovery will not be possible when a breakdown occurs.

The restart testing must be carefully done to ensure that the restart points within the system are carefully and accurately defined. If they are not, then it will not be possible to restart the system with all files in tact. Although in a batch system there are not a great deal of restart problems, this problem can become more difficult when the programs processing the data are very long running programs and checkpoints are used, or when an ''on-line'' application is used and there is not an accurate record of transactions received or updates processed. If a file is destroyed, then restarting is difficult and it is in these areas that a good deal of testing and organizing must take place if a system is to be restarted properly.

The important point is, however, that no matter what the backup and restart procedures are for the system, and no matter how simple or difficult they may be, they must be thoroughly tested. Therefore, in addition to testing the entire system as a whole, the analyst must be sure that all backup and restart procedures are adequate. If they are not, and if the analyst does not fullfill his responsibility in this area, there could at some future time be catastrophic results because of a failure within the system. It is very important that these backup and restart procedures be thoroughly reliable before the system is implemented.

Management and User Approval

After the final systems testing is completed, the analyst should once again obtain management and user approval. The purpose of the management and user approval at this point is to show them what they are getting out of the system and to ensure that such operations as the calculations on the reports are being performed properly. In addition, the procedures for processing the data through the system should be exhibited so that all concerned are aware of what is necessary in order to efficiently operate the system.

It must be emphasized that there is no way that changes in file or report formats can be accommodated at this time without substantially delaying the project. These changes should have been made when the presentation was given after the system design phase. The only changes which will be made at this time are corrections of errors which are discovered when the users and management review the output of the system. If there are errors in programming then the errors will have to be corrected and the system retested. Again, however, it is imperative that no changes, other than corrections, be accepted by the analyst after the system has been extensively tested and before it has been implemented.

Changes which must be made to the programs in order to add additional features or restructure files, reports, etc. can be made to the system after it is in production and when there is time to properly make changes. In most systems, after the systems testing has been completed, the remaining time left in the schedule is quite short and will be devoted to one of several methods of converting from the current system to the new system. In addition, the pressure to get the system "on the air" is quite heavy and any changes made in this time period are very susceptible to error. Thus, if format changes, additional reports, or other changes are to be made to the system, they will have to wait until after the system is in production.

It is important, however, to have the user's and management's approval of the system which is shown to them. As with previous encounters with management and the users, it is quite helpful to the analyst to have approvals on everything which is going to be produced from the system. In this way, if there is a problem at a later date with some of the output, the analyst can point out that the calculations or whatever else is the problem were approved at the conclusion of the testing of the system. There is some protection for the analyst and it many times causes the user to very closely examine the output because he knows that "what he sees is what he gets."

DOCUMENTATION

A tremendous amount of time, resources and knowledge must be expended in the programming and testing of the system. The results of this effort must be carefully organized so that the myriad of details related to the programs within the system can be recorded in a clear and orderly fashion. This means that significant information about the programs and the system must be written down and stored in a clear and concise fashion. The creation and maintenance of this information is called DOCUMENTATION. Documentation is a vital part of any system.

Computer programmers are challenged by new applications to program. Thus, all too frequently after coding, debugging and testing a program, the ID is written with a felt pen upon the top of the source deck and the programmer moves on to the next assignment. After all, the program works whether it is documented or not! As a result, there have been far too many installations where documentation has been neglected, completely ignored, or written in a haphazard manner. In these installations, when it becomes necessary to maintain or make a minor modification to the program, many hours of effort are required to reconstruct the input, output, and processing operations performed by the program.

Thus, it is essential that the analyst establish "documentation standards" which are firmly adhered to by all programmers.

There are three important areas to consider in final systems documentation. These include:

1. Program Documentation
2. Operations Documentation
3. User Documentation

In planning a system of documentation it is also important to determine how much documentation is needed. Insufficient documentation can result in inefficiencies in maintaining and operating the system. Excessive documentation becomes burdensome, inefficient and difficult to use.

Program Documentation

Adequate program documentation is essential for the continued operation and maintenance of the system. Although program documentation takes time, almost invariably the lack of adequate program documentation leads to much greater efforts in program maintenance in the future than any savings resulting from neglecting documentation. The consequences are even greater if the original programmer has left the company and a new employee must maintain a system with little or no documentation.

The following chart summarizes the contents of a program documentation manual.

PROGRAM DOCUMENTATION	
1. Title Page	The title page should contain the program name, a program number or some unique identification, the name of the original programmer, and the date.
2. Revision Page	The revision page should be a form to document the history of the program. Contents should include the name of the original programmer or programmers, the date the program became operational, and the time required to program the project. In a chart form provision should be made to document subsequent revisions, a brief description of the revisions, the names of the programmers making the revisions, and the date the revised program was operational.
3. Abstract or General Description	The program abstract should contain a general description of the program, its basic purpose and features, frequency of use, input, output, and processing performed. The abstract provides an overview of the function of the program.
4. Record Layouts	Record layouts include multiple card layout forms, printer spacing charts, and/or disk and tape record layouts. These forms should specifically define the format of the input and output, the blocking factors and record sizes, an estimated number of records to be processed by the programmer, and the requirements for disk and tape storage space.
5. Systems Flowchart	The systems flowchart documents the flow of data through the system, and provides a visual means of identifying the input and output files being used and the required steps in the total processing cycle.
6. Detailed Description	This part of the documentation manual should provide a detailed description of the program including special editing (such as checks for valid numeric fields), sequence checking, checking for maximum field size, reasonableness checks, any table used within the program, special forms used, special carriage control tape required, special operating instructions, and size of storage required for the program.
7. Program Logic	The detailed program logic must be illustrated. Program flowcharts and/or decision tables should be used.
8. Program Listing	The program listing should list all source codes, storage maps, and job control.
9. Test Data	A listing of the test data used to test the program, as well as the output of the program, should be provided in the program documentation manual. This test data should include sufficient data to test out all routines within the program.

Figure 12-3 Program Documentation

OPERATIONS DOCUMENTATION

Another very important area in which documentation must be accurately prepared is for computer operations, that is, instructions to the computer operator on how to run the system and the programs within the system. It is said that a system and programs that are perfectly designed are useless if they cannot be run on the computer; therefore, it is critical that complete and correct operation instructions be prepared for the computer operators.

The instructions for the operator must be concerned with several areas of information—the operator must know what programs the system consists of and the general flow and function of each of the programs; he must be made aware of what files are used, when they are used, where they come from, and what to do with them when processing is completed; he must be informed of any messages which require operator intervention within the system, including restart procedures; and, he must be given instructions concerning any set-up operations which must take place such as switch settings or special control card handling. The operator must also be given step-by-step procedures to be followed if there is any special processing to be done either within the entire system or within one program.

The Operations Documentation normally consists of two elements: a Master Run Book and a Console Run Book. The master run book provides detailed information relative to the entire system with emphasis on the factors necessary to provide for the efficient operation of the system. This document is primarily a reference tool for the operations department. The console run book, on the other hand, is designed to be used as the actual operating document used on a day-to-day basis by the computer operator. This operations documentation is discussed in detail in the following paragraphs.

Master Run Book

A separate "run" manual should be prepared for each major computer program or group of programs. It should contain all the information necessary to process the data to produce the current output.

Materials which should be contained in the run book include:

1. GENERAL FLOWCHART AND DESCRIPTION—A general input/output flowchart showing the input, main flow of the data, and the output of the program is needed. Along with this should be a narrative describing the run, giving its purpose, input and output requirements, and other pertinent information. In addition, the relationship of the particular program to the whole application should be indicated. This could be accomplished by including a portion of the overall flowchart of the application and outlining the specific run.

2. RUN DESCRIPTION—A section should be provided giving the program identification, its function, the name of the programmer, input description and sequence, output description and sequence, controls, machine setup instructions, restart procedures, disposition of input and output, and any other information pertinent to the overall operating picture of the program.

3. DETAILED FLOWCHARTS—Current charts incorporating all changes and revisions should be included. The degree of detail should be commensurate with the need for adequate understanding and with the requirements defined by the installation's programming standards.

4. RECORD LAYOUTS—Detailed layouts of all records used in the program should be included. These would include input and output cards, master disk and tape files, printed and typed reports.

5. SAMPLE FORMS—Card and report forms used in the program should be shown, along with form numbers, approximate quantities, number of carbons, etc. Sample reports can have actual information printed on them for the purpose of clarity. The carriage tape needed for the printer can also be included here.

6. PROGRAM LISTING—A current machine listing of the source deck should be included. These listings are often bound in a separate folder.

7. PROGRAM HISTORY—This page should contain the record of program maintenance and provide an audit trail by listing all modifications to the program. These should be documented by showing date, authorization, description of the change made, and the initials of the programmer.

8. FILE DESCRIPTIONS—This section should contain a description of every file used in the system. It lists all volumes, uses, retention periods, etc.

Console Run Book

A run book should be set up for use at the console to assist the computer operator in running the programs. It should contain complete operating instructions and other information required for normal operation of the runs. Many of the items included in the master run book described above can simply be duplicated and inserted in the console run book.

Some items that might be included are: a brief program narrative, program setup information, disposition of files, sample carriage control tapes, and printed forms, etc., and any information needed by the computer operator to run the program.

The data in the console book should be organized in a form suited primarily for operator convenience and can be arranged either in a separate binder for each run, or in a binder for all runs in a particular application. For any occasions when more information than that contained in the console run book is needed, the operator should have access to the master run book.

An example of a page from a console run book is illustrated in Figure 12-4.

CONSOLE RUN FORM

PROGRAM NAME		PROGRAM NUMBER	ANALYST	PROGRAMMER
SYSTEM	FREQUENCY	DEADLINE	DATE	

RUN DESCRIPTION:

INPUT/OUTPUT INSTRUCTIONS

DEVICE TYPE	UNIT	USE	FILE DESCRIPTION	SOURCE	DISPOSITION	RETENTION PERIOD

PRINTER INSTRUCTIONS

FORM	NO. OF PARTS	VERTICAL SPACING	CARRIAGE CONTROL TAPE	EST. SIZE	SPECIAL INSTRUCTIONS

RESTART INSTRUCTIONS:

MESSAGES:

SPECIAL INSTRUCTIONS:

Figure 12-4 Example of Console Run Form

As can be seen from Figure 12-4, the Console Run Form contains all of the information required by the operator to run the program, including the input/output instructions, the printer instructions, restart instructions, and any other special instructions which may be necessary. In most applications the information contained on the form in Figure 12-4 will be sufficient, but if there is additional information required for a special application, the analyst must ensure that it will be available to the operator.

MANAGER AND USER DOCUMENTATION

As has been noted previously, it is not only necessary to have documentation explaining the operation of the system, what each of the programs in the system accomplishes, and how, it is also necessary to have documentation for both management and the users of the system to indicate how the system should be used, what information is available on the reports, and how the information should be handled at the user's site.

Management documentation consists primarily of examples of reports which are generated from the system and how to read the information provided. It normally is not extensive documentation and in many systems and installations, is never prepared. For those members of high-level management who are using the output of the system, however, it could prove quite beneficial. Thus, if high-level management is to make use of the output of the system, then explanations of reports should be prepared and distributed to those members using the output.

The documentation which is prepared for the users of the system is more important than the management documentation and should be prepared for virtually every system which is developed. This documentation not only describes the output from the system, it also would illustrate any data preparation which the user must perform for entry into the system. Thus, the source documents and their use must be thoroughly covered in this documentation. Additionally, any timing considerations, such as deadlines for preparing source documents or other material which must be submitted to the data processing department, must be included because without these firm schedules, the system stands little chance of being processed on time.

SUMMARY

One of the most important aspects of the entire systems analysis and design spectrum is the efficient testing of the programs within the system and the testing of the system itself. A competent systems analyst will not allow his system to be implemented on the computer until he is sure that all features of the system will function without error. This, as described previously, entails a great deal of planning and attention to detail when the system is tested. A system which is implemented to meet a deadline before it is properly and completely tested will cause many more problems than one which misses its deadline but is reliable once it is implemented. Too much emphasis cannot be placed on the requirement that a system be adequately tested prior to implementation.

Without proper documentation, it is likely that even if reliable, the system will not be operated either properly or with the best efficiency. In addition, the actual task of implementation will be more difficult and if any changes must be made to the system subsequent to implementation, it will be very difficult to make the changes without the proper documentation.

CASE STUDY - JAMES TOOL COMPANY

Once the programming of the payroll system had begun, the analysts Mard and Coswell set about designing the testing which they would perform to ensure that the payroll system would operate properly. As noted previously, each program will be individually tested by the programmer. The analysts determined that after each program had been tested, a series of link tests were called for in order to test the compatibility of the programs within the system.

Case Study - Testing Edit and File Creation Programs

The first set of programs to be tested are the edit program and the file creation program which are used to create the Employee Master File and the Terminated Employees Master File. The following sequence of testing is to be performed:

1. Properly prepared data will be input to the Edit program and will then be input to the file creation program. The data will be prepared in such a way that no errors should be found in the edit program and no errors should be found in the file creation program. Therefore, the output of the programs should be a properly formatted Employee Master File containing the information concerning each employee and his year-to-date earnings.

2. After all "good" data has been loaded properly, data with errors which should be detected by the edit program will be input to the edit program, together with some good data. The output from this sequence of testing should be entries on the edit exception report, no entries on the load exception report, and only the good data on the loaded Employee Master File. Particular attention must be paid in this sequence of testing to be sure that only the records with good data are loaded on the master file and that all of the input records with errors are detected by the edit program.

3. After the edit program is checked, data will be input to the system which will pass the edit checks but will not be loaded because of problems found in the file creation program. Typical of the type of error which will be tested is where there are not two addition cards for an employee but rather one of the two required cards is missing. The output from this testing should be entries on the file creation exception report for the errors which were found, an Employee Master File containing only the input records with valid data, and the listing of the employees loaded onto the Employee Master File.

4. After the link testing for loading the Employee Master File is completed, good data will be used to create both the Employee Master File and the Terminated Employee File. After it is verified that this processing works correctly, data containing errors will be introduced for both the edit program and the processing in the file creation program which creates the Terminated Employee File.

Note from the processing indicated above that the link testing takes place a step at a time, first using data containing no errors and then incorporating errors which will occur in the input data. Each step of the testing will be closely monitored so that any errors which occur can be corrected and retested prior to any further testing. It should be noted also that any corrections which are incorporated into a program should be retested prior to further testing. It should never be assumed that a correction will work properly and therefore need not be tested.

Another set of programs which must be "link-tested" are the edit program and the update program which are used in the weekly processing of the payroll system. The analysts would assemble the "job stream" to include the job control statements which would, for the most part, be used in the actual production run. In addition, they would develop the test data and produce an Employee Master File and a Terminated Employees File to be used in this testing. It should be noted that these files would be created for testing purposes only with special data designed specifically to test the various conditions determined by the analysts. It is quite important that the testing at this stage use specifically designed test data rather than some "live" data since the analyst will define those conditions which must be tested. After the two files have been created, the analysts determined that the following were some of the tests which must be performed.

1. Data without any errors will be used to test additions to the Employee Master File and terminations from the Employee Master file. The data will not include any errors which will be detected by either the edit program or the update program. It should be noted that the first and last record conditions will be specifically tested.

2. Changes to the master file will next be tested. Again, the data in this test will not have any errors which will be detected by either the edit or the update programs.

3. Data to be used to add records to the master file but with errors which should be found by the edit program will next be tested. The results of this testing should be that errors are reported on the edit program exception report and the records in error should not be added to the master file.

4. Next, data to be used to add records which contain errors that will be found in the update program will be input to the programs. Such errors as no sequence 2 record, no sequence 1 record, and attempting to add a record to the file which is already on the file will be incorporated into the data. The result of this testing should be the update exception report with the errors listed.

5. Invalid change data will then be tested. Change data with errors which can be caught in both the edit program and the update program will be used.

6. The terminations will then be tested by inputting data which indicates that employees should be terminated. Checking must be performed on both the Employee Master File to ensure that the employees are deleted and on the Terminated Employees File to ensure that the records are added to that file. Particular attention will be paid to first and last record processing on both files.

7. Throughout each testing procedure, the contents of the Report File will be checked to ensure that the data for producing the reports in the system is properly recorded.

8. Limits which are established in the update program and the edit program, such as the number of hours worked and the number of overtime hours, will be checked as well as all other editing functions. Particular attention must be paid to data which is just under and just over limits.

As can be seen from the partial list of tests which are to be performed in the link testing, the analyst must very carefully determine all of the types of processing which is to take place in the programs within the system and then design tests which adequately check all of the conditions which can be found once the system is in production. It is very important that the analysts define each type of error which must be checked and all of the combinations which can occur. If this is not done, experience has shown that the errors not checked will be the errors which occur when the system is put into production.

Case Study - System Test

After all of the programs within the system have been individually tested and then tested in the link tests and all have performed satisfactorily, the analysts must test the entire payroll system. This "system test" is used to not only further check the programs within the system, but to also check all of the procedures which have been developed to process the data completely through the system. Since these procedures must be known by the persons performing them, most training sessions should be completed by the time the system testing commences.

The following are some of the elements which will take place when the system test is performed for the payroll system:

1. In order to run the system test, the analysts decided to place all the data for employees from one department on the Employee Master File and the Terminated Employees File. Therefore, the personnel department will be instructed to transcribe the personal and payroll information for employees from one department from the manual forms to the new source documents designed for the new payroll system.

2. The source documents will be transmitted from the personnel department to the data entry department in the same manner as will be done when the actual implementation of the system begins.

3. The data entry department will prepare the input records from the source documents supplied by the personnel department. These input records will be used to create the Employee Master File and the Terminated Employees File which will be used in the system test.

4. The computer processing will be performed to create the master files. If there are any errors which are detected in the processing by either the edit program or the file creation program, these corrections will be made in the same manner as when the system is implemented. It should be noted that the analysts will purposely include some data with errors so that the error procedures can be checked.

5. Time cards will be prepared for the employees in the department which is being tested. These time cards will be made by the analysts in the form which will be received by the payroll department when the payroll system is implemented, that is, they will be the ones prepared from the system itself. The payroll department will perform the work which must be done in the payroll department with the time cards.

6. Changes, additions, and terminations will be made available to the personnel department so that this department can prepare the source documents. Again, the analysts will purposely include errors so that all of the error recovery procedures can be tested.

.7. The source documents from the payroll department and the personnel department are to be transported to the data entry area of the data processing department according to the methods and schedules established in the design of the system. Thus, the documents will be sent through intracompany mail on the same schedule as when the system is implemented.

8. Data preparation in the data entry area of the data processing department will take place when the source documents are delivered. Again, this will occur according to the schedule which has been previously designed. Any improperly completed source documents will be returned to the originating department for correction as required by procedure.

9. The input data will be transported to the staging area in the computer room. In addition, the master tapes which have been previously created will be removed from the tape library and made available the same as if the job were in production.

10. Computer processing will take place using the input data and master tapes made available. The operators will follow the instructions which have been prepared by the systems analyst and all output will be produced as if this was a normal production run.

11. The output reports will be processed and distributed according to instructions. The decollating and bursting of the payroll register and any other work which must be accomplished will be done by the operations staff.

12. The time cards which are produced will be transported to the payroll department the same as will take place when the system is in production.

13. All of the input tapes, the output tapes, the input cards, and the source documents will be placed in their proper places according to the instructions supplied by the systems analysts.

As can be seen from the previous steps, a simulation of the actual system in production will be carried out. Everything which is to take place when the system is in production will take place in the system testing. If the processing can be carried out successfully in the system testing, then there is a strong likelihood that the implementation of the system will also be successful.

An important consideration in the system testing, as in all testing, is the preparation of the test data. When conducting a system test, the analyst should design all of the test data so that the conditions which can arise within the processing will be tested. This data must be very meticulously developed so that not only will the conditions be tested, but the results will be easily interpreted and any incorrect processing will be spotted. This normally means that a large volume of data is not required.

Case Study - Successful Criteria

It does little good to test a system if it is not known what constitutes a successful test. In addition, a general "everything seems to work O.K." will not usually suffice for a criteria. Therefore, as a part of the design of the system test, Mard and Coswell will have to specify the exact results which they expect as a result of the testing. If these results are not met, corrections must be made to the system, either in the programs or the system procedures, and the system testing will have to be repeated. It should be noted that the system should never be allowed to go into production until the system test has been successfully completed.

Among the elements which must be checked by the analysts in the payroll system are:

1. All functions, including the preparation of the source documents, the preparation of the input data, the computer processing, and the distribution of the output should be accomplished per the schedule designed by the analysts.

2. The results of the computer processing must be correct. This includes not only the contents of the reports produced and the files updated, but should include such things as proper labels on the tape files, correctly punched time cards, proper marking of void checks or any other checks which are not used in the processing, and accurate correlation between the contents of all of the reports produced.

3. Distribution must be accomplished properly. The reports must be correctly prepared for distribution by decollation and bursting, the source documents, input data, and tapes must be filed in the proper areas, and all external labeling for tape files must be done correctly.

4. The operations staff must be questioned concerning their reactions to the system and how it is designed. If they have any suggestions which would lead to a more efficiently operated system, the analyst must consider these suggestions.

5. The user departments must also be questioned concerning any problems which they encountered in the system. It is quite important that the user departments be satisfied with the operation of the system during this testing phase so that there will be a smooth implementation.

Case Study - Scheduling of Tests

In scheduling the system testing, the analysts must consider both the needs of the user department and the data processing department. If either of these departments is unable to perform their functions in the system test, then the test will not provide the assurance of reliability which is desired from this testing. Therefore, the analyst must contact all interested parties to be sure that a convenient testing time and schedule can be arranged.

In addition, it is likely that four or five full system tests will be required before the system can be certified ready to go into production. Thus, the schedule must be flexible enough to allow all of this testing. Time must also be allocated to allow for corrections to be made to the system prior to the next system test. Obviously, if there are program errors, changes and corrections must be made to the programs but time must also be allocated for changes which must be made in procedures and documentation in case there are difficulties in this area.

In the payroll system, the analyst contacted the payroll department, the personnel department, the data entry department, and the computer operations staff in order to coordinate the entire system test. In addition, they planned on four system tests, each four days apart. In this manner, it was felt that adequate time was allocated for any corrections which must be made and that at the conclusion of system testing, the payroll system would be ready for conversion and implementation.

Case Study - Management and User Approval

At the conclusion of the system testing, Mard and Coswell planned a series of meetings with the management of the personnel department, the payroll department, and top-level management to present the results of the testing. It should be noted that this will be the first opportunity for the management and user personnel to see the actual reports which will be produced from the system as previously they had only viewed the report layouts on the printer spacing charts. Therefore, if any adjustments must be made to reports, they can be made at this time because this is the final review prior to implementation of the system. As has been noted previously, however, it is costly and time-consuming to make changes at this point in the system project, but if necessary they can be made prior to the implementation of the system.

These meetings also offer the opportunity for management personnel to comment upon the operation of the system in the system tests and to offer any suggestions concerning improvements in the procedures to be followed when the system is implemented. If changes are suggested which are accepted, then the documentation concerning these procedures should be changed and the procedures should be tested prior to implementation.

Case Study - Documentation

The documentation for a system consists of the program documentation, the computer operations documentation, and the documentation for the procedures which are to be followed in the various areas of the company using the system. The program documentation is largely the responsibility of the programmer writing the program, but the analyst should oversee the preparation of this documentation so that the standards required are adhered to.

The preparation of the operations documentation is also largely the responsibility of the programmer although the analyst will normally take a larger part in its preparation because the flow through the system was designed by the analyst. One area in which the analyst is particularly interested is the console run form because any special operating instructions which are necessary must be contained on this form. The console run form for the edit program in the payroll system is illustrated in Figure 12-5.

CONSOLE RUN FORM

PROGRAM NAME		PROGRAM NUMBER	ANALYST	PROGRAMMER
PAYROLL EDIT		PAYRL04	MARD; COSWELL	KOHLER

SYSTEM	FREQUENCY	DEADLINE	DATE
PAYROLL	WEEKLY	WEDNESDAY, 8 A.M.	6/5/76

RUN DESCRIPTION:

Using the time cards, additions, terminations, and changes as input, this program performs all the editing functions of the payroll system to ensure valid data entering the payroll system.

INPUT/OUTPUT INSTRUCTIONS

DEVICE TYPE	UNIT	USE	FILE DESCRIPTION	SOURCE	DISPOSITION	RETENTION PERIOD
Cards	00C	I	Additions, Changes, Terminations, Time Cards	Staging Area	File Storage	6 Months
Printer	00E	O	Exception Report		Control Desk	
Tape	182	O	PYRLT03- Edited Transactions	Scratch	To PAYRL05	1 Day

PRINTER INSTRUCTIONS

FORM	NO. OF PARTS	VERTICAL SPACING	CARRIAGE CONTROL TAPE	EST. SIZE	SPECIAL INSTRUCTIONS
Standard	1	6/inch	Standard	2 Pages	Return To Control Immediately

RESTART INSTRUCTIONS:

Restart From Beginning

MESSAGES:

None

SPECIAL INSTRUCTIONS:

None

Figure 12-5 Example of Console Run Form

Note from the example that the input and output files are specified together with the devices on which they are to be processed. In addition, all information required to run the program is contained on the form. It is quite important that the console run form be properly prepared because without accurate information, the computer operator will not be able to run the payroll system programs.

The other area of concern for the analyst in preparing the documentation for the system is the procedures which must be followed in all other departments in order for the system to operate properly. In the payroll system, some of the procedures which must be documented are:

1. The procedure for distributing and retrieving the time cards and preparing them for the data entry department.

2. The procedure for distributing reports generated from the payroll system.

3. The procedure for issuing paychecks to the employees, including the signing of the paychecks and the distribution of the paychecks.

4. The procedure for controlling the paychecks and providing the necessary security.

As can be seen from the partial list above, the procedures to be documented can vary a great deal. It should be noted that all of the procedures to be documented should have previously been designed and will be tested in the system testing. The format of the documentation is normally a narrative with step by step instructions. It cannot be overemphasized that this documentation must be clear and concise for without good instructions, there is every likelihood that there will be errors in the processing of the system.

SUMMARY

The testing which takes place within a system can many times determine whether the system will ever be a successful system once implemented. With adequate planning and thorough testing, most systems will perform to the level desired. Without this complete testing, there is little chance of a trouble-free system operating to the expectations of the user and management.

STUDENT ACTIVITIES—CHAPTER 12

1. List the tasks which must be completed after the programmer has received the programming specifications and prior to systems implementation.

2. What subtasks must be performed to produce programs from a set of programming specifications?

3. Discuss the problems associated with program testing.

4. Summarize the basic rules for program testing and debugging.

5. What is meant by "Unit testing"?

6. Discuss some of the factors that must be considered when creating test data.

7. What is meant by ''link'' or ''string'' testing?

8. What is meant by ''systems testing''?

9. List the main objectives of system tests.

10. Discuss the testing which must be done on the backup system.

11. List three factors that should comprise the final systems documentation.

    1)

    2)

    3)

12. List the contents of a program documentation manual.

    1)
    2)
    3)
    4)
    5)
    6)
    7)
    8)
    9)

13. What information should be contained in a Master Run Book?

14. What information should be contained in a console run book used for operations documentation?

15. What should be included in the documentation for management?

16. What should be included as a part of user documentation?

1. "We insist that our programmers thoroughly document each program. This documentation must include a systems flowchart, a detailed narrative description of the overall function of the program, a narrative description of all main routines within each program, a program flowchart, and a sample of the input and output. Without this type of documentation our installation couldn't operate as over 50% of our programmers' time is involved with program maintenance."

"Most of our programs run between 5,000 to 20,000 lines of coding. The only documentation we require is the program listing and an example of the input and output. If we required program flowcharts they would be 50 pages or more in length. If our programmers had to redraw a program flowchart every time a change was made to one of our programs, and if they had to rewrite a narrative description of the program and related routines we would never get anything done! I've been in companies that have entire staffs that do nothing but maintain the documentation for programs. This is a waste of money! Many companies today over-document."

Which position do you support? Can a system be over-documented?

2. One analyst recently asserted that "link tests are a waste of time. If each program has been adequately tested and if the programming specifications were prepared properly, there is no reason for link testing. All that is required is a system test to check out the system procedures."

Is this a responsible attitude for a systems analyst to take?

3. A senior analyst remarked that "taking all the time and trouble to design and make up system test data is largely a waste of time if the programs have been debugged. Instead, live data from previous cycles of the system being replaced should be used. After all, if any errors are going to occur, they most certainly will show up in the live data which has been previously used in the system."

As data processing manager, would you let this analyst implement a large system in your installation? Why?

CASE STUDY PROJECT—CHAPTER 12
The Ridgeway Company

INTRODUCTiON

After programming specifications for the billing system of the Ridgeway Country Club have been completed it is then necessary to:

1. Program the system
2. Complete the program testing
3. Prepare the final documentation

STUDENT ASSIGNMENT 1
SYSTEM TESTING

1. Design the testing which will be required for the billing system. Both link testing and system testing should be considered.

2. Design the test data which will be used for all of the testing of the system. This should include the data for all of the tests which will be performed on the system in all phases of the testing.

3. Specify the criteria which will constitute successful testing of the billing system.

STUDENT ASSIGNMENT 2
DOCUMENTATION

1. Prepare the operations documentation which will be required in order to operate the billing system.

2. Prepare the procedural documentation for all departments which will be concerned with the billing system.

PHASE V

SYSTEMS IMPLEMENTATION AND EVALUATION

SYSTEMS IMPLEMENTATION AND EVALUATION

CHAPTER 13

SYSTEMS IMPLEMENTATION AND EVALUATION

13

INTRODUCTION

After the testing has been completed and the documentation prepared, the system is ready to be implemented on the computer and run in a production environment. Assuming that the testing and documentation have been properly completed, the implementation should be a relatively smooth transition from the old system to the new system.

The basic steps to be taken in the Systems Implementation and Evaluation phase of the systems project include the following:

1. Convert existing files.
2. Determine the conversion method to be used.
3. Perform the "change-over" to the new system.
4. Perform a post-implementation evaluation of the new system.

FILE CONVERSION

Prior to system implementation, the data which is to comprise the master files must be converted from the form in which it is used in the manual system to the form which will be used for the new computer system. The two foremost elements which must be observed and strived for when converting data from one system to another are:

1. File Integrity
2. File Security

File Integrity

File integrity refers to the accurate transfer of all records and fields from the old file to the new file. Although it would seem that this is a simple matter of copying from one file to another, in actual practice some problems may arise. First, the new file is not always identical in format to the file used in the old system. In fact, many times there may be several files in the old system which are to be combined and changed to make up the new file. Therefore, it is mandatory that the analyst carefully define all sources of information for the new files.

In addition, he must gather the data on schedule and define the exact procedure which is to be used to transfer the data on the original source documents to the new file. This may involve the transferring of information from the old source documents to the new source documents and then transferring the data from the new source documents to the new file storage media such as magnetic tape or magnetic disks. This transferring of data is subject to error and tight control and efficient checking procedures must be developed to ensure the data is transferred accurately.

File Security

Whenever currently existing master data is transferred to a new form to build a master file, there is always the danger of a "security leak," that is, that confidential data which is contained on these files may inadvertently be placed where it is available to persons who should not see it. For example, the payroll information which is to be contained upon an Employee Master File is normally confidential data and should not be disclosed to anyone except those persons who are directly involved in the file conversion process or those persons working in the payroll department. Thus, care must be taken so that the documents from the old systems are returned to their proper place and any new source documents are carefully controlled and processed through the proper channels.

File security is extremely important in data processing applications. During file conversion, because a variety of people may be working on various aspects of the conversion, there is a good chance for file security to break down. Although file security may not seem important, consider the file conversion problem when thousands of credit records for persons throughout the country are transferred from one system to another. A leak of this information to unauthorized people could cause a great deal of difficulty, so it is important that the analyst consider the problem of file security when file conversions are taking place.

File Conversion Timing

The time within the system plan when file conversion should take place will vary, dependent primarily upon the availability of the data to be converted and the requirement for current data. In many systems, the data which must be converted will not be available and current until the time the system is to be put into production. For example, in a payroll system, the information in the manual system is updated each week in order to reflect the weekly and year-to-date pay of an employee. The most current information must be placed on the master file when the system is implemented or the system will not reflect the correct information when it is run. Therefore, the file conversion, including the preparation of the source documents and the actual loading of the file, must take place after one cycle of the payroll has been completed and before another is scheduled to be run. Thus, for example, if the payroll is done manually on Wednesday, the week's payroll would be completed and then the preparation of the source documents would begin. The entire conversion of the manual data to the master file for use in the computer system would have to be completed by the following Wednesday.

Another factor which enters into the determination of the timing for file conversion is the method of testing the system which is to be employed and the type of conversion to the computer system which will be utilized. When the systems testing is conducted, there is always a requirement that some information be contained in a master file because the processing of the programs requires a master file. When testing the system, the normal procedure is to build a partial master file which contains enough data to test the programs but which does not contain all of the data which will be on the master. The source of this data may be current information within the system or may be data which is made up by the analyst to test certain conditions within the system. Thus, for the systems testing, it is not normally necessary to convert the entire manual file to a format to be used in the new system.

If parallel conversion and testing is to be used, however, it is necessary to convert the entire file. This method of converting from the old system to the new system, which is discussed in the next section, requires that a current file of information be available for processing.

Thus, it can be seen that the analyst must consider a number of factors when converting the data from the old system to the new system. Without these considerations, the analyst runs the danger of not having data properly prepared when the system is placed into production. If this occurs, it will be impossible to meet the scheduled implementation date.

DETERMINING THE CONVERSION METHOD

Conversion refers to the process of making the change from the old system and procedures to the new data processing procedures. The difficulty of conversion is directly dependent upon the complexity of the system being implemented.

There are three common methods which are used when converting and implementing a newly designed data processing system. These methods are:

1. Direct Conversion
2. Partial Conversion
3. Parallel Conversion

It is the decision of the analyst to determine which of the three methods of providing for the actual implementation of the system is to be used.

Direct Conversion

With direct conversion, at a given date, the old system merely ceases to be operational and the new system is placed into use. To those not familiar with the implementation of large system projects, this may seem to be the logical method for implementing a system, as many months of effort have gone into the design and testing of the new system. In reality, however, direct conversion is extremely ''risky,'' for in spite of the great care that has been taken by the analyst in the design and testing of the system, it is likely that the results will not be exactly as planned because of the hundreds of variables that are possible in most systems.

Although direct conversion is the least expensive method of converting to a new system, this method is normally not used except on the simplest of systems and then only with those systems in which a failure would not create a ''disaster'' for the company.

Partial Conversion

Another approach which sometimes proves useful is the partial or the step-by-step approach, where either a portion of the system is implemented without requiring that the whole new system go ''on the air'' at once; or the entire system is implemented, but only for limited users until the system is proven reliable. An example of this would be to implement a payroll system in total but only for several departments within the company. The remainder of the departments would use the old system when the new system is first implemented. As the bugs are ironed out, more departments would be implemented until finally the entire company is using the new system.

This method has the advantage of allowing the system to be implemented relatively quickly without the requirements of additional personnel and efforts as is needed with parallel runs, and the entire company is not placed on a ''do-or-die'' basis. Obviously, the department which is implemented is dependent upon the new system, but it is quite likely that with the proper choosing of the areas of the system to be implemented, there will be a minimum of disruption if the system, for some reason, fails. In many cases, this is the most efficient method of implementing a system.

Parallel Conversion

In many complex systems, it is desirable to provide for further testing of the system prior to implementation, even though each of the programs have been tested individually, the programs have been tested as a group, and a complete systems test has been performed. To further ensure the accuracy of the system, an approach called Parallel Conversion or parallel testing is commonly used prior to the actual ''stand-alone'' implementation of the new system.

Parallel Conversion consists of processing the **current** input data in both the old and new system simultaneously and then comparing the results of the two. If the output from both systems is identical it indicates that the new system is functioning in its proper manner.

The advantages of Parallel Conversion include:

1. The entire system may be checked for accuracy prior to the first production run.

2. Parallel Conversion provides for operational training and a check on the completeness of operational documentation, as the entire application is run under the same condition as will be encountered in actual production runs.

The disadvantages of Parallel Conversion are:

1. Cross-checking the results of the new system with the results of the old system may prove difficult because the new system may contain reports in different formats then the old system and additional data may be generated by the new system.

2. The company must pay the cost of operating both systems during the conversion period.

Parallel Conversion, although being an extremely good test to determine if a system is working properly and to check the time the run takes with normal production loads, does create some operational problems. Obviously, if both an old and a new system are to be run at the same time, input for both the old system and the new system must be prepared and this will require an increase in the number of people who must work on the system that is being parallel-tested. This increased workload creates several problems—first, the help which is working on the system being tested will be temporary because when the parallel testing is completed, there will be no need for additional people. Secondly, since temporary help is required, there will be a training problem since the temporary personnel will have to be trained to operate either the old or the new system.

In most instances, the persons who will be permanently assigned to working on the new system will work with it in the parallel testing state also. This has the benefit of allowing them to become familiar with the system prior to actual implementation. Thus, in addition to testing the programs and the system in general, there is valuable experience gained by the personnel who will be working with the system. The drawback, as mentioned, is the need for additional people during the parallel testing.

The time period during which parallel testing takes place will vary depending upon the complexity of the system, the confidence of the systems department in the correctness of the system, and the need for reliability. The more complex systems will normally require a longer period to be totally checked out. The need for reliability will vary greatly between different systems. In general it can be said that where money is involved in a system, such as a payroll system, an accounts payable system, or an accounts receivable system, there is a very great need for complete realiability and management will not usually allow a system to become the only system being used until they are completely satisfied that the system is totally reliable. As with most situations, however, parallel testing can be carried to the extreme, so at some point it will be necessary for the analysts and management to approve the system and allow it to go into production.

SUMMARY - SYSTEMS CONVERSION METHODS

The first step in the systems implementation phase of the systems project requires the systems analyst to determine the procedure for the actual implementation of the system.

For relatively small systems that are not critical to the operation of the company, direct conversion may be attempted in which the analyst determines a time in which the old system ceases to be operational and the new system is implemented. A similar approach is the implementation of one segment of the system at a given time. This technique allows difficulties to be worked out of the system prior to implementing the system throughout the entire business organization.

In many cases further testing of the system is desirable prior to implementation. Perhaps the most common approach to implementation is to provide for the parallel operation of both the old and the new system, comparing the results to assure that the proper output is being obtained, prior to running the new system by itself.

It must again be emphasized, however, that none of these conversion plans are to be undertaken without the most thorough program and systems testing. As noted previously, the analyst should never allow a system to be even partially implemented until it is felt that the system works perfectly and there is no chance for failure.

IMPLEMENTATION OF THE SYSTEM

At some point in time it is necessary for the analyst to give final approval for the system "cut-over," that is, the newly designed system must be placed into production.

When the implementation date arrives, hopefully all of the pieces fit together and the system goes "on the air" with little or no trouble. In actual practice, however, the implementation of a large system is normally a hectic time. Even with the best preparations, there always seem to be last minute things which must be done and which are not foreseen. In addition, the use of a new system and unfamilliar procedures will many times cause minor errors with which the analyst will have to contend.

An important consideration in the implementation of a system is when to put it into production. Most systems are somewhat "cyclic," that is, they are run in cycles of a week, month, quarter, and so on. In addition, there are normally some types of accumulative periods when the system has alternate processing which it performs. For example, a payroll system will be run each week. At the end of each quarter, however, there will be tax statements due for the company, so additional processing will be run each quarter. It is not normally a good idea to implement a system in the middle of one of these cycles because the accumulative processing would have to be performed partially from the old manual system and partially from the new computer system. Thus, instead of merely picking a date on which to implement the new system, it behooves the analyst to determine a point in the cycle of the system when implementation will not work a hardship on any subsequent processing which must be performed with the system. In a payroll system, this could be at the end of a quarter when all of the reports for the previous quarter would be prepared manually and the new quarter would start out with the computer system.

The timing for implementation depends primarily upon the system itself but typical implementation times are the beginning of a quarter and at the beginning of a fiscal or calendar year. If the system is not dependent upon processing such as quarter or year-end, then the system could be implemented as soon as the system is fully tested and all files converted.

As can be seen, if the implementation date is the beginning of a quarter or the beginning of a year, there will be a great deal of pressure to have the system fully operative at this time. For example, if it is determined that the system should be implemented on the first day of the new fiscal year and the system is not ready for implementation, theoretically there is another year to wait until it can be implemented again. In actual practice, of course, this would not occur; however, it can be seen that by missing the scheduled date of implementation there can be some real difficulties in performing the required processing. Thus, it is essential that realistic schedules are established and followed throughout the systems project.

Problem Areas During Implementation

Assuming that the programs have been properly tested and that they run as designed, the most likely area for problems when the system is first implemented is in the handling of the data itself. This includes the transporting of the source documents within the user department and to the data processing department, the handling of the source documents and the preparation of computer input in the data processing department, and the distribution of the output reports at the conclusion of processing. Most analysts and certainly the programmers involved in a system are extremely concerned about the programs working, and rightfully so, because the system will not function if the programs do not work as designed. Equally important, however, are the logistics involved in getting data to and from the computer, the preparation of the input data, and the distribution of the output data. These are areas where, in some cases, too little attention is paid. No matter how good the programs are, the system just will not be useful if data does not flow smoothly.

Another area of potential problems when the system is first implemented is in the running of the programs on the computer. When a strange system and unfamiliar programs are first run, operators will not be as comfortable with them as at a later time after the system has been processed a number of times. Thus, any flaw in the operations documentation will appear quite quickly when the operator follows the instructions in the documentation. It is therefore important that these operating procedures be thoroughly checked and the validity of the documentation ensured by testing prior to the implementation date.

POST-INSTALLATION EVALUATION

After the new system has been installed and is in operation, there is a need for the systems analyst to provide for systematic follow-up to ensure that the system is operating as intended and that the original objectives of the system are being accomplished. Specific questions that should be raised by the analyst during this post-installation evaluation include:

1. Have the objectives of the system been accomplished?
2. How do the actual results compare to the anticipated results?
3. What are the actual operating costs compared to the anticipated costs?
4. What modifications or revisions to the system are necessary?

To assist in performing the post-installation evaluation the analyst should establish some type of "feedback" procedure, that is, a procedure whereby both the users of the system and high-level management can indicate when some portion of the system is not operating the way it was designed. This feedback is quite important because without it, there is no way for the dissatisfied user to communicate that something is wrong and to get it corrected. As has been noted previously, the satisfaction of the user and the preparation of useful data is the primary reason for a computer system and if the system is not producing, then steps must be taken to correct the problems.

Once feedback has been received which indicates there is some type of dissatisfaction, then it is mandatory that some corrective action be taken. In some instances, there is merely a misunderstanding on the part of the user or management and an explanation, either orally or in the form of written instructions, will suffice to clear up the problem. In other cases, however, there will have to be changes made to the system to accommodate the requests from users or management. These changes can vary from different spacing on a printed report to major changes in master files.

For the most part, in a newly installed system there will be no requirements for wholesale changes because all of the contingencies will have been covered when the system was originally designed. There is no guarantee of this, however, and if some type of major change must be made, like adding a number of fields to a master file or adding reports and programs to the system, then these changes are essentially handled in the same manner as if a new system were being developed; that is, an investigation will have to be done to determine exactly what is wanted and how the new requirements will fit with the system currently in existence, the system will have to be approved, designed, and then implemented in much the same manner as the system which it is modifying. As can be imagined, this is a large task and should, in most imaginable circumstances, be uncalled for because these major areas of the system should be considered when the system is designed in the first place. About the only time when major changes will have to be made to a system which has been recently implemented is if there is a major change or adjustment within the company and new company policy requires something different from what is in the system.

Minor changes, however, such as changes in a report, are not at all uncommon. The major reason for these changes is that the users of the system did not critically evaluate their needs and how the data from the system would relate to them when they were asked to approve the systems design prior to implementation. Although the typical changes which are requested are not of such a magnitude as to cause large programming changes, they are many times quite time consuming because the programmer must change a program which has been thoroughly checked out and proven reliable. Minor changes made to programs are notorious in data processing for causing unforeseen problems in other portions of a program and there are many programs that once changed have never been properly debugged and periodically fail to process data properly. Thus, as discussed previously, changes to programs and to the system in general can be costly, both in terms of the time required to make the changes and any problems which may arise as a result of the changes.

It must be noted again, however, that the user must have the ability to request changes and the procedure for implementing these changes must be determined. A system, no matter how well designed, will not be satisfactory if those who use it are not satisfied. The analyst must, therefore, plan for and expect changes to be requested and made to the system which is implemented.

Although the term ''post-installation'' evaluation suggests that an evaluation is made only immediately after the system has been put into operation, this evaluation should not be considered a ''one-time'' activity. In fact, as long as the system is in operation a continuing evaluation should be made of the system to determine if the system is functioning as planned and meeting the needs of the company.

FINAL REPORT TO MANAGEMENT

After the system has become operational and a post-installation evaluation has been conducted, the systems analyst should prepare a final report to management relative to the systems project. At this time the analyst should review the history of the project, the projected costs and time for the systems project, the actual costs and time of the project, and the success or failure of the entire systems project.

The final report to management provides an excellent opportunity for the systems analyst to point out the advantages that the implementation of the systems project has brought about. The analyst, however, should be very frank and honest in his appraisal of the system, for it is certain that if a system is not operating as intended management will be informed and aware of the difficulties.

The project completion report can serve several useful purposes. First, the report serves as an opportunity to honestly evaluate the success of the project; and secondly, the report serves as an official sign-off from the systems department indicating that the analysts are turning over the operation of the new system to line personnel who are responsible for the day-to-day operation of the system.

SUMMARY

The implementation of a data processing system requires a great deal of planning and attention to detail. If done properly, it is a relatively painless experience but if all contingencies are not planned for, the implementation of a system and the subsequent changes and corrections which must be made can be costly in terms of money and in terms of peace of mind. The whole secret to a successful implementation is sufficient planning to ensure that there are no surprises.

CASE STUDY - JAMES TOOL COMPANY

Once the system testing has been completed and the results approved by management and the users, the analysts Mard and Coswell must plan for the conversion and implementation of the payroll system. They are concerned with two primary areas: File Conversion and Conversion from the manual system to the computer system.

Case Study - File Conversion

As has been noted, the Employee Master File and the Terminated Employees File must both be created from the data in the manual payroll system. Since all of the forms and procedures have been designed for this preparation, it remains for the exact schedules to be formulated and the personnel involved to be defined.

Since sensitive payroll data is to be dealt with in this file conversion effort, Mard and Coswell determined that the file conversion effort should be handled by the same people who will be involved in the actual processing once the system is in production. In this manner, the potential for unauthorized disclosures of sensitive data is minimized.

In scheduling the file conversion, the analysts were aware that the year-to-date pay information for each employee had to be current, that is, this data could not be prepared until the previous week's processing with the manual system had been completed. On the other hand, the constant data such as employee number and name, address, pay rate, social security number, etc. could be prepared prior to the current week because this data will not usually be changed and if it is changed during the last week, it can be updated on the master file through the use of change transactions. Thus, the file conversion effort was scheduled for a three week period prior to the day the first computer payroll processing will take place. The first two weeks of this period will be devoted to preparing the constant data and the last week will allow for the preparation of the current pay data.

Since the payroll was manually processed on Tuesday, the conversion of the current pay data will begin Wednesday morning. The analysts further decided that the file conversion programs which will create the new Employee Master File and the Terminated Employees File should be run on the following Sunday so that the results of the processing could be checked prior to beginning the processing of the computer payroll system on Tuesday evening. Any difficulties which occur in the creation of these files should be detected and corrected so that when the initial processing run takes place, all data in the master files is correct.

The analysts further determined that the payroll system would be implemented on August 4, which was the start of a new month and presented the minimum problems concerning any overlap between the manual system and the payroll system.

Case Study - Conversion

After analyzing the various options concerning the conversion method to be used, the analysts decided to use a parallel conversion because this method afforded the opportunity to evaluate the output of the computer system in comparison to the output from the manual system. This was deemed quite important because of the need for complete accuracy when dealing with the company payroll.

Mard and Coswell decided that a three week period was sufficient for the parallel processing, provided that the output from the computer system was satisfactory. They made the further determination that no additional personnel should be hired for the parallel processing because the additional training which would have to be done would be time-consuming and expensive and would subject the payroll system to the possibility of security problems. After checking with user department management, it was agreed that whatever overtime was required in order to run the parallel payroll processing would be authorized in the user departments.

The parallel processing would continue for a three week period, with careful scrutiny by the analysts after each run. It should be noted that the output from the computer processing will not be the only element of the payroll system which will be checked during the parallel processing. All procedures of the system will be examined to be sure that every procedure within the system works properly.

If errors are found in either the computer processing or within the procedures established for the system, then obviously these errors will have to be corrected. As has been noted previously, corrections to a system should not be implemented without further testing. Therefore, if corrections must be made to the system, the analysts determined that further parallel runs should be made in increments of one week until the entire system functions satisfactorily. At that time, the operation of the manual system will cease and the entire payroll will be processed using the new computer system.

Case Study - Report to Management

After solving several problems with the system, the cutover from the manual system to the computer system was made on September 1. After an evaluation period of one month in which no significant complaints were received, the following report was made to management.

MEMORANDUM

DATE: October 3, 1976
TO: Management
FROM: Systems - Mard and Coswell
SUBJECT: Payroll System

The computer payroll system for the James Tool Company was implemented on September 1. As of this date, there have been no significant complaints or criticisms and it would appear that all user departments are satisfied with the system.

The system was implemented approximately one month past the scheduled implementation date. This was due to some unforeseen programming problems in handling overtime calculations. The cost overrun for the project was $3,850.00, which was primarily accounted for by the excess programming time required to solve the overtime problems.

A continuing analysis of the system will be carried on for a period of four months to monitor any problems which appear. At this time, there are no plans for additional processing to be added to the system.

The payroll system is installed and running and should process the payroll in an error-free manner.

D. Mard
H. Coswell

sjm

Figure 13-1 Example of Memorandum

In the example above it can be seen that the memorandum summarizes the completion of the project and the fact that it is fully implemented. Note also that the time and cost overruns are indicated. In some systems it may be necessary to specify in more detail the results of the system, but the material contained would be a further explanation of the information contained in the memorandum in Figure 13-1.

Case Study - Summary

After the presentation has been made to management and there is satisfaction as to the results of the systems project, the analysts Mard and Coswell would move on to another project.

STUDENT ACTIVITIES—CHAPTER 13

1. List the steps that must be taken in the systems implementation and evaluation phase of the systems project.

    1)
    2)
    3)
    4)

2. What is meant by the term "file integrity"?

3. Discuss the problems associated with file security during file conversion.

4. List the three commonly used conversion methods when implementing a newly designed data processing system.

    1)
    2)
    3)

5. Explain the term "direct conversion." When is direct conversion normally used?

6. What is meant by "partial conversion"?

7. What is meant by "parallel conversion"? Discuss the advantages and disadvantages of parallel conversion.

8. Discuss the problem areas that are likely to occur during systems implementation.

9. What questions should be asked by the systems analyst during the post-installation evaluation?

DISCUSSION QUESTIONS

1. An efficiency expert recently noted that "a properly designed, programmed and tested system should be able to be implemented using direct conversion. It is much better to spend more time in the testing phase of a system in order to eliminate the costly parallel conversion method than to use parallel conversion."

   Do you agree with this position? Why?

2. It has been stated that "system evaluation is a continuing effort. A system must be continually monitored in order to be sure it is producing accurate results."

   How would you establish a continuing evaluation program for a large data processing system?

3. After implementing a system using some new hardware, a systems analyst was called in to the vice-president's office and informed that "The new system will have to be abandoned and we are going back to the old manual system. The reason is the new hardware is too expensive and there have been numerous complaints about the new system."

   What arguments would you present to convince the vice-president that this is not a good business move?

CASE STUDY PROJECT—CHAPTER 13
The Ridgeway Company

INTRODUCTION

After management approval has been given to the results of the system testing, it remains to implement the billing system.

STUDENT ASSIGNMENT 1
FILE CONVERSION

1. Prepare a plan and a schedule for converting the currently used master files in the billing system to the files which will be used in the new computer system. Document this plan in a narrative form.

STUDENT ASSIGNMENT 2
CONVERSION AND IMPLEMENTATION

1. Determine the method to be used for system conversion. Justify your selection in a memo to the data processing department manager.

2. Prepare a schedule for the implementation of the billing system and document the schedule in a memo to the data processing manager.

STUDENT ASSIGNMENT 3
EVALUATION AND MANAGEMENT REPORT

1. Prepare an evaluation plan for the billing system and submit this plan to the manager of the Country Club.

2. Prepare a management report on the billing system for presentation to top management in the Ridgeway Company.

CURRENT TRENDS IN SYSTEMS ANALYSIS AND DESIGN

CHAPTER 14

CURRENT TRENDS IN SYSTEMS ANALYSIS AND DESIGN 14

INTRODUCTION

The methods and techniques of Systems Analysis and Design presented in the preceding chapters are applicable to all systems projects. There are, however, a number of developments in business data processing which are presenting additional alternatives in the design of systems and the manner in which data is processed. The future systems analyst should be aware of these developments and understand how these changes are influencing hardware acquisition decisions and systems design projects within business organizations throughout the country. The two major factors which are influencing systems design are:

1. The increased use of data communication facilities.
2. The trend toward the use of ''data base'' oriented systems.

DATA COMMUNICATIONS

In increasingly great numbers, data systems are being developed in which the input and output data is entered and produced from terminals of some type which are remotely located from the central computer facility. These terminals are connected to the central computer by means of data communications equipment. The data to be processed by the computer is transmitted via the terminals and related data communications equipment through communication channels, such as telephone lines, to the computer.

Systems that provide for this type of processing are known by such names as:

1. On-Line Information Systems
2. Real-Time Information Systems
3. Teleprocessing Systems
4. Data Communication systems

Regardless of the title these types of systems have the basic characteristic of transmitting information from a remotely located terminal to the computer, the processing of that information by the computer, and the transmission of the processed information back to the terminal.

DATA COMMUNICATION SYSTEMS

Data Communication systems provide the ability to access a computer from terminals which are located at a site remote from the computer. This remote site may be a few feet from the computer or many thousands of miles.

The basic components of a data communication system consist of the following:

1. Some type of remote terminal.

2. A "modem" on the sending and receiving end of the communications network. Modems are merely electronic devices whose function is to provide signal compatibility between the communication lines and the terminals and/or computer.

3. A communications channel such as telephone lines which provide for the transmission of data between the remote terminal and the computer.

4. A communications control unit for interfacing the communication system and the central processing unit of the computer. These units contain the electronics that provide the remote terminals with access to the storage unit of the computer.

5. The central processing unit of the computer.

DATA COMMUNICATION SYSTEMS APPLICATIONS

Although there are many types of data communication systems, there are four common application areas which make use of data communication networks. These applications include:

1. REMOTE BATCH PROCESSING—In remote batch processing systems the terminal acts essentially in the same manner as a card reader and printer at the computer site. The primary difference is that the card reader and printer transmit and receive data via a communication line from a location separated from the central processing unit.

2. DATA COLLECTION SYSTEMS—A data collection system is one in which data is transmitted from a remote site to a computer and the data is stored on some medium, either disk or tape, for later processing. Data collection systems are used in such applications as payroll, where the employee clocks in and this data is transmitted directly to a computer for subsequent processing.

3. MESSAGE SWITCHING—A message switching system is one in which a number of terminals are connected to a computer and a message is transmitted from one of the terminals to the computer and then transmitted to another terminal. For example, a company with an office in Los Angeles could transmit data to the computer in Chicago. After the data is processed the output would be transmitted to a different office in New York.

4. INQUIRY SYSTEMS—An inquiry system is one in which there is a direct interaction between the operator at the remote terminal and the computer. Inquiry systems are commonly used in air line reservation systems, and banks and savings and loan institutions, as well as many other applications.

DATA BASE

In addition to the use of data communication systems, many users of large scale computers are organizing data into a "data base," wherein all data relevant to a number of applications is linked together to provide access to the data by multiple users.

Traditionally, data files were designed to serve individual applications areas, such as personnel and payroll. For example, the personnel department would have a file containing the names and addresses of the employees, job classifications, pay rates, job skills, test scores, etc. The payroll department would have a separate file containing the names and addresses, job classifications, pay rates, YTD earnings, and taxes of each employee.

Normally, each data file, such as the personnel file and the payroll file, is specifically designed with its own storage area on a direct access device or each file is stored on its own reel of magnetic tape. Thus, when the personnel department needs a list of employees the personnel file is used; and when the weekly payroll is to be processed, the payroll file is used. As a result of this approach two problems are apparent:

1. There is often duplicate or redundant information in each of the files.

2. There is no integration of information.

In the example of the personnel file and the payroll file it can be seen that much of the data is duplicated in each of the files and, in addition, when it is necessary to delete an employee or change a pay rate both files would have to be updated. These undesirable attributes of data files have been eliminated by the use of DATA BASE SYSTEMS. A data base is defined as "a nonredundant collection of interrelated data items processable by one or more applications."

DATA BASE DESIGN

The basic approach to data base design is to have the systems analyst review the data requirements of all related applications within a company and then define a data base that services those applications. One data base organization method is to conceptually store the data base in a hierarchical manner, that is, the most significant data resides on a "high" level while less significant, but related data, appears on a subordinate level. Figure 14-1 illustrates this concept.

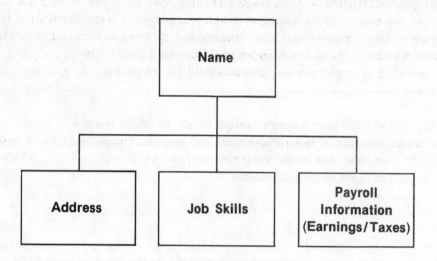

Figure 14-1 Personnel/Payroll Data Base

Note in the example in Figure 14-1 that the data base consists of the employee name, address, job skills, and payroll information. Thus, conceptually a single "file" contains all relevant information related to payroll and personnel systems.

In a "data base system" if one user requires access to the employee name and address and job skills this information may be accessed. If another user requires the employee name and payroll information this data may be accessed through the same data base.

DATA SECURITY

One of the most important problems in the use of a data base system is "file security." In most business organizations a certain portion of the information related to the company is confidential. By definition and design data base systems may be accessed by multiple users. Thus, the problem arises as to how to design the data base so that all users do not have access to all the information in the data base.

In a data base system, each user can only access the data in the structure to which it has been authorized. Assume one user requires the name and address and a second user requires name and payroll information. Even though the applications share common data (NAME) the first user could only access the subordinate data item "address" and the second user could only access the subordinate data items containing the payroll information. Control over the structure of the data base and data to be accessed is handled by the Data Base Management System.

DATA BASE MANAGEMENT SYSTEMS

An integral part of every data base is an associated DATA BASE MANAGEMENT SYSTEM. The data base management system consists of some type of programming system and "language" that provides for the creation of the data base, provides for inquiries into the data base, and provides for the maintenance and production of reports from the elements within the data base.

Data base management system languages are commonly divided into three categories:

1. Languages that describe and generate the data base.
2. Languages which provide the maintenance and updating of the data base.
3. Languages which provide for the inquiry and retrieval from the data base.

In some data base management systems special languages have been created to access the data base. These systems are available from a number of software companies specializing in this area.

In other systems the approach is to use a host language, such as COBOL, and incorporate into the language data base management functions.

The particular advantage to a data base management system is that such systems permit very high-level statements of a generalized nature to access the data base. Statements entered via a terminal such as: "RETRIEVE RECORD FOR EMPLOYEE - 1275" or "CHANGE RATE TO 375," are possible. Thus, it can be seen that the ability to access a data base by an individual within an organization offers great potential for the efficient management of a business organization.

SUMMARY—THE FUTURE

Data communications and data base management systems will play an increasingly greater role in systems design in the future because of the advantages they offer in terms of cost, control, and ability to store and access data. In addition, with the continuing developments in small-scale computer hardware, data communications equipment, and peripheral devices, it can be anticipated that many organizations which, in the past, could not afford any type of data processing system will now be able to install and use data processing systems to their advantage.

Index